T0257838

Olive Oil Handbook

Volume II

Olive Oil Handbook
Volume II

Edited by **Thelma Bosso**

New York

Published by Callisto Reference,
106 Park Avenue, Suite 200,
New York, NY 10016, USA
www.callistoreference.com

Olive Oil Handbook
Volume II
Edited by Thelma Bosso

International Standard Book Number: 978-1-63239-492-7 (Hardback)

Printed in the United States of America.

Contents

Preface

Over the recent decade, advancements and applications have progressed exponentially. This has led to the increased interest in this field and projects are being conducted to enhance knowledge. The main objective of this book is to present some of the critical challenges and provide insights into possible solutions. This book will answer the varied questions that arise in the field and also provide an increased scope for furthering studies.

The health-promoting impacts associated with olive oil and the growth of the olive oil industry has reinforced the quest for novel information, stimulating a broad spectrum of research. This book is a source of currently compiled information. It incorporates a wide spectrum of topics organized under the sections: Olive Oil Extraction & Waste Water Treatment, Bioavailability & Biological Properties of Olive Oil Constituents and Innovative techniques for the production of olive oil based products. It will serve as useful source of information for a broad audience, mainly food scientists, nutritionists, biotechnologists, researchers, pharmacologists, olive oil producers and consumers.

I hope that this book, with its visionary approach, will be a valuable addition and will promote interest among readers. Each of the authors has provided their extraordinary competence in their specific fields by providing different perspectives as they come from diverse nations and regions. I thank them for their contributions.

Editor

Part 1

Olive Oil Extraction and Waste Water Treatment – Biotechnological and Other Applications

New Olive-Pomace Oil Improved by Hydrothermal Pre-Treatments

G. Rodríguez-Gutiérrez, A. Lama-Muñoz, M.V. Ruiz-Méndez,
F. Rubio-Senent and J. Fernández-Bolaños
Departamento de Biotecnología de los Alimentos,
Instituto de la Grasa (CSIC), Avda, Seville,
Instituto al Campus de Excelencia Internacional Agroalimentario, ceiA3
Spain

1. Introduction

The health properties of virgin olive oil (VOO) are well known in the Mediterranean Diet, in which VOO is the main source of fat (Boskou, 2000). The Mediterranean area provides 97% of the total olive production of the world and represents a major industry in the region (Aragon & Palancar, 2001). The fatty acid composition is not the only healthy component of olive oil; in addition, minor components have high biological activities (Pérez-Jiménez et al., 2007). From the olive oil by-product, the olive-pomace oil (OPO) is obtained. Recent studies have demonstrated the positive benefits of OPO on health, and these effects are due mainly to the presence of minor components (Ruiz-Gutiérrez et al., 2009). The new olive oil extraction processes in the olive mills make the extraction of OPO and the general utilisation of wastes more difficult. New thermal systems are proposed to pre-treat the olive oil wastes to facilitate their utilisation and OPO extraction.

1.1 Olive oil extraction systems

The manufacturing process of olive oil has undergone evolutionary changes. The traditional discontinuous pressing process was initially replaced by continuous centrifugation, using a three-phase system and later a two-phase system. Depending on the different olive oil production method, there are different kinds of wastes. The classic production of olive oil generates three phases and two wastes: olive oil (20 %), solid waste (30 %) and aqueous liquor (50 %). The solid waste (olive cake or "orujo") is a combination of olive pulp and stones. The aqueous liquor comes from the vegetation water and the soft tissues of the olive fruits, with water added during processing (so-called "alpechin" or "olive-mill waste water"). The presence of large amounts of organic substances (oil, polyphenols, protein, polysaccharides, etc.) and mineral salts represents a significant problem for the treatment of wastewater (Borja et al., 1997).

The use of a modern two-phase processing technique to which no water is added generates oil and a new by-product that is a combination of liquid and solid waste, called "alperujo", "alpeorujo" or "two-phase olive mill waste". This by-product is a high-humidity residue with a thick sludge consistency that contains 80 % of the olive fruit, including skin, seed, pulp and pieces of stones, which is later separated and usually used as solid fuel (Vlyssides et al., 2004). In Spain, over 90 % of olive oil mills use this system, which means that the annual production of this by-product is approximately 2,5-6 million tons, depending on the season (Aragon & Palancar, 2001).

1.2 Utilisation of olive oil wastes

Alperujo presents many environmental problems due to its high organic content and the presence of phytotoxic components that make its use in further bioprocesses difficult (Rodríguez et al., 2007a). Most of these components mainly phenolic compounds, confer bioactive properties, to olive oil. The extraction of the phenolic compounds has a double benefit: the detoxification of wastes and the potential utilisation as functional ingredients in foods or cosmetics, or for pharmacological applications (Rodríguez et al., 2007a). Although olive mill wastes represent a major disposal problem and potentially a severe pollution problem for the industry, they are also a promising source of substances of high value. In the olive fruits, there is a large amount of bioactive compounds, many of them known to have beneficial health properties. During olive oil processing, most of the bioactive compounds remain in the wastes or alperujo (Lesage- Meessen et al., 2001).Therefore, new strategies are needed for the utilisation of this by-product to make possible the bioprocess applications and the phase separation of alperujo.

Until now, efforts focussed on detoxifying these wastes prior to disposal, feeding, or fertilisation/composting, because they are not easy degradable by natural processes, or even used in combustion as biomass or fuel (Vlyssides et al., 2004). However, the recovery of high value compounds or the utilisation of these wastes as raw matter for new products is a particularly attractive way to reuse them, provided that the recovery process is of economic and practical interest. This, added to the alternative proposals to diminish the environmental impact, will allow the placement of the olive market in a highly competitive position, and these wastes should be considered as by-products (Niaounakis & Halvadakis, 2004).

1.3 Olive-pomace oil

After VOO extraction, the residual oil, or crude olive-pomace oil (COPO), is extracted by organic solvent extraction or centrifugation from olive oil wastes. After the COPO refining step, the refined olive-pomace oil (ROPO) is blended with VOO, obtaining OPO for human consumption. Currently, the growing interest in OPO is due to its biological active minor constituents (Ruiz-Gutiérrez et al., 2009). The concentration of these components in OPO is higher than the concentration in VOO, with the exception of polar phenols (Perez-Camino & Cert, 1999). Today, new processes for COPO refining are studied in order to diminish the loss of minor components (Antonopoulos et al., 2006). Some of these components are recovered in the refining steps.

Alperujo is treated by the OPO extractors for crude olive-pomace oil extraction (**Figure 1**). First, the major part of the stone present is removed, with the initial stone concentration of about 45% and, after the stone extraction, less than 15%. The stone is easily commercialised for numerous uses, such as in combustion materials, activated carbon, liquid and gas production from stone pyrolysis, an abrasive for surface preparation or for cosmetics, in addition to others (Rodríguez et al., 2008). The pitted alperujo is frequently centrifuged in the OPO extractor because the new decanter technology allows treating low-fat material for oil extraction, through which crude olive-pomace oil is obtained. After this mechanical extraction, a partially defatted and pitted alperujo is obtained, with a humidity close to 50%. This material is dried to no more than 10% humidity for both solvent extraction and combustion. Drying consumes much energy, therefore attempts are continuously to reduce energy costs and to avoid the appearance of undesirable compounds in pomace-olive oil formed by the high temperatures (up to 500 °C) such as polycyclic aromatic hydrocarbons (PAHs) (León -Camacho et al., 2003) or oxidised compounds (Gomes and Caponio, 1997).

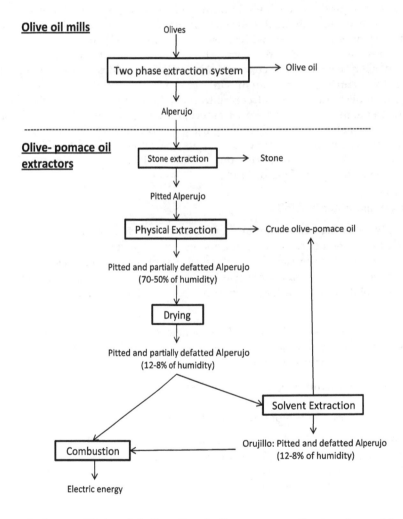

Fig. 1. General scheme of industrial olive oil and olive-pomace oil extraction and by-product processing.

The drying process is usually carried out in rotary heat dryers (Espínola-Lozano, 2003) in which alperujo and hot gases obtained from orujillo, olive stones or exhausted gases from co-generation systems (Sánchez & Ruiz, 2006) are introduced at 400 to 800 °C. The high temperatures have negative effects on the final composition of COPO. After drying, the pitted and partially de-fatted alperujo, with humidity close to 10%, is extracted with organic solvents. After the extraction, the organic solvent is removed for COPO production. The COPO obtained by physical or chemical methods has to be refined for human consumption. The final solid, called orujillo, is commonly used as a biomass for energy production.

The apparition of alperujo was supposed to be a great advantage for olive oil mills because the liquid waste (alpechin) was removed, but it was a serious inconvenience for COPO extractions with regard to the high humidity of the new semi-solid waste, or alperujo. Previous to the two-phase extraction system, the solid waste, or orujo, from the three-phase extraction system was treated with lower humidity (50%) than the alperujo (70%). Nowadays, despite the use of the final solid as biomass, the extraction of olive-pomace oil does not, in many cases, have economic advantages. In addition, the olive oil mills are improving the centrifugation systems in order to increase the quantity of olive oil, producing alperujo with lower oil concentrations. Consequently, the higher humidity in addition to the high organic content of alperujo complicate the COPO extraction, higher temperatures in heat dryers or alperujo with lower oil content. Therefore, pre-treatment alternatives are necessary to properly process the alperujo in the OPO extractor and improve the oil extraction balance and quality, while at the same time obtain new components and add value to the product.

1.4 Minor components in OPO

Interest in olive-pomace oil is growing due to its economic advantages. It is cheaper than olive oil, and contains many minor components with bioactivities. OPO contains all of the functional compounds found in virgin olive oil, except for the polyphenols, in addition to other biologically active components (De la Puerta et al., 2009; Ruiz-Gutiérrez et al., 2009) that could be solubilised from leaves, skin or seeds of olives, depending on the extraction systems.

Phytosterols, tocopherols, aliphatic alcohols, squalene and triterpenic acid are some of the most important compounds that make the minor components an interesting fraction from the point of view of bioactive compounds that are agents for disease prevention.

The phytosterol´s structure is similar to cholesterol, and they are a powerful agent in the cholesterol-lowering effects in human blood (Jiménez-Escrig et al., 2006) and as a cytostatic agent in inflammatory and tumoral diseases (Sáenz et al., 1998).

Tocopherols (α-, β-, and γ-form) are present in high concentrations in OPO. α-tocopherol is an essential micronutrient involved in several oxidative stress processes related to atherosclerosis, Alzheimer's disease, accelerated aging and cancer (Mardones & Rigotti, 2004). Recently, biological activities against diseases like cancer in animal models have been also attributed to γ-form (Ju et al., 2010).

There is also squalene in olive-pomace oil. This compound has a beneficial effect on atherosclerotic lesions (Guillén et al., 2008, Bullon et al., 2009), dermatitis (Kelly et al., 1999) and cellular proliferation and apoptosis in skin and intestinal cancers (Rao, 1998). After being absorbed by the human skin surface, squalene acts as a defence against oxidative stress due to the exposure to ultraviolet (UV) radiation from sunlight.

Aliphatic alcohols with long-chain fatty alcohols (C26 or hexacosanol, C28 or octacosanol and C30 or triacontanol) obtained from OPO have shown activity in reducing the release of different inflammatory mediators (Fernández-Arche et al., 2009), reducing platelet aggregation and lowering cholesterol (Taylor et al., 2003, Singh et al., 2006).

Uvaol and erythrodiol are the triterpenic alcohol fraction present in OPO. They are active antioxidant agents in the microsomal membranes of rat liver (Perona et al., 2005), with positive effects on the inflammatory process (Márquez-Martín et al., 2006), or in the prevention and treatment of brain tumours and other cancers (Martín et al., 2009).

1.5 Thermal pre-treatments

Alperujo is a high humidity solid that needs special pre-treatments to obtain a viable utilisation of all its phases. Only a few pre-treatments have been proposed for the total utilisation of alperujo, extracting the main interesting fractions. One of the more attractive processes is based on thermal pre-treatments that allow the recovery of all of the bioactive compounds and valuable fractions, making possible the utilisation of alperujo (Fernández-Bolaños et al., 2004). Thermal treatments produce the solubilisation of bioactive compounds to the liquid phase, leaving a final solid enriched in oil, cellulose and proteins. From the liquid, it is possible to extract and purify the bioactive compounds that confer healthy properties to olive oil, mainly phenols such as hydroxytyrosol (HT). HT is one of the more important phenols in the olive oil and fruit because it has excellent activities as a pharmacological and antioxidant agent (Fernández-Bolaños et al., 2002). HT has been recently commercialised by a patented system (Fernández-Bolaños et al., 2005). In addition to other important compounds, a novel phenol has been isolated and purified for the first time: 3,4-dihydroxyphenylglycol (DHPG). DHPG has never been studied as a natural antioxidant or functional compound with a higher antiradical activity and reducing power than the potent HT (Rodríguez et al., 2007b). After the thermal treatment and the solid-liquid separation, a solid that is rich in cellulose and oil is obtained. The cellulose is easy to extract and use as a source of fermentable sugar, animal feed or fertiliser (Rodríguez et al., 2007a). The thermal reaction improves the concentration in oil of minor components with functional activities. In addition, phenols increase in the liquid due to the solubilisation, with this fraction a rich source of interesting phenols, sugar and oligosaccharides, all of them with a potential use in the food or nutraceutical industry.

This alternative pre-treatment not only increases the concentration of oil in the final solid, but also the content of minor components in COPO prior to the refining process. The thermal treatment improves the functional profile, enhancing the quality and healthy properties of this oil (Lama-Muñoz et al., 2011). The application of thermal pre-treatment to alperujo makes the extraction of olive-pomace oil easier, improving its functional composition. On the other hand, it is important to note that all chemical changes of fats and oils at elevated temperatures result in oxidation, hydrolysis, polymerisation, isomerisation or cyclisation reactions (Quiles et al., 2002, Valavanidis et al., 2004). All of these reactions may be promoted by oxygen, moisture, traces of metal and free radicals (Quiles et al., 2002). Several factors, such as contact with the air, the temperature and the length of heating, the type of vessel, the degree of oil unsaturation and the presence of pro-oxidants or antioxidants, affect the overall performance of oil (Andrikopoulos et al., 2002). In this work, the effect of two different thermal pre-treatments on COPO composition has been individually studied to balance the positive and negative factors in the final COPO.

1.5.1 Steam explosion system (SES)

The SES is commonly used as a hydrolytic process for lignocellulosic material utilisation (McMillan, 1994). This process (**Figure 2**) combines chemical and physical effects on lignocellulosic materials. The material is treated with high-pressure saturated steam for a few minutes and then the pressure is swiftly reduced, causing the materials to undergo an explosive decompression. The process causes hemicellulose degradation and lignin transformation due to high temperature, increasing the solubilisation of interesting compounds not only into the aqueous phase but also into the oil fraction. It is used mainly for the treatment of bagasse, such as wheat or rice straw, sugar cane, etc. The pre-treatment can enhance the bio-digestibility of the wastes for bioprocess applications to obtain, for instance, ethanol or biogas, and to increase the accessibility of the enzymes to the materials (De Bari et al., 2004; Palmarola-Adrados et al., 2004; Kurabi et al., 2005). High pressures (10-40 Kg/cm²) and temperatures (180-240 °C) are applied with or without the addition of acid in a short period of time, followed by explosive depressurisation. The SES makes it possible to obtain a final solid that is rich in COPO from alperujo. The thermal treatment solubilises a high proportion of solid, leaving behind components such as oil, proteins and cellulose. All these components are concentrated in the final solid.

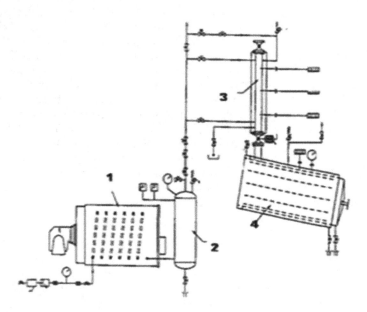

Fig. 2. Steam Explosion System scheme. Laboratory pilot unit designed in the Instituto de la Grasa (Seville, Spain), equipped with: 1) steam generator, 2) steam accumulator, 3) 2 L reactor stainless-steel and 4) expansion chamber.

1.5.2 New steam treatment (ST)

The system scheme is shown in **Figure 3**. A lower range of pressure and temperatures (3-9 Kg/cm² and 140-180 °C) than SES is applied for a longer period of time (15-90 min) in the novel system. The conditions and the contact of steam with the sample have been successfully improved, avoiding the technical complications and the high costs of the SES. This treatment has been recently patented to treat olive oil wastes, and the first tests have been carried out to assess its industrial viability for alperujo utilisation (Fernández-Bolaños et al., 2011).

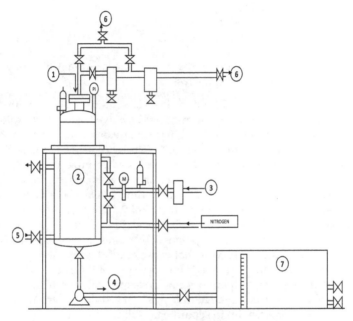

Fig. 3. New steam treatment (ST) reactor scheme designed in the Instituto de la Grasa (Seville, Spain) with: 1) sample entrance, 2) reactor chamber (100 L), 3) water steam, 4) sample exit, 5) cold water for refrigeration system, 6) vacuum and 7) solid-liquid separation.

The sample is introduced into the reaction chamber together with water steam. The sample temperature is increased up to 190°C for 30-60 minutes. After the reaction time, the sample is cooled with indirect water as a refrigerant. The liquid and solid phases of the treated sample are separated and the solid is finally extracted to obtain the crude olive-pomace oil.

The advantages of both systems are based on the important solubilisation of the initial solid to the liquid phase that occurs during the thermal treatment, leaving a final solid in which several components like oil, cellulose and protein are concentrated. The humidity of the final solid is also easier to remove by centrifugation or filtration, simplifying the drying process and the undesirable compounds that are formed at high temperatures.

In addition, the liquid phase is rich in bioactive compounds that are easy to extract. All these factors make possible the total utilisation of olive oil wastes, diminishing their environmental impact (Rodríguez et al., 2007a).

First, the application of SES on alperujo in order to obtain pomace olive oil was studied with or without acid as a catalytic agent. Due to the technical disadvantages of SES and to make the use of thermal pre-treatment in OPO mills easier and more convenient low severities, no depressurisation or acid addition were used in the new system (ST).

2. Experimental procedures

Samples of stored olive pomace or alperujo were collected from the COPO extraction factory Oleícola El Tejar (Córdoba, Spain) with 70% humidity. This by-product is generated as a waste from the two-phase olive oil extraction system.

2.1 Thermal treatments

The thermal treatments were carried out in the Instituto de la Grasa (CSIC) pilot plant by the steam explosion system and a new thermal system:

a. The SES was carried out using a flash hydrolysis laboratory pilot unit with a 2 L reactor. The reactor was equipped with a quick-opening ball valve for the final explosion into the expansion chamber. Alperujo samples of 250 g were treated with saturated steam in a 2 L reactor with a maximum operating pressure of 40 Kg/cm^2. The reactor was equipped with a quick-opening ball valve and an electronic device programmed for the accurate control of steam time and temperature for the final steam explosion. Prior to the treatment, some of the samples were acidified with H_3PO_4. The acid was added to the moist sample so as to reach a final concentration of 2,5 % (v/v). After the treatment, the samples were collected and filtered in vacuo through filter paper using a Buchner funnel.
b. The new ST reactor has recently been patented (Fernández-Bolaños et al., 2011). It has a 100 L capacity stainless steel reservoir that can operate at temperatures between 50 and 190 °C by direct heating and at a maximum pressure of 9 Kg/cm^2. The system allows the appropriate treatment of alperujo without steam explosion or high pressures and temperatures. The wet treated material was filtered by centrifugation at 4700 g (Comteifa, S.L., Barcelona, Spain) to separate the solids and liquids.

After solid separation, the solid phase was dried in a stove at 50 °C, and the reduction (%) in the mass of the solid phase was determined.

2.2 Analytical determinations

Oil was extracted from the treated solid obtained by SES and ST with n-hexane using a Soxhlet apparatus. The obtained oils were filtered and stored at -20 °C until analysis. Oil content and fat enrichment (pitted dry matter) were determined and compared with the values for untreated alperujo samples.

Determination of the concentrations of aliphatic alcohols, sterols and triterpenic dialcohols (erythrodiol and uvaol) was performed according to the Commission Regulation (EEC) No 2568/91 for olive oil and pomace oil. After the silylation reaction, 1 mL of heptane was added to the mixture, and 1 µL of the solution was injected into an Agilent 7890A gas chromatograph system (Agilent Technologies, Palo Alto, USA) equipped with an FID detector. The analytical column was an HP-5 (5%-phenyl)-methylpolysiloxane column (30 m x 0,32 mm i.d., 0,25 µm film thickness). The results were expressed as mg/kg of oil.

Tocopherols were evaluated using the IUPAC 2.432 method. Results were expressed as mg/kg of oil.

The wax and squalene compositions were determined according to the European Regulation EEC/183/93, by separation on a silica gel 60 (70-230 mesh ASTM) chromatographic column (Merck KGaA, Darmstadt, Germany) using hexane/ether (98:2) as the eluent with a few drops of Sudan I as a colorant. Dodecyl arachidate (Sigma) and squalane (Fluka) were added as internal standards. The results were expressed as mg/kg of oil.

Polar compounds, triglycerides and their derivatives oxidise and hydrolyse were prepared using solid-phase extraction and size-exclusion chromatography and monostearin as internal standard (Márquez -Ruiz et al., 1996). An aliquot (20 μL) of the final solution was injected into a Hewlett Packard Series 1050 HPLC system equipped with a refractive index detector (LaChrom L-7490 Merck) and a 100-Å PL gel column (5 μm) (Agilent). Elution was performed at 0,6 mL/min, with tetrahydrofuran as the mobile phase.

Determination of fatty acid, free acidity and peroxide value (PV) was carried out according to the Official Methods described in the European Community Regulation EEC/2568/91. The results were expressed as the percentage of oleic acid. The peroxide value was expressed in milliequivalents of active oxygen per kilogram of oil (mequiv O_2/kg oil).

The indexes K_{232}, K_{270} and ΔK were determined using the European Communities official methods (European Union Commission, 1991). Oil samples were diluted in isooctane and placed into a 1 cm quartz cuvette; for values calculation, each solution was analysed at 270 and 232 nm, with isooctane as a blank.

3. Results and discussion

Both systems allow the utilisation of the final solid for crude olive-pomace oil extraction. These COPOs have been characterised to assess the positive and negative effects of both treatments on its composition. The SES was studied as a commonly used method for lignocellulosic materials, and the ST was designed to simplify the first system and to diminish the negative effects of SES on crude olive-pomace oil. The lipid fraction of POO extracted from solids treated with either treatment was evaluated, and the minor components were also characterised, in the case of the ST.

3.1 Steam explosion system (SES)

An average temperature of 200 °C and a time of 5 minutes were used with or without acid impregnation of alperujo. The acid increases the severity of the treatment, enhancing the oxidation of the samples. A vacuum was applied to one of the treatments, with acid addition in order to diminish the possible oxidative effect of oxygen at high temperatures and pressures. The results showed (**Table 1**) an important solubilisation of the solid in all treatment. In addition, the oil was concentrated in the final solid from 8,3 up to 19,9 % with respect to the dry final solid. Despite the high level of acidity in the initial sample after the treatment, these values decreased.

K_{232} and K_{270} are simple and useful parameters for assessing the state of oxidation of olive oil. The coefficient of specific extinction at 232 nm is related to the presence of products of the primary stage of oxidation (hydroperoxides) and conjugated dienes, which are formed by a shift in one of the double bonds. The extinction coefficient at 270 nm is related to the presence of products of secondary oxidation (carbonylic compounds) and conjugated trienes (the primary oxidation products of linolenic acid).

The K_{232} values of all treated samples were lower than the untreated alperujo, unlike the K_{270} values in which only the sample treated with vacuum and without acid presented a similar absorbance at 270 nm. Only when vacuum and acid were applied to the SES did the value of K_{270} exceed the maximum concentration in ROPO (2,0), with all the ΔK values lower than the maximum in ROPO (0,2). All these oxidised compounds diminished after the refining process.

The polar compound values that show the alteration level by the non-volatile compounds of oil are practically the same in all treatments, except when the acid and the vacuum are used simultaneously. Polar compounds provide an idea not only of oxidative reactions, but also of hydrolytic degradation, because they are partial constituents of FFA and glycerides.

Curiously, the concentration of oxidised triglycerides and polymers are lower after the SES treatments. This result could be explained by their partial solubilisation during the thermal treatment in the liquid phases that are previously separated by the oil extraction.

	Untreated sample	SES (200° C, 5 min)	SES (200° C, 5 min, 2,5% H_3PO_4)	SES (200° C, 5 min, vacuum)	SES (200° C, 5 min, 2,5% H_3PO_4, vacuum)
% of solid reduction	-	51,7	54,6	52,1	58,5
% of oil in final solid	8,3	17,2	18,0	16,9	19,9
Acidity (% oleic acid)	6,76	4,97	4,44	5,08	5,48
K_{232}	5,72	4,57	3,23	5,18	5,21
K_{270}	1,27	1,74	1,44	1,24	2,37
ΔK	-0,03	0,00	0,04	0,06	0,07
Polar compounds mg/g	114,3	116,59	112,81	111,37	129,87
Oxidised Triglycerides [a]	1,61	1,25	1,34	1,41	1,36
Diglycerides[a]	3,10	4,47	4,83	4,37	5,13
Monoglycerides[a]	0,42	0,53	0,40	0,49	0,56
FFA (% as oleic acid)	6,10	5,28	4,49	4,66	5,73
Polymers[a]	0,23	0,14	0,23	0,21	0,21

[a] % with regard to the oil sample.

Table 1. Solid reduction, oil concentration in final solid and chemical characteristics of crude olive-pomace oil treated or untreated by SES in several conditions.

The quantity of triglycerides decreased after the SES treatment, with an increased presence of diglycerides and monoglycerides as an unmistakable sign of hydrolytic degradation.

The oxidative effects do not seem to be the main cause of the COPO alteration during the SES treatment, while hydrolysis seems to be an important effect on the triglyceride loss.

The composition of triglycerides was determined and the results are shown in the **Table 2**. The main triglyceride peaks in all samples were oleic-oleic-oleic (OOO), oleic-oleic-palmitic (OOP) and linoleic-oleic-oleic (LOO).

Despite the low quality of the initial oil, the relation of triglycerides was not altered by SES. As expected, in the crude olive-pomace oil obtained after SES treatment, the total content of triglycerides decreased up to 22% compared to the oil obtained from the untreated alperujo. Despite the high severity, only 22% of triglyceride composition was lost, with the rest susceptible for refining. The oxidative conditions of SES treatment were minimised using a vacuum or avoiding the acid addition.

The great advantages of the SES application on alperujo are based, in addition to other reasons, on the solid reduction (up to 58%) and oil concentration in the final solid (up to 20%). Because the triglycerides (TG) are concentrated, their loss is not a significant or negative factor to limit the use of this system. These reasons make technically possible the extraction of the crude olive-pomace oil from one sample of alperujo treated by SES. Therefore, the application of SES to this kind of alperujo allows for the obtaining of a final solid rich in oil in a high concentration that is susceptible for further refining processes for olive-pomace oil production. However, the technical inconveniences of SES such as high temperatures and pressures or the explosive decompression make it an inadequate system for olive-pomace oil extractors.

Triglycerides	Untreated sample		SES (200° C, 5 min) treated sample	
	Peak area	% of total glycerides Mean ± SD	Peak area	% of total glycerides Mean ± SD
LLL	18039	0,7 ± 0,05	11076	0,6 ± 0,02
LnLO	15271	0,6 ± 0,03	5886	0,3 ± 0,01
OLL+PoLO	99759	4,0 ± 0,10	75243	3,9 ± 0,77
PLL+LnOO	62860	2,5 ± 0,15	32818	1,7 ± 0,26
POLn	23935	1,0 ± 0,07	0	0,0 ± 0,00
LOO	404446	16,1 ± 1,02	321011	16,6 ± 1,23
LOP	154422	6,2 ± 0,81	126669	6,5 ± 0,87
LPP	4000	0,2 ± 0,01	6064	0,3 ± 0,02
OOO	959632	38,3 ± 1,67	788016	40,7 ± 2,03
OOP	448959	17,9 ± 1,11	364371	18,8 ± 1,30
POP	57959	2,3 ± 0,90	39691	2,0 ± 0,11
SOO	157822	6,3 ± 1,43	108877	5,6 ± 0,95
POS	28350	1,1 ± 0,03	21116	1,1 ± 0,05
AOO	49158	2,0 ± 0,16	26398	1,4 ± 0,06
Área total	2504443		1938316	

P, palmitic, Po, palmitoleic, M, margaric, S, stearic, O, oleic, L, linoleic, Ln, linolenic, and A, arachidic acids

Table 2. Triglycerides composition of crude olive-pomace oil obtained from alperujo untreated and treated by SES.

3.2 Effects of the new steam treatment

The ST effect on POO composition was determined by characterisation of the fatty acid fraction. After the thermal treatment in the range of 150-170°C for 60 min (**Table 3**), the final treated solid had an increase in oil yield up to 97%, with a reduction in solids up to 35,6-47,6% by solubilisation. The oxidative damage was lower in the new treatment. The analysis of the polar fraction showed that oxidised triglycerides and peroxide values increased slightly and that no polymerisation reactions occurred. The hydrolytic process is shown in the diglycerides increasing from 2,5 to 6,6%, with the FFA and the unsaponifiable matter for all treatments remaining constant.

	Untreated sample	ST (150° C, 60 min)	ST (160° C, 60 min)	ST (170° C, 60 min)
% of solid reduction	-	35,6	47,1	47,6
% of oil in final solid	8,1	11,8	14,3	16,0
Acidity (% oleic acid)	3,6	4,7	4,9	5,1
Peroxide Values (meq/Kg)	8,7	9,4	10,9	12,3
Oxidised Triglycerides [a]	0,7	1,1	1,1	1,6
Diglycerides[a]	2,5	5,2	6,6	6,6
Monoglycerides[a]				
FFA (% as oleic acid)	3,3	2,8	2,9	2,8
Unsaponifiable matter (%)	2,53	3,02	2,50	2,54

Table 3. Solid reduction, oil concentration in final solid and chemical characteristics of crude olive-pomace oil treated or untreated by ST at 150, 160 and 170°C for one hour. [a] % with regard to the oil sample.

The concentration of minor components (**Table 4**) was significantly increased by ST. Sterols, aliphatic alcohols, triterpenic alcohols, and squalene increased up to 33%, 57%, 23% and 43%, respectively. In addition, the content of tocopherols increased up to 57% compared to untreated POO. This increase is due to solubilisation during the thermal treatment. The waxes level is also increased because of the high solubilisation from the external cuticle of the olive fruit and the leaves. Waxes are easily removed by the refining of COPO.

The samples of alperujo had been stored for a long time and the oil was partially extracted by centrifugation in OPO extractors just before pitting. The alperujo was chosen because its low fat concentration makes the COPO extraction economically unviable. In this condition, the initial oil has a very low quality, as previously shown in the tables. The analysed oils showed, despite the low quality of initial oil present in the alperujo studied, that the effect of thermal treatment increases slightly the values of oxidised components and hydrolytic degradation. All COPOs obtained after the thermal treatments are susceptible for a posterior refining process for OPO obtention.

Components	Untreated sample	ST (150°C, 60 min)	ST (160°C, 60 min)	ST (170°C, 60 min)
Sterols	4927 (104)	5687 (291)	6546 (216)	6555 (298)
Aliphatic alcohols	4065 (77)	5532 (603)	5880 (283)	6389 (68)
Triterpenic alcohols	992 (58)	1054 (34)	1220 (124)	1189 (107)
Waxes	1535 (3)	2971 (5)	3124 (94)	3461 (110)
Squalene	2404 (36)	2472 (11)	2729 (109)	3439 (171)
Tocopherols	425 (33)	460 (6)	668 (14)	533 (20)

Table 4. Total minor component composition (mg/kg ± SD) of oils from steam-treated and untreated alperujo. Numbers between parentheses indicate the standard deviation of three replicates.

For human consumption, the refining process of OPO is necessary. The refining (physical or chemical) process eliminates undesirable compounds (peroxides, degradation products, volatile compounds responsible for off-flavours, free fatty acids, etc.) but also results in the loss of valuable bioactive compounds and natural antioxidants (Ruiz-Méndez et al., 2008). The new trends of refining systems involve losing as few minor components as possible to obtain a final OPO that is rich in the minor components. The thermal treatments increase the minor component of COPO that help to obtain a final olive-pomace oil rich in interesting compounds, whose concentrations might be higher after the refining process, mainly using the new physical systems. Moreover, some of these minor components are recuperated during the refining, such as squalene, the concentration of which is increased up to 43 % after the pre-treatment. After extraction, the defatted solid is lacking in phenols and then in phytoxic compounds for further bio-utilisation and rich in components like cellulose and protein.

Figure 4 shows the main aspects of both thermal systems. The high difference between temperatures and pressures together with the absence of explosive decompression makes ST more appropriate and economically viable for industrial applications. A longer period of reaction time is necessary to treat with ST, but is easily applicable in an industrial continuous reactor. The percentage of solid reduction and, consequently, the final oil concentration show that despite the high severity difference between both treatments, there is not a significant difference in the results. Then, the new ST provides the major advantages of SES without technical complications.

Thus, the positive effect of a novel thermal treatment for the extraction of crude olive-pomace oil that could improve the commercial value of OPO and its bioactivities by increasing the concentrations of minor components concentration has been demonstrated. This treatment also significantly reduces the cost of oil extraction by centrifugation or solvent extraction because the starting solid is more concentrated in oil and is drier than untreated alperujo.

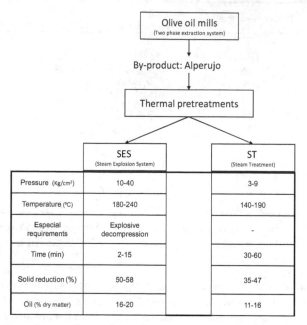

	SES (Steam Explosion System)		ST (Steam Treatment)
Pressure (Kg/cm²)	10-40		3-9
Temperature (°C)	180-240		140-190
Especial requirements	Explosive decompression		-
Time (min)	2-15		30-60
Solid reduction (%)	50-58		35-47
Oil (% dry matter)	16-20		11-16

Fig. 4. Comparative scheme of two thermal pre-treatments used for alperujo utilisation.

4. Conclusion

The new treatment ST not only maintains the advantages of the SES with regard to the concentration of oil in the final solid and phase separation, but also diminishes the oxidation and significantly improves the concentration of the most interesting components of the minor fractions of POO. Thus, the application of ST enhances the functional properties of this new POO, increasing the oil extraction yield and the total recovery of bioactive compounds from the refining process.

The steam treatment offers not only serious advantages in terms of the oil but also in terms of the total recovery of alperujo as described above. The application of ST to treat olive oil wastes allows the phase separation and the concentration of interesting compounds and components in each phase. In the liquid phase, bioactive compounds like phenols and oligosaccharides are solubilised and are easy to extract. In the solid fraction, the oil and cellulose are concentrated. After the oil extraction, the solid has a low content of phytotoxics that are in the liquid phase, and therefore, it is susceptible for bio-treatment for the total utilisation of this fraction.

5. Acknowledgment

Financial support for this study was provided by Ministerio de Ciencia e Innovación and Junta de Andalucía through projects AGL 2009-12352 (ALI) and P08-AGR-03751 respectively.

6. References

Andrikopoulos, N. K.; Kalogeropoulus; N.; Falirea; A. & Barbagianni, M. N. (2002). Performance of virgin olive oil and vegetable shortening during domestic deep-frying and pan-frying of potatoes. *International Journal of Food Science and Technology*, Vol. 37, No.2, (February 2002), pp. 177–190, ISSN 0950-5423

Antonopoulos, K.; Valet, N.; Spiratos, D. & Siragakis, G. (2006). Olive oil and pomace olive oil processing. *Grasas y Aceites*, Vol.57, No.1, (January-March 2006), pp.56-67, ISSN 0017-3495

Aragón, J.M. & Palancar, M.C. (2001). Improlive 2000. Present and future of Alpeorujo, pp. 242-300, ISBN 84-7491-593-7, Editorial Complutense S. A., Madrid.

Borja, R.; Alba, J. & Banks, C.J. (1997). Impact of the main phenolic compounds of olive mill wastewater (OMW) on the kinetics of acetoclastic methanogenesis. *Process Biochemistry*. Vol.32, No.2, (February 1997), pp. 121-133, ISSN 1359-5113

Boskou, D. (2000). Olive oil. *World Review of Nutrition & Dietetics*. Vol.87, pp.56-77, ISSN 0084-2230

Bullon, P.; Quiles, J.L.; Morillo, J.M.; Rubini, C.; Goteri, G.; Granados-Principal, S.; Battino, M. & Ramirez-Tortosa, M.C. (2009). Gingival vascular damage in atherosclerotic rabbits: hydroxytyrosol and squalene benefits. *Food and Chemical Toxicology*, Vol 47, No.9, (September 2009), pp. 2327-2331, ISSN: 0278-6915

Commission Regulation (EEC) No 2568/91 of 11 July 1991 on the 454 characteristics of olive oil and olive-residue oil and the relevant methods of 455 analysis. *Off. J. Eur. Communities*. 1991. 248.

Commission Regulation (EEC) No 183/93 L 28 of the 29 January 1993 on the 462 characteristics of olive oil and olive-residue oil and the relevant methods of 463 analysis. *Off. J. Eur. Communities*. 1993.

De Bari, I.; Cuna, D.; Nanna, F. & Braccio, G. (2004). Ethanol production in immobilized-cell bioreactors from mixed sugar and enzymatic hydrolyzates of steam-exploded biomass. *Biotechnology and Applied Biochemistry*, Vol.114, No.1-3, (April 2004), pp. 539-557, ISSN 0273-2289

De la Puerta, R.; Márquez-Martín, A.; Fernández-Arche, A. & Ruiz-Gutiérrez, V. (2009). Influence of dietary fat on oxidative stress and inflammation in murine macrophages. *Nutrition*, Vol. 25, No.5, (February 2009), pp. 548-554, ISSN: 0899-9007

Jiménez-Escrig, A.; Santos-Hidalgo; A. B. & Saura-Calixto; F. (2006). Common sources and estimated intake of plant sterols in the Spanish diet. *Journal of Agriculture and Food Chemistry*. Vol. 54, No.9, (May 2006), p.p.3462-3471, ISSN 0021-8561

Espínola-Lozano, F. (2003). Technological change in the extraction of olive pomace oils. *Alimentación Equipos y Tecnología, Vol.22*, pp. 60-64, ISSN 0212- 1689

European Union Commission (1991). Regulation 2568/91. Official Journal of the European Communities.

Fernandez-Arche; A.; Marquez-Martin, A.; de la Puerta Vazquez; R.; Perona, J.; Terencio; C.; Pérez-Camino, C. & Ruiz-Gutierrez, V. (2009). Long-chain fatty alcohols from pomace olive oil modulate the release of proinflammatory mediators. *Journal of Nutritional Biochemistry*, Vol.20, No.3, (March 2009), pp. 155-162, ISSN: 0955-2863

Fernández-Bolaños, J.; Rodríguez, G.; Rodríguez, R.; Heredia, A.; Guillén, R. & Jiménez, A. (2002). Production in large quantities of highly purified hydroxytyrosol from liquid-solid waste of two-phase olive oil processing or "alperujo". *Journal of*

Agriculture and Food Chemistry. Vol.50, No.22, (October 2002), pp. 6804–6811, ISSN 0021-8561

Fernández-Bolaños; J.; Rodríguez; G.; Gómez; E.; Guillén; R.; Jiménez; A. & Heredia, A. (2004). Total recovery of the waste of two-phase olive oil processing: isolation of added-value compounds. *Journal of Agricultural and Food Chemistry,* Vol.52, No.19, (August 2004), pp. 5849–5855, ISSN 0021-8561

Fernández-Bolaños, J.; Heredia, A.; Rodríguez, G.; Rodríguez, R.; Jiménez, A. & Guillén, R. (2005). Methods for obtaining purified hydroxytyrosol from products and by-products derived from the olive tree, patent No. US 6849770 B2

Fernández-Bolaños; J.; Rodríguez; G.; Lama, A. & Sanchez, P. (2011). Dispositivo y procedimiento para el tratamiento de los subproductos de la obtención de aceite de oliva, patent No. PCT/ES2011/070583.

Guillén, N.; Acín; S.; Navarro; M. A.; Perona; J. S.; Arbonés-Mainar; J. M.; Arnal; C.; Sarría; A. J.; Surra; J. C.; Carnicer, R.; Orman, I.; Segovia, J. C.; Ruiz-Gutiérrez, V. & Osada, J. (2008). Squalene in a sex-dependent manner modulates atherosclerotic lesion which correlates with hepatic fat content in apoE-knockout male mice. *Atherosclerosis,* Vol.197, No.1, (March 2008), pp. 72-83, ISSN 0021-9150

Gomes, T. & Caponio, F. (1997). Evaluation of the state of oxidation of crude olive-pomace oils. Influence of olive-pomace drying and oil extraction with solvent. *Journal of Agriculture and Food Chemistry,* Vol.45, No.4, (April 1997), pp. 1381-1384, ISSN 0021-8561

International Union of Pure and Applied Chemistry. (1992). Determination of tocopherols and tocotrienols in vegetable oils and fats by high performance liquid chromatography. *Standard Methods for the Analysis of Oils, Fats and Derivatives 1st Supplement to the 7th Edition.* Blackwell Scientific Publications, Osney Mead, Oxford.

Ju, J.; Picinich, S. C.; Yang, Z.; Zhao, Y.; Suh, N.; Kong, Ah-Ng & Yang, C. S. (2009). Cancer-preventive activities of tocopherols and tocotrienols. *Carcinogenesis,* Vol.31, No.4, (September 2009), pp.533-542, ISSN 0143-3334

Kelly, G.S. (1999). Squalene and its potential clinical uses. *Alternative Medicine Review,* Vol. 4, No.1, (February 1999), pp. 29-36, ISSN 1089-5159

Kurabi, A.; Berlin, A.; Gilkes, N.; Kilburn, D.; Bura, R.; Robinson, J.; Markov, A; Skomarovsky, A.; GustaKov, A.; Okunev, O.; Sinitsyn, A.; Gregg, D.; Xie, D. & Saddler, L. (2005). Enzymatic hydrolysis of steam-exploded and ethanol organosolv-pretrated Douglas-Fir by novel and commercial fungal-cellulase. *Applied Biochemistry and Biotechnology,* Vol.212, No.121-124, (March 2005), pp. 219-230, ISSN 0885-4513

Lama-Muñoz, A.; Rodríguez-Gutiérrez, G.; Rubio-Senent, F.; Gómez-Carretero, A. & Fernández-Bolaños, J. (2011). New Hydrothermal Treatment of Alperujo Enhances the Content on Bioactive Minor Components in Crude Pomace Olive Oil. *Journal of Agriculture and Food Chemistry,* Vol.59, No.4, (February 2011), pp. 1115-1123, ISSN 0021-8561

Lesage-Meessen, L.; Navarro, D.; Maunier, S.; Sigoillot, J.C.; Lorquin, J.; Delattre, M.; Simon, J.L.; Asther, M. & Labat, M. (2001). Simple phenolic content in olive oil residues as a function of extraction systems. *Food Chemistry,* Vol.75, No.4, (December 2001), pp. 501-507, ISSN 0308-8146

León-Camacho, M.; Viera-Alcaide, I. & Ruiz-Méndez, M.V. (2003). Elimination of polycyclic aromatic hydrocarbons by bleaching of olive pomace oil. *European Journal of Lipid Science and Technology*, Vol.105, No.1, (January 2003), pp. 9-16, ISSN 1438-7697

McMillan, J.D. (1994). Pretreatment of lignocellulosic biomass. In: M.E. Himmel, J.O. Baker and R.P. Overend, Editors. *Enzymatic Conversion of Biomass for Fuels Production*, American Chemical Society, Washington, DC (1994); pp. 292–324. ISBN 0841229562

Mardones, P. & Rigotti, A. (2004). Cellular mechanisms of vitamin E uptake: relevance in α-tocopherol metabolism and potential implications for disease. *Journal of Nutritional Biochemistry*, Vol.15, No.5, (May 2005), pp. 252-260, ISSN 0955-2863

Márquez-Martín, A.; De la Puerta, R.; Fernández-Arche, A.; Ruiz-Gutiérrez, V. & Yaqoob, P. (2006). Modulation of cytokine secretion by pentacyclic triterpenes from olive pomace oil human mononuclear cells. *Cytokine*, Vol.36, No.5-6, (December 2006), pp. 211-217, ISSN 1043-4666

Márquez-Ruiz, G.; Jorge, N.; Martin-Polvillo, M. & Dobarganes, M.C. (1996). Rapid, quantitative determination of polar compounds in fats and oils by solid-phase extraction and size-exclusion chromatography using monostearin as internal standard. *Journal of Chromatography. A*, Vol.749, No.1-2, (October 1996), pp. 55-60, ISSN 0021-9673

Martín, R.; Ibeas, E.; Carvalho-Tavares, J.; Hernández, M.; Ruiz-Gutiérrez, V. & Nieto; M. L. (2009). Natural triterpenic diols promote apoptosis in astrocytoma cells through ROS-mediated mitochondrial depolarization and JNK activation. *PLoS ONE*, Vol.4, No.6, (June 2009), pp. e5975, ISSN 1932-6203

Niaounakis, M. & Halvadakis, P. (2004). Olive-mill waste management: literature review and patent survey, 1st Ed; Typothito-George Dardanos Publications, Athens. ISBN 960-402-123-0

Palmarola-Adrados, B.; Galbe, M. & Zacchi, G. (2004). Combined steam pretreatment and enzymatic hydrolysis of starch-free wheat fiber. *Applied Biochemistry and Biotechnology*, Vol.113, No.1-3, (Spring 2004), pp. 989-1002, ISSN 0885-4513

Pérez-Camino, M. C. & Cert, A. (1999). Quantitative determination of hydroxy pentacyclic triterpene acids in vegetable oils. *Journal of Agricultural and Food Chemistry*, Vol.47, No.4, (April 1999), pp. 1558-1562, ISSN 0021-8561

Pérez-Jiménez, F.; Ruano, J.; Pérez-Martínez, P.; López-Segura, F. & López-Miranda, J. (2007). The influence of olive oil on human health: not a question of fat alone. *Molecular Nutrition and Food Research*, Vol.51, No.10, (October 2007), pp. 1199-1208, ISSN 1613-4125

Perona, J. S.; Arcemis, C.; Ruiz-Gutiérrez, V. & Catalá, A. (2005). Effect of dietary high-oleic-acid oils that are rich in antioxidants on microsomal lipid peroxidation in rats. *Journal of Agriculture and Food Chemistry*, Vol.53, No.3, (February 2005), pp. 730-735, ISSN 0021-8561

Quiles, J. L.; Ramírez-Tortosa, M. C.; Gómez, J. A.; Huertas, J. R.; & Mataix, J. (2002). Role of vitamin E and phenolic compounds in the antioxidant capacity; measured by ESR; of virgin olive; olive and sunflower oils after frying. Food Chemistry, Vol.76, No.4, (April 2002), pp. 461–468, ISSN 0308-8146

Rao, C. V.; Newmark, H. L. & Reddy, B. S. (1998). Chemopreventive effect of squalene on colon cancer. *Carcinogenesis*, Vol.19, No.2, (March 1998), pp. 287-290, ISSN 0143-3334

Rodríguez, G.; Rodríguez, R.; Guillén, R.; Jiménez, A. & Fernández-Bolaños, J. (2007a). Effect of steam treatment of alperujo on the enzymatic saccharification and in vitro digestibility. *Journal of Agriculture and Food Chemistry*, Vol.55, No.1, (January 2007), pp. 136-142, ISSN 0021-8561

Rodríguez, G.; Rodríguez, R.; Fernández-Bolaños, J.; Guillén, R.; & Jiménez, A. (2007b). Antioxidant activity of effluents during the purification of hydroxytyrosol and 3,4-dihydroxyphenyl glycol from olive oil waste. *European Food Research and Technology*, Vol.224, No.6, (April 2007), pp. 733-741, ISSN 1438-2377

Rodríguez, G.; Lama, A.; Rodríguez, R.; Jiménez, A.; Guillén, R. & Fernández-Bolaños, J. (2008). Olive stone an attractive source of bioactive and valuable compounds. *Bioresource Technology*. Vol.99, No.13, (September 2008), pp. 5261-5269, ISSN 0960-8524

Ruiz-Gutiérrez, V.; Sanchez-Perona, J. & Osada, J. (2009). Using refined oil from olive pressings for retarding development of atherosclerosis; reduces levels of triglycerides and specific lipoproteins; and the number of leucocytes carrying the Mac-1 integrin , patent No. WO2005092354-A1

Ruiz-Méndez, M.V.; López-López, A. & Garrido-Fernández, A. (2008). Characterization and chemometric study of crude and refined oils from table olive by-products. *European Journal of Lipid Science and Technology*, Vol.110, No.6, (June 2008), pp. 537-546, ISSN 1438-7697

Taylor, J. C.; Rapport, L. & Lockwood, G. B. (2003). Octacosanol in human health. *Nutrition*, Vol.19, No.2, (February 2003), pp. 192-195, ISSN 0899-9007

Sáenz, M.T.; García, M.D.; Ahumada, M.C. & Ruiz, V. (1998). Cytostatic activity of some compounds from the unsaponifiable fraction obtained from virgin olive oil. *Il Farmaco*, Vol.53, No.6, (June 1998), pp. 448-449, ISSN 0014-827X

Sánchez, P.; Ruiz, M.V. (2006). Production of pomace olive oil. *Grasas y Aceites*, Vol.57, No.1, (March 2006), pp. 47-55, ISSN 0017-3495

Singh, D. K.; Li, L. & Porter, T. D. (2006). Policosanol inhibits cholesterol synthesis in hepatoma cells by activation of AMP-kinase. *Journal of Pharmacology and Experimental Therapeutics*, Vol.318, No.6, (May 2006), pp. 1020-1026, ISSN 0022-3565

Valavanidis, A.; Nisiotou, C.; Papageorghiou, Y.; Kremli, I.; Satravelas, N.; Zinieris, N. & Zygalaki, H. (2004). Comparison of the radical scavenging potential of polar and lipidic fractions of olive oil and other vegetable oils under normal conditions and after thermal treatment. *Journal of Agriculture and Food Chemistry*, Vol.52, No.8, (April 2004), pp. 2358-2365, ISSN 0021-8561

Vlyssides, A.G.; Loizides, M. & Karlis, P.K. (2004). Integrated strategic approach for reusing olive oil extraction. *Journal of Cleaner Production*, Vol.12, pp. 603-611, ISSN 0959-6526

2

Olive Oil Mill Waste Treatment: Improving the Sustainability of the Olive Oil Industry with Anaerobic Digestion Technology

Bárbara Rincón, Fernando G. Fermoso and Rafael Borja

Instituto de la Grasa (Consejo Superior de Investigaciones Científicas), Sevilla, Spain

1. Introduction

The processes used to treat waste streams are chosen according to technical feasibility, simplicity, economics, societal needs and political priorities. However, the needs and priorities of a sustainable society undergo pressure which means a shift in the focus of wastewater treatment from pollution control to resource exploitation (Angenent et al., 2004). The organic fraction of agro-wastes (e.g. olive oil wastes, sugar beet pulp, potato pulp, potato thick stillage or brewer´s grains) has been recognized as a valuable resource that can be converted into useful products via microbially mediated transformations. Organic waste can be treated in various ways, of which bio-processing strategies resulting in the production of bioenergy (methane, hydrogen, electricity) are promising (Khalid et al., 2011). The aim of the present chapter is to discuss: firstly the quantities, characteristics and current treatments of the solid wastes and wastewaters from the continuous olive oil extraction industry for their exploitation and recovery; secondly, anaerobic digestion processes as an alternative for waste treatment and valuable energy recovery

2. Olive mill extraction wastes

2.1 Olive oil extraction technology

The olive mill elaboration system has evolved over time for economic and environmental reasons. The traditional system or "pressing system" was replaced in the 70s by the three-phase continuous centrifugation system. The centrifugation of the milled and beaten olives to obtain olive oil by the three-phase system produces 20% oil, 50% three-phase olive mill wastewaters (3POMWW) and 30% three-phase olive mill solid wastes (3POMSW). This three-phase system led to an increase in the processing capacity and consequently to an increase in the yield of the mills and the growth of the average mill size. However, large quantities of water needed for carrying out the three-phase process generate a high volume of olive mill wastewaters. The uncontrolled discharge of 3POMWW brought environmental problems. In some countries, technology manufacturers developed the "ecological" two-phase process. This system enables reduced fresh water consumption in the centrifugation phase. The two-phase process has attracted special interest in countries where water supplies are restricted. The quantity of water required to carry out the two-phase process is much lower than in the three-phase process and a considerable reduction in generated two-phase olive mill wastewaters (2POMWW) is achieved. However, the two-phase process led to a slight increase in solid wastes. The quantities of two-phase olive mill solid wastes (2POMSW) are 60% higher than those generated in the three-phase system (3POMSW).

Over 2.9 million tonnes of virgin olive oil are produced annually worldwide, of which 2.4-2.6 million tonnes are produced in the European Union (IOOC, 2009). Currently, both elaboration systems, three- and two-phase, coexist in the Mediterranean area (Niaounakis & Halvadakis, 2004). Spain, the largest producer of olive oil in the world, currently uses the two-phase system in 98% of its olive mills. Over the past few years, Spain has produced between 1,412,000 tonnes (2003/2004 season) and 1,028,000 tonnes (2008/2009 season) of olive oil, which meant 57.7% and 53% of European production (IOOC, 2009). Croatia uses the two-phase system in 55% of its mills and produces 4,500 tonnes of olive oil (2008/2009 season). In olive oil producing countries such as Cyprus, Portugal and Italy, only around 5% of the mills use the two-phase system (Roig et al., 2006). Other large producers such as Greece or Malta have continued using mainly the three-phase system although the two-phase system is being introduced slowly. The high quantities of wastes produced in both systems makes sustainable treatments necessary.

2.2 Waste quantities and characteristics

3POMSW are produced in a proportion of 500 kg per ton of olives and are basically made up of dry pulp and stones.

3POMWW are the main wastes generated in the three-phase olive mill system (1,200 L ton^{-1} milled olives). The annual 3POMWW production of Mediterranean olive-growing countries is estimated between 7-30 million m^3. The chemical composition of 3POMWW is complex due to the water from the milled olives (vegetation water) and the soft tissues from the olive fruit. Typical composition of three-phase olive mill wastewaters is: pH 5.04, COD 43.0 g L^{-1} (COD: Chemical Oxigen Demand), total sugars 17.4 g L^{-1}, total phenols 2.5 g L^{-1} and lipids 0.75 g L^{-1} (D'Annibale et al., 2006).

The change from the three-phase to the two-phase elaboration system reduces the high generation of wastewaters produced in the three-phase process. The two-phase elaboration process generates 800 kg of 2POMSW per ton of olives processed. 2POMSW have a 60%-70% humidity content, 13%-15% lignin, 18%-20% cellulose and hemicellulose and 2.5%-3% oil (Borja et al., 2002). In a similar way to 3POMWW, the composition of 2POMSW is complex due to the vegetation water. Consequently, 2POMSW and 3POMWW are the main problematic streams.

2POMWW are a mixture between the water used for olive washing before the milling process and the water coming from washing the oil in a vertical centrifuge. Initial studies gave volumes of 2POMWW of around 250 L ton^{-1} of olives in total. However, current studies have suggested a significant reduction in the amount of water to be added to the vertical centrifuge.

3. Olive oil mill wastes - Current treatments

3.1 Three-phase olive mill wastes

3.1.1 Three-phase olive mill solid waste (3POMSW)

The traditional use of 3POMSW is to extract the residual olive oil. 3POMSW have around 4% residual oil on wet basis, which can be extracted by mechanical and/or chemical treatments. There are several extraction methods, the most usual being a first centrifugation where 40%-50% of the residual oil is extracted (Sánchez & Ruiz, 2006), followed by a drying process from 60-70% to 8% moisture (400°C-800°C) and extraction with solvents (hexane). Finally, the oil extracted 3POMSW is used for co-generation of heat and electricity in combustion-turbine cycles or a gas-turbine cycle. Oil extraction factories usually use this energy for their own drying process before extraction.

3.1.2 Three-phase olive mill wastewater (3POMWW)

Due to their high organic load and problematic disposal, the depuration of 3POMWW has been the subject of a great number of studies over the years. Initial treatments in the 60s focused on the use of 3POMWW as a soil conditioner if previously neutralized with lime (Albi Romero & Fiestas Ros de Ursinos, 1960). The addition of 3POMWW to the soil seems beneficial, as it produces an increase in nitrogen-fixing organisms (Garcia-Barrionuevo et al., 1992).

3POMWW are a potential source of biophenols, some being studied for potential industrial exploitation (Cardoso et al., 2011). The extraction of polyphenols provides a double opportunity to obtain high added value biomolecules and to reduce the phytotoxicity of the effluent (Bertín et al., 2011). López and Ramos-Cormenzana (1996) showed the possibility of obtaining 4.4 g L^{-1} of Xanthan with 3POMWW diluted to 30%-40%. The 3POMWW have also been studied as a source of natural pigments (anthocyanins) and different exopolysaccharides, and as a growth medium for algae (Ramos-Cormenzana et al., 1995). 3POMWW have been used as a growth media for the microbial production of extra-cellular lipase (D'Annibale et al., 2006) and for composting with olive leaves (Michailides et al., 2011).

Most of these studies, although very interesting, do not solve the problem because the quantities required for these studies are very small in contrast to the high quantity generated annually. The final destination of these wastewaters is mainly evaporation ponds. In the Mediterranean countries the summers are very hot, the evaporation ponds are large pools built with waterproof materials where the wastewaters can be stored for their evaporation in the summer period. After solar drying, the remaining solids can be used as fertilizer (Rozzi & Malpei, 1996). Although the evaporation ponds are very simple constructions, failure in the insulation of the basin can contaminate the ground water. Another disadvantage of these ponds is the production of putrid odors and insects during the decomposition processes (Khoufi et al., 2006).

3.2 Two-phase olive mill wastes

3.2.1 Two-phase olive mill solid wastes (2POMSW)

2POMSW have around 3.5% residual oil in wet basis. Like 3POMSW, this waste is also used for residual olive oil extraction. However, the humidity of 2POMSW is higher than

3POMSW. In order to obtain 8% humidity before extraction, the intensity and the length of the drying process are higher for 2POMSW than for 3POMSW. Furthermore, the vegetation water fraction of the olives gives 2POMSW a complex composition generating a high number of problems during the drying process. The high concentration in reducing sugars gives 2POMSW a doughy consistency in the continuous rotary dryer. This consistency creates dead areas which cannot be dried properly in the drying place, making residual oil extraction more difficult. Although the extraction process is more expensive and less profitable for 2POMSW than 3POMSW, this residual oil extraction is still applied. The extracted 2POMSW have 30%-45% stones, 15%-30% olive skin and 30%-50% pulp (Cruz et al., 2006). They are used for the co-generation of heat and electricity in combustion-turbine cycles or a gas-turbine cycle in the same way as 3POMSW. The oil extraction factory usually uses this type of energy for its own drying process before extraction.

Composts of 2POMSW is another alternative. The initial 2POMSW is phytotoxic, but Alburquerque et al. (2006) found the mixture with grape stalk and olive leaves as bulking agents free of phytotoxicity and suitable as soil conditioners.

Currently there are several experimental treatments for 2POMSW using it as a source of pharmaceutical compounds. A new process based on the hydrothermal treatment of 2POMSW led to a final solid enriched in minor components with functional activities (Lama-Muñoz et al., 2011). Other studies have been carried out using the bacteria *Penibacillus Jamila* for the production of exo-polysaccharides with 2POMSW as growth media (Ramos-Cormenzana & Monteoliva-Sánchez, 2000). There are two patented products extracted from 2POMSW: oleanoic acid and maslinic acid. Maslinic acid is being used for a treatment against the human immunodeficiency virus (HIV-1) (Parra et al., 2009). The walls of the olives are rich in polysaccharides such as L-arabinose. These polysaccharides are part of 2POMSW and can also be extracted and exploited (Cardoso et al., 2003).

2POMSW have also been used as feeding for animals. There are several studies about the digestibility of the protein content in 2POMSW used as sheep and goat feed (Martín et al., 2003; Molina Alcaide et al., 2003). Maslinic acid obtained from 2POMSW added to the diet of rainbow trout increased growth and protein-turnover rates (Fernández-Navarro et al., 2008).

The application of 2POMSW as a fertilizer has also been considered. Although the vegetation water gives a phytotoxic effect similar to 3POMWW, it has been observed that the fertilizer effect prevails over the phytotoxic effect when the dosage is not very high (Sierra et al., 2000).

An extremely low quantity of 2POMSW is used in these treatments, so none could be used as an integral treatment for this problematic waste.

3.2.2 Two-phase olive mill wastewaters (2POMWW)

Different options have been studied for the treatment of the wastewaters generated during the purification of olive oil. The use of oxidative methods for the treatment of 2POMWW has been reported in literature (Martínez-Nieto et al., 2011). These methods are based on the use of chemical oxidants such as permanganate, hydrogen peroxide (H_2O_2) or Fenton-like reaction. Aerobic treatment using a completely mixed activated sludge reactor was also reported (Borja et al., 1995a). The results obtained with the aerobic treatment indicated that more than 93% of the input COD concentration can be removed. The most commonly used treatment of both 2POMWW and 3POMWW is storage in evaporation ponds (section 3.1.2).

4. Anaerobic treatments

Anaerobic wastewater treatment has evolved into a competitive treatment technology in the past few decades. Many different types of organically polluted wastewaters, even those that were previously believed not to be suitable for anaerobic wastewater treatment, are now treated by anaerobic high-rate conversion processes (Van Lier, 2008).

Similar to anaerobic wastewater treatment, since the introduction of anaerobic digestion of solid waste in the beginning of the 1990s, adoption of the technology has been increasing (De Baere & Mattheeuws, 2010). European energy output from solid waste digestion plants rose to 5.3 Mtoe in 2009, which is 236 ktoe more than in 2008 (EurObserv´ER, 2010).

This section focuses on the principles of bioenergy production through anaerobic processes. Methanogenic anaerobic digestion (methane), biological hydrogen production (hydrogen) and microbial fuel cell technology (electricity) will be explained and discussed.

4.1 Methanogenic anaerobic digestion

The anaerobic digestion process is a biological process carried out by three different groups of microorganisms (hydrolytics, acetogenics and methanogenics) (Gujer & Zehnder, 1983)) which transform organic matter to obtain 90% biogas [a mixture CH_4/CO_2 ($\approx65\%$-35%)] and only 10% excess sludge. Biogas has a high calorific value (5000-6000 kcal m^{-3}) and can be used for electricity or heat.

The main advantages of the anaerobic process compared with other types of treatment are (Van Lier, 2007):

- High applicable chemical oxygen demand (COD).
- No use of fossil fuels for treatment
- No use of or very little use of chemicals; simple technology (Figure 1) with high treatment efficiencies.
- Anaerobic sludge can be stored unfed, so the reactors can be operated during the harvesting seasons only (e.g. 4 months per year in the olive oil mill industry).
- Effective pathogen removal.
- High degree of compliance with many national waste strategies implemented to reduce the amount of biodegradable waste entering landfills.
- The slurry produced (digestate) is an improved fertiliser.

Methanogenic anaerobic digestion of organic material has been performed for about a century. Therefore, the food web of anaerobic digestion is reasonably well understood (Figure 2).

Anaerobic digestion of biodegradable wastes involves a large spectrum of bacteria of which three main groups can be distinguished. The first group comprises fermenting bacteria which perform **hydrolysis and acidogenesis** (e.g. *Clostridium butyricum*, *Propionibacterium*). This involves the action of exo-enzymes to hydrolyze matter such as proteins, fats and carbohydrates into smaller units which can then enter the cells to undergo an oxidation-reduction process resulting in the formation of volatile fatty acids (VFA) and some carbon dioxide and hydrogen. The fermenting bacteria are usually designated as an acidifying or acidogenic population because they produce VFA.

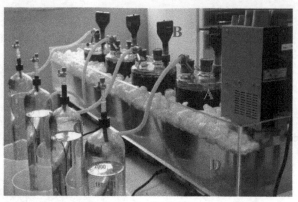

Fig. 1. Experimental laboratory scale anaerobic reactors (A: reactor, B: pH-meter, C: gasometer with NaOH for methane measurement, D: water bath for temperature control).

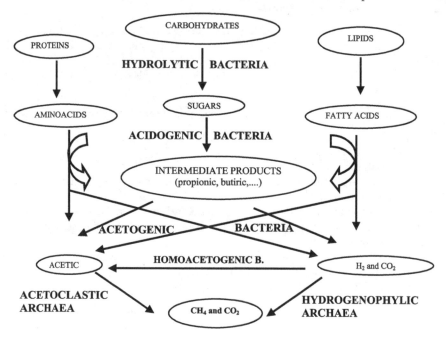

Fig. 2. Diagram of the different steps of anaerobic digestion

Acetogenic bacteria (e.g. *Clostridium aceticum, Acetobacterium woodii)* constitute the second group and are responsible for breaking down the products of the acidification step to acetic acid. In addition, hydrogen and carbon dioxide are also produced during acetogenesis.

The third group involves methanogenic Archaea (e.g. *Methanobrevibacter smithii, Methanobacterium thermoautotrophicum, Methanosarcina barkerii, Methanotrix soebugenii)* convert acetic acid or carbon dioxide and hydrogen into methane. Other possible methanogenic substrates such as formic acid, methanol, carbon monoxide, and methylamines are of minor importance in most anaerobic digestion processes. In addition to these three main groups, hydrogen consuming acetogenic bacteria are always present in small numbers in an anaerobic digester. They produce acetic acid from carbon dioxide and hydrogen and, therefore, compete for hydrogen with the methanogenic archaea.

The synthesis of propionic acid from acetic acid, as well as the production of longer chain VFA, occur to a limited extent in anaerobic digestion. Competition for hydrogen can also be expected from **sulfate reducing bacteria.**

In conventional applications, the acid- and methane-forming microorganisms are kept together inside the reactor system with a delicate balance between these two groups of organisms, because they differ greatly in terms of physiology, nutritional needs, growth kinetics and their sensitivity to environmental conditions. Problems encountered with stability and control in conventional design applications have led researchers to new solutions such as the physical separation of acid-formers and methane-formers in two separate reactors. Optimum environmental conditions for each group of organism is provided separately to enhance the overall process stability and control (Cha & Noike, 1997).

4.2 Biological hydrogen production

Hydrogen is a clean, recyclable, and efficient energy carrier. The possibility of converting hydrogen into electricity via fuel cells makes the application of hydrogen energy very promising (Chang et al., 2002).

Hydrogen production via dark fermentation is a special type of anaerobic digestion consisting of only hydrolysis and acidogenesis. It leads to the production of hydrogen, carbon dioxide and some simple organic compounds [VFA and alcohols]. These readily degradable organic compounds can be used for further methane production. (Bartacek et al., 2007)

Much interest has recently been expressed in the biological production of hydrogen from waste streams by dark fermentation. Biological hydrogen production shares many common features with methanogenic anaerobic digestion, especially the relative ease with which the two gaseous products can be separated from the treated waste.

From hydrogen-producing mixed cultures, a wide range of species have been isolated, more specifically from the genera Clostridium (*Clostridium pasteurianum, Clostridium saccharobutylicum, C. butyricum*), Enterobacter (*E. aerogenes*) and Bacillus under mesophilic conditions; and from the genera Thermoanaerobacterium (*Thermoanaerobacterium thermosaccharolyticum*), and Caldicellulosiruptor (*Caldicellulosiruptor saccharolyticus, C. thermocellum, Bacillus thermozeamaize*) under thermophilic or extremophilic conditions.

However, the low efficiency of the hydrogen production process remains the main limiting factor. Much research will be needed to be carried out to reach hydrogen yields comparable with the theoretical efficiency maximum. Although a relatively high efficiency has been reached using pure substrates, the low hydrogen yield with complex (real) substrates remains a great challenge.

4.3 Microbial fuel cell technology

The microbial fuel cell (MFC) is a new energy technology in which microorganisms produce electricity directly from renewable biodegradable materials (Logan et al., 2006). During microbial oxidation of biodegradable matter, not only are protons and oxidized products formed but electrons are remarkably transferred from the bacteria towards a solid electrode. The electrons flow through an electrical circuit towards the cathode where a final electron acceptor is reduced resulting in generation of electrical power (Figure 3).

Although interest in microbial fuel cells was relatively high in the 1960s, research has been limited as the cost of other energy sources remained low and the available microbial fuel cells lacked efficiency and long term stability. However, in the past seven to eight years there has been a resurgence in microbial fuel cell research. In fact, the efficiency of this energy conversion is potentially higher than the actual waste treatment technology for energy recovery, such as anaerobic digestion or incineration (Logan et al., 2006).

Microbial fuel cells have been validated at lab-scale with simple organic substrates, pure culture and highly controlled experimental conditions. Organic substrates as volatile fatty acids and more recently wastewater have generated high energy production (Catal et al., 2008; Clauwaert et al., 2007; Clauwaert et al., 2008; Rabaey et al., 2003; Rabaey & Verstraete, 2005). Over the last 10 years, the improvement in the design of microbial fuel cells has increased electrical generation 10,000 times (Debabov, 2008). However, full scale application has not yet been developed.

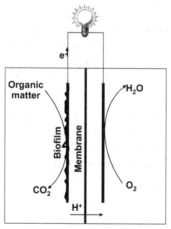

Fig. 3. Microbial Fuel Cell (MFC) set up

5. Anaerobic treatment of olive oil mill wastes

5.1 Three-phase olive mill wastes

5.1.1 Three-phase olive mill solid wastes (3POMSW)

The heterogeneous cellulosic and lignocellulosic structures of the husk make the anaerobic digestion of this waste impossible because the microorganisms are unable to attack these complex structures. Therefore, anaerobic digestion is not a suitable technology for the treatment of 3POMSW.

5.1.2 Three-phase olive mill wastewaters (3POMWW)

Anaerobic digestion is a promising alternative for the treatment of 3POMWW. It allows for the disposal of these wastewaters achieving considerable organic material removals and producing renewable energy in the form of biogas, which could be used as an energy source in the olive oil mill itself.

Certain components of 3POMWW such as poly-phenols, pH, oil, etc. may inhibit the AD process. Martín et al., (1991) obtained methane yields of 260 mL CH_4 g^{-1} COD for 3POMWW. Borja et al. (1995b) improved the methane production using a pre-treatment stage with *Geotrichum candidum*, *Azotobacter Chroococcum* and *Aspergillus terreus*. The latest study reported methane yield coefficients of 300 (*Geotrichum*-pretreated 3POMWW), 315 (*Azotobacter*-pretreated 3POMWW) and 350 (*Aspergillus*-pretreated 3POMWW) mL CH_4 g^{-1} COD against the 260 mL CH_4 g^{-1} COD obtained for the untreated 3POMWW.

3POMWW have a low nitrogen content which limits the AD process due to the fact that the microorganisms need this element for their metabolism. In this way, co-digestion with rich nitrogen substrates may improve the biodegradability of the mixture. Azbar et al., (2008) studied the co-digestion of 3POMWW with laying hen litter obtaining a significant improvement in the biodegradability of 3POMWW. Co-digestion with liquid cow manure [20% 3POMWW, 80% liquid cow manure (v:v)] also showed good results in terms of COD and volatile solids removal (Dareioti et al., 2010).

Another option is the combination of catalytically oxidized olive mill wastewaters (by Fenton's reagent) plus anaerobic digestion. El-Gohary et al. (2009) found that the digestion of catalytically oxidized 3POMWW followed by a classical upflow anaerobic sludge blanket reactor (UASB) and a hybrid UASB as a post-treatment step is a promising alternative.

Other treatments envisage the combination of an initial liquid-liquid extraction with ethyl acetate for exploitation of the phenol content, followed by aerobic or anaerobic digestion of the phenolic extracted 3POMWW (Khoufi et al., 2006).

The use of sand filtration and subsequent treatment with powered activated carbon in batch systems has also been studied as a pre-treatment. This pre-treatment allowed COD removal efficiencies of 80%-85% for an HRT of 5 days and at an OLR of 8 g COD L^{-1} d^{-1}. A methane yield of 300 mL biogas g^{-1} COD removed was achieved (Sabbah et al., 2004).

The separation of the digestion phases, hydrolytic-acidogenic reactor and methanogenic reactor, in two completely independent reactors can also be considered as a way to improve the AD digestion of these wastes. Bertín et al. (2010) studied different acidogenic configurations of biofilm reactors using ceramic filters or granular activated carbon with good results.

The latest research studies report AD as a promising technology for the treatment of 3POMWW, leading to sustainable waste treatment and an environmentally friendly solution.

5.2 Two-phase olive mill wastes

5.2.1 Two-phase olive mill solid wastes (2POMSW)

Borja et al. (2002) carried out an initial anaerobic digestibility study with four different dilutions of 2POMSW (20%, 40%, 60% and 80 %). The main findings showed that the performance of the reactor [in terms of COD removal (%)] is practically independent of the feed COD concentration. Studies with no-diluted 2POMSW were carried out by the same authors with similar results (Rincón et al., 2007). The methane yields obtained in these studies ranged between 200-300 mL CH_4 g^{-1} COD removed. The 2POMSW is easily biodegradable by mesophilic anaerobic digestion and COD removal efficiencies up to 97 % may be achieved (Rincón et al., 2007).

Rincón et al. (2006, 2008a) studied the different microorganisms participating in the 2POMSW anaerobic digestion. For the determination of the microorganism population polymerase chain reaction (PCR), denaturing gradient gel electrophoresis (DGGE), cloning and sequencing techniques were employed. The results showed differences in the microbial communities, both bacterial and archaeal, with varying OLRs. Analysis of the microbial communities may be decisive in understanding the microbial processes taking place during 2POMSW decomposition in anaerobic reactors and optimizing their performance. During this experimental study the most frequently encountered microbial group were the Firmicutes (53.3% of analyzed sequences), represented mostly by members of the Clostridiales (Figure 4). Chloroflexi also represented an important bacterial group in the study (23.4% of sequences) and has been reported as a major constituent in anaerobic systems (Rincón, 2006). The Gamma-Proteobacteria (8.5% sequences, represented mainly by the genus *Pseudomonas*), Actinobacteria (6.4%) and Bacteroidetes (4.3%) are also significant components of the microbial communities during the anaerobic decomposition of 2POMSW (Figure 4). The major archaeal component detected for the 2POMSW anaerobic digestion was *Methanosaeta concilii* (formerly *Methanothrix soehngenii*) (Figure 5). Furthermore, results showed the existence of molecular diversity within the genus *Methanosaeta* in the anaerobic process under study (Rincón et al., 2006).

As has been explained in section 4.1, the anaerobic digestion process could be more stable if the hydrolytic-acidogenic and the methanogenic stages were physically separated. The microorganisms participating in this kind of biological treatment (bacteria and methanogenic archaea) have different requirements in terms of growing kinetic, optimal working conditions and sensitivity to environmental conditions. Studies in two stages allow for the enrichment of the different populations of microorganisms (Cha & Noike, 1997). The separation in the hydrolytic-acidogenic and the methanogenic steps showed improved results as compared to one simple AD stage. The acidification of the 2POMSW in an initial hydrolytic-acidogenic step achieved a high concentration of total volatile fatty acids 14.5 g L^{-1} (expressed as acetic acid) at an OLR as high as 12.9 g COD L^{-1} d^{-1} (Rincón et al., 2008b). After this initial acidification, the OLRs achieved in the methanogenic reactor were in the order of 22.0 g COD L^{-1} d^{-1} with COD and volatile solid removals of 94.3%-61.3% and 92.8%-56.1%, respectively, for OLRs between 0.8 and 20.0 g COD L^{-1}d^{-1}. Methane yields of 268 mL CH_4 g^{-1} COD removed were achieved (Rincón et al., 2009, 2010).

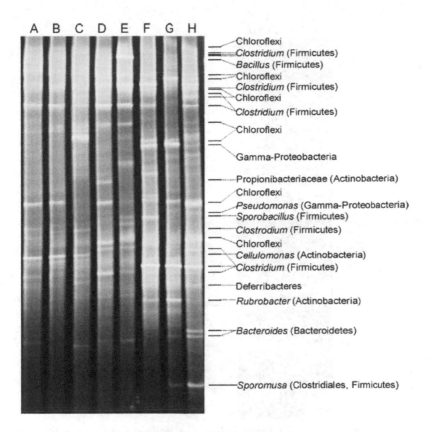

Fig. 4. DGGE analysis of the diversity of bacterial communities at different OLRs studied in one stage anaerobic digestion of 2POMSW (Rincón et al., 2008a). The position of the major electrophoretic bands corresponding to the 16S rRNA gene of the identified bacteria are indicated. A, B, C, D, E, F, G and H are the increasing OLRs studied in g COD L^{-1} d^{-1}: 2.3 (A), 3.0 (B), 4.5 (C), 5.8 (D), 6.8 (E), 8.3 (F), 9.2 (G) and 11.0 (H).

Other studies in two-stages at thermophilic scale reported 2POMSW as an ideal substrate for biohydrogen and methane production. These studies used diluted 2POMSW (1:4) with tap water achieving 18.5 ± 0.4 mmol CH$_4$ g^{-1} total solid added (TS). Experiments for biohydrogen production followed by methane production, generated 1.6 mmol H$_2$ g^{-1} TS added and 19.0 mmol CH$_4$ g^{-1} TS in the methanogenic stage (Gavala et al., 2005). Mesophilic bio-hydrogen production from 3POMSW has shown to be feasible at mesophilic temperature resulting in 2.8-4.5 mmol H$_2$ per gram of carbohydrates consumed in the reactor (Koutrouli et al., 2006). Methane production in these assays achieved a maximum value of 1.13 L CH$_4$ L^{-1} d^{-1} at 10 days of HRT. Hydrogen is a renewable energy source and one of the most attractive applications is the conversion of hydrogen to electricity via fuel cells (Koutrouli et al., 2006)

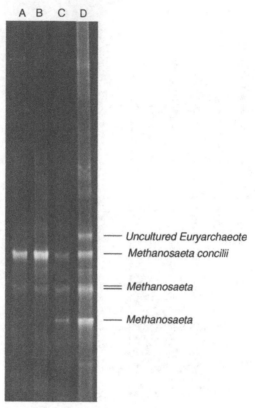

Fig. 5. DGGE analysis of the diversity of archaeal communities at different OLRs studied in one stage anaerobic digestion of 2POMSW (Rincón et al., 2006). The position of the major electrophoretic bands corresponding to the 16S rRNA gene of the identified archaea are indicated. A, B, C and D are the increasing OLRs studied in g COD L^{-1} d^{-1}): 0.75 (A), 1.5 (B), 2.25 (C) and 3.0 (D).

5.2.2 Two-phase olive mill wastewaters

Anaerobic treatment of this wastewater is very promising and beneficial. The production of biogas enables the process to generate or recover energy instead of just saving energy. This reduces operational costs as compared with other processes such as the physical, physicochemical or biological aerobic treatments previously mentioned (Wheatley, 1990). A kinetic study of the anaerobic digestion of wastewaters derived from the washing of virgin olive oil was previously reported (Borja et al., 1993). The study was carried out in a completely mixed reactor with biomass immobilized on sepiolite at a concentration of 10.8 g VSS L^{-1} operating at 35°C. COD removal efficiencies of more than 89% were achieved. Olive oil mills are usually small enterprises. Therefore, complex waste treatment systems are usually difficult to implement. Energy recovery from the generated wastewater with a single unit like the previously explained MFC (section 4.3) is very promising. Preliminary studies of the treatment of 2POMWW have been reported by Fermoso et al. (2011).

6. Conclusion

Three-phase olive mill wastewaters (3POMWW) and two-phase olive mill solid wastes (2POMSW) are the main wastes generated in the olive mill industry (1,200 L of 3POMWW per ton of milled olives and 800 kg of 2POMSW per ton of milled olives, respectively). The composition of 3POMWW and 2POMSW is very complex due to the vegetation water. Currently, the final destination of 3POMWW is mainly evaporation ponds and the final destination of 2POMSW is evaporation ponds and co-generation. Although the evaporation ponds are very simple constructions, failure in the insulation of the basin can contaminate the ground water and they generate putrid odors and insects during the decomposition processes. The co-generation processes have a high number of environmental disadvantages: nitrogen oxides production, emission of suspended ashes, etc.

Anaerobic digestion is already successfully used for many agro-industrial residues, such as sugar beet pulp, potato pulp, potato thick stillage or brewer´s grains. This technology allows an efficient solids stabilisation and energy recovery. Both 2POMSW and 3POMWW have been shown to be promising substrates for anaerobic digestion, however full scale application is not a reality yet.

7. Acknowledgment

The authors wish to express their gratitude to the Spanish "Ministerio de Educación y Ciencia" (Project REN 2001-0472/TECNO) and to the contracts JAE-Doc from "Junta para la Ampliación de Estudios del CSIC" co-financed by the European Social Funds.

8. References

Albi Romero, M. A. & Fiestas Ros De Ursinos, J. A. (1960). Estudio del alpechin para su aprovechamiento industrial. Ensayos efectuados para su posible utilización como fertilizante. *Grasas y Aceites*, Vol. 11, pp. 123-124, ISSN: 0017-3495

Alburquerque, J. A.; Gonzálvez, J.; García, D. & Cegarra, J. (2006). Effects of bulking agent on the composting of "alperujo", the solid by-product of the two-phase centrifugation method for olive oil extraction. *Process Biochemistry*, Vol. 41, pp. 127-132, ISSN: 1359-5113.

Angenent, L. T.; Karim, K.; Al-Dahhan, M. H.; Wrenn, B. A. & Dominguez-Espinosa, R. (2004). Production of bioenergy and biochemicals from industrial and agricultural wastewater. *Trends in Biotechnology*, Vol. 22, pp. 477-485, ISSN: 0167-7799

Azbar, N.; Keskin, T. & Yuruyen, A. (2008). Enhancement of biogas production from olive mill effluent (OME) by co-digestion. *Biomass and Bioenergy*, Vol. 32, pp. 1195-1201, ISSN: 0961-9534

Bartacek, J.; Zabranska, J. & Lens, P. N. L. (2007). Developments and constraints in fermentative hydrogen production. *Biofuels, Bioproducts and Biorefining*, Vol. 1, pp. 201-214, ISSN: 1932-104X

Bertín, L.; Lampis, S.; Todaro, D.; Scoma, A.; Vallini, G.; Marchetti, L.; Majone, M. & Fava, F. (2010). Anaerobic acidogenic digestion of olive mill wastewaters in biofilm reactors packed with ceramic filters or granular activated carbon. *Water Research*, Vol. 44, pp. 4537-4549, ISSN: 0043-1354

Bertín, L.; Ferri, F.; Scoma, A.; Marchetti, L. & Fava, F. (2011). Recovery of high added value natural polyphenols from actual olive mill wastewater through solid phase extraction. *Chemical Engineering Journal*, In Press, Corrected Proof, ISSN: 1385-8947

Borja, R.; Alba, J.; Martín, A.; Ruiz, A. & Hidalgo, F. (1993). Caracterización y digestión anaerobia de las aguas de lavado de aceite de oliva virgen. *Grasas y Aceites*, Vol. 44, pp. 85-90, ISSN: 0017-3495

Borja, R.; Alba, J. & Banks, C. J. (1995a). Activated sludge treatment of wash waters derived from the purification of virgin olive oil in a new manufacturing process. *Journal of Chemical Technology and Biotechnology*, Vol. 64, pp. 25-30, ISSN: 0268-2575

Borja, R.; Martín, A.; Alonso, V.; García, I. & Banks, C. J. (1995b). Influence of different aerobic pretreatments on the kinetics of anaerobic digestion of Olive Mill Wastewater. *Water Research*, Vol. 29, pp. 489-495, ISSN: 0043-1354

Borja, R.; Rincón, B.; Raposo, F.; Alba, J. & Martín, A. (2002). A study of anaerobic digestibility of two-phase olive mill solid waste (OMSW) at mesophilic temperature. *Process Biochemistry*, Vol. 38, pp. 733-742, ISSN: 1359-5113

Cardoso, S. M.; Coimbra, M. A. & Lopes Da Silva, J. A. (2003). Calcium-mediated gelation of an olive pomace pectic extract. *Carbohydrate Polymers*, Vol. 52, pp. 125-133, ISSN: 0144-8617

Cardoso, S. M.; Falcão, S. I.; Peres, A. M. & Domingues, M. R. M. (2011). Oleuropein/ligstroside isomers and their derivatives in Portuguese olive mill wastewaters. *Food Chemistry*, In Press, Corrected Proof, ISSN: 0308-8146

Catal, T.; Xu, S.; Li, K.; Bermek, H. & Liu, H. (2008). Electricity generation from polyalcohols in single-chamber microbial fuel cells. *Biosensors and Bioelectronics*, Vol. 24, pp. 849-854, ISSN: 0956-5663

Clauwaert, P.;Rabaey, K.;Aelterman, P.;De Schamphelaire, L.;Pham, T. H.;Boeckx, P.;Boon, N. & Verstraete, W. (2007). Biological denitrification in microbial fuel cells. *Environmental Science and Technology*, Vol. 41, pp. 3354-3360, ISSN: 0013-936X

Clauwaert, P.; Aelterman, P.; Pham, T. H.; De Schamphelaire, L.; Carballa, M.; Rabaey, K. & Verstraete, W. (2008). Minimizing losses in bio-electrochemical systems: The road to applications. *Applied Microbiology and Biotechnology*, Vol. 79, pp. 901-913, ISSN: 0175-7598

Cruz, P.;Palomar, J. M. & Ortega, A. (2006). Integral energy cycle for the olive oil sector in the province of Jaén (Spain). *Grasas y Aceites*, Vol. 57 (2), 219-228. ISSN: 0017-3495

Cha, G. C. & Noike, T. (1997). Effect of rapid temperature change and HRT on anaerobic acidogenesis. *Water Science and Technology*, Vol. 36, pp. 247-253, ISSN: 0273-1223

Chang, J.-S.; Lee, K.-S. & Lin, P.-J. (2002). Biohydrogen production with fixed-bed bioreactors. *International Journal of Hydrogen Energy*, Vol. 27, pp. 1167-1174, ISSN: 0360-3199

D'annibale, A.;Sermanni, G. G.;Federici, F. & Petruccioli, M. (2006). Olive-mill wastewaters: a promising substrate for microbial lipase production. *Bioresource Technology*, Vol. 97, pp. 1828-1833, ISSN: 0960-8524

Dareioti, M. A.; Dokianakis, S. N.; Stamatelatou, K.;Zafiri, C. & Kornaros, M. (2010). Exploitation of olive mill wastewater and liquid cow manure for biogas production. *Waste Management*, Vol. 30, pp. 1841-1848, ISSN: 0956-053X

Debabov, V. G. (2008). Electricity from microorganisms. *Microbiology*, Vol. 77, pp. 123-131, ISSN: 1350-0872

El-Gohary, F.; Tawfik, A.; Badawy, M. & El-Khateeb, M. A. (2009). Potentials of anaerobic treatment for catalytically oxidized olive mill wastewater (OMW). *Bioresource Technology*, Vol. 100, pp. 2147-2154, ISSN: 0960-8524

EurObserv'ER (2010). Biogas barometer. Retrieved from EurObserv'ER: http://www.eurobserv-er.org/pdf/baro200b.pdf

Fermoso, F. G.; Ter Heijne, A.; Rincón, B.; Borja, R. & Hamelers, H. V. M. (2011). Feasibility of olive oil mill wastewaters as electron donor in a boelectrical system. *The 8th Intrational IWA Syposium on Waste Management Problems in Agro-industries. (22-24 June 2011, Izmir, Turquey.*

Fernández-Navarro, M.; Peragón, J.; Amores, V.; De la Higera, M. & Lupiañez, J. A. (2008). Maslinic acid added to the diet increases growth and proteína-turnover rates in the White muscle of rainbow trour (*Oncorhynchus mykiss*). *Comparative Biochemistry and Physiology, Part C*, Vol. 147, pp. 158-167, ISSN: 1532-0456

Garcia-Barrionuevo, A.; Moreno, E.; Quevedo-Sarmiento, J.; Gonzalez-Lopez, J. & Ramos-Cormenzana, A. (1992). Effect of wastewaters from olive oil mills (alpechin) on Azotobacter nitrogen fixation in soil. *Soil Biology and Biochemistry*, Vol. 24, pp. 281-283, ISSN: 0038-0717

Gavala, H. N.;Skiadas, I. V.; Ahring, B. K. & Lyberatos, G. (2005). Potential for biohydrogen and methane production from olive pulp. *Water Science and Technology*, Vol. 52, pp. 209-215, ISSN: 0273-1223

Gujer, W. & Zehnder, A. J. B. (1983). Conversion processes in anaerobic digestion. *Water Science and Technology*, Vol. 15, pp. 127-167, ISSN: 0273-1223

IOOC (2009). (http://www.internationaloliveoil.org/downloads/production2_ang.PDF).

Khalid, A.; Arshad, M.; Anjum, M.; Mahmood, T. & Dawson, L. (2011). The anaerobic digestion of solid organic waste. *Waste Management*, Vol. 31, pp. 1737-1744, ISSN: 0956-053X

Khoufi, S.;Aloui, F. & Sayadi, S. (2006). Treatment of olive oil mill wastewater by combined process electro-Fenton reaction and anaerobic digestion. *Water Research*, Vol. 40, pp. 2007-2016, ISSN: 0043-1354

Koutrouli, E. C.; Gavala, H. N.; Skiadas, I. V. & Lyberatos, G. (2006). Mesophilic biohydrogen production from olive pulp. *Process Safety and Environmental Protection,* Vol. 84, pp. 285-289, ISSN: 0957-5820

Lama-Muñoz, A.;Rodríguez-Gutiérrez, G.;Rubio-Senent, F.;Gómez-Carretero, A.;Fernández-Bolaños, J., (2011). New hydrothermal treatment of alperujo enhances the content of bioactive minor components in crude pomace olive oil. Journal of Agricultural and Food Chemistry, Vol. 59, pp. 1115-1123, ISSN: 0021-8561

Logan, B. E.;Hamelers, B.; Rozendal, R.; Schroder, U.; Keller, J.; Freguia, S.; Aelterman, P.; Verstraete, W. & Rabaey, K. (2006). Microbial fuel cells: Methodology and technology. *Environmental Science and Technology,* Vol. 40, pp. 5181-5192, ISSN: 0013-936X

López, M. J. & Ramos-Cormenzana, A. (1996). Xanthan production from olive-mill wastewaters. *International Biodeterioration & Biodegradation,* Vol. 38, pp. 263-270, ISSN: 0964-8305

De Baere, L. & Mattheeuws, B. (2010). Anaerobic digestion of MSW in Europe. *Biocycle,* Vol., pp. 24-26, ISSN: 0276-5055

Martín, A.; Borja, R.; García, I. & Fiestas, J. A. (1991). Kinetics of methane production from olive mill wastewater. *Process Biochemistry,* Vol. 26, pp. 101-107, ISSN: 1359-5113

Martín, A. I.; Moumen, A.; Yánez, D. R. & Molina, E. (2003). Chemical composition and nutrients availability for goat as and sheep of two-stage olive cake and olive leaves. *Animal Feed Science and Technology,* Vol. 107, pp. 61-74, ISSN: 0377-8401

Martínez-Nieto, L.; Hodaifa, G.; Rodríguez, S.; Giménez, J. A. & Ochando, J. (2011). Degradation of organic matter in olive-oil mill wastewater through homogeneous Fenton-like reaction. *Chemical Engineering Journal,* In Press, Corrected Proof, ISSN: 1385-8947

Michailides, M.; Christou, G.; Akratos, C. S.; Tekerlekopoulou, A. G. & Vayenas, D. V. (2011). Composting of olive leaves and pomace from a three-phase olive mill plant. *International Biodeterioration & Biodegradation,* Vol. 65, pp. 560-564, ISSN: 0964-8305

Molina Alcaide, E.; Yáñez Ruiz, D. R.; Moumen, A. & Martín García, I. (2003). Chemical composition and nitrogen availability for goats and sheep of some olive by-products. *Small Ruminant Research,* Vol. 49, pp. 329-336, ISSN: 0921-4488

Niaounakis, M. & Halvadakis, C. P. (2004). Olive-Mill Waste Management: Literature review and Patent Survey. *Typothito – George Dardanos Publications, Athens, Greece.*

Parra, A.; Rivas, F.; Lopez, P.E.; Garcia-Granados, A.; Martinez, A.; Albericio, F.; Marquez, N. & Muñoz, E. (2009). Solution- and solid-phase synthesis and anti-HIV activity of maslinic acid derivatives containing amino acids and peptides. *Bioorganic & Medicinal Chemistry.* Vol. 17, pp. 1139-1145, ISSN: 0968-0896

Rabaey, K.; Lissens, G.; Siciliano, S. D. & Verstraete, W. (2003). A microbial fuel cell capable of converting glucose to electricity at high rate and efficiency. *Biotechnology Letters,* Vol. 25, pp. 1531-1535, ISSN: 0141-5492

Rabaey, K. & Verstraete, W. (2005). Microbial fuel cells: Novel biotechnology for energy generation. *Trends in Biotechnology,* Vol. 23, pp. 291-298, ISSN: 0167-7799

Ramos-Cormenzana, A.; Monteoliva-Sanchez, M. & Lopez, M. J. (1995). Bioremediation of Alpechin. *International Biodeterioration and Biodegradation*, Vol. 35, pp. 249-268, ISSN: 0964-8305

Ramos-Cormenzana, A. & Monteoliva-Sánchez, M. (2000). Potencial biofarmacéutico de los residuos de la industria oleícola. *Ars Pharmaceutica*, Vol. 41, pp. 129-136, ISSN: 0004-2927

Rincón, B.; Raposo, F.; Borja, R.;Gonzalez, J. M.; Portillo, M. C. & Saiz-Jimenez, C. (2006). Performance and microbial communities of a continuous stirred tank anaerobic reactor treating two-phases olive mill solid wastes at low organic loading rates. *Journal of Biotechnology*, Vol. 121, pp. 534-543, ISSN: 0168-1656

Rincón, B.;Travieso, L.;Sánchez, E.;De Los Ángeles Martín, M.;Martín, A.;Raposo, F. & Borja, R. (2007). The effect of organic loading rate on the anaerobic digestion of two-phase olive mill solid residue derived from fruits with low ripening index. *Journal of Chemical Technology and Biotechnology*, Vol. 82, pp. 259-266, ISSN: 0268-2575

Rincón, B.; Borja, R.; González, J. M.; Portillo, M. C. & Sáiz-Jiménez, C. (2008a). Influence of organic loading rate and hydraulic retention time on the performance, stability and microbial communities of one-stage anaerobic digestion of two-phase olive mill solid residue. *Biochemical Engineering Journal*, Vol. 40, pp. 253-261, ISSN: 1369-703X

Rincón, B.; Sánchez, E.; Raposo, F.; Borja, R.; Travieso, L.; Martín, M. A. & Martín, A. (2008b). Effect of the organic loading rate on the performance of anaerobic acidogenic fermentation of two-phase olive mill solid residue. *Waste Management*, Vol. 28, pp. 870-877, ISSN: 0956-053X

Rincón, B.; Borja, R.; Martín, M. A. & Martín, A. (2009). Evaluation of the methanogenic step of a two-stage anaerobic digestion process of acidified olive mill solid residue from a previous hydrolytic-acidogenic step. *Waste Management*, Vol. 29, pp. 2566-2573, ISSN: 0956-053X

Rincón, B.; Borja, R.; Martín, M. A. & Martín, A. (2010). Kinetic study of the methanogenic step of a two-stage anaerobic digestion process treating olive mill solid residue. *Chemical Engineering Journal*, Vol. 160, pp. 215-219, ISSN: 1385-8947

Roig, A.; Cayuela, M. L. & Sánchez-Monedero, M. A. (2006). An overview on olive mill wastes and their valorisation methods. *Waste Management*, Vol. 26, pp. 960-969, ISSN: 0956-053X

Rozzi, A. & Malpei, F. (1996). Treatment and disposal of olive mill effluents. *International Biodeterioration & Biodegradation*, Vol. 38, pp. 135-144, ISSN: 0964-8305

Sabbah, I.; Marsook, T. & Basheer, S. (2004). The effect of pretreatment on anaerobic activity of olive mill wastewater using batch and continuous systems. *Process Biochemistry*, Vol. 39, pp. 1947-1951, ISSN: 1359-5113

Sánchez, P. & Ruiz, M. V. (2006). Production of pomace olive oil. *Grasas y Aceites*, Vol. Enero-Marzo, 47-55, ISSN: 0017-3495

Sierra, J.; Martí, E.; Garau, M.A. & Crueñas, R. (2007). Effects of the agronomic use of olive oil mill wastewater: Field experiment. *Science of the Total Environment*, Vol. 378, pp. 90-94, ISSN: 0048-9697

Van Lier, J. B. (2007) Current and future trends in anaerobic digestion: diversifying from waste(water) treatment to resource oriented conversion techniques. *In: Proc. of the 11th IWA-International Conference on Anaerobic Digestion.*

Van Lier, J. B. (2008). High-rate anaerobic wastewater treatment: Diversifying from end-of-the-pipe treatment to resource-oriented conversion techniques. *Water Science and Technology,* Vol. 57, pp. 1137-1148, ISSN: 0273-1223

Wheatley, A. (1990) *Anaerobic digestion: a waste treatment technology,* Elsevier, ISBN: 1851665269 ,London, United Kingdom.

3

Microbial Biotechnology in Olive Oil Industry

Farshad Darvishi

Department of Microbiology, Faculty of Science, University of Maragheh,
Iran

1. Introduction

Microbial biotechnology is defined as any technological application that uses microbiological systems, microbial organisms, or derivatives thereof, to make or modify products or processes for specific use (Okafor 2007). Current agricultural and industrial practices have led to the generation of large amounts of various low-value or negative cost crude wastes, which are difficult to treat and valorize. Production of agro-industrial waste pollutants has become a major problem for many industries. The olive oil industry generates large amounts of olive mill wastes (OMWs) as by-products that are harmful to the environment (Roig et al. 2006).

However, OMWs have simple and complex carbohydrates that represent a possible carbon resource for fermentation processes. In addition, OMWs generally contain variable quantities of residual oil, the amount of which mainly depends on the extraction process (D'Annibale et al. 2006). Therefore, OMWs could be used as substrate for the synthesis of biotechnological high-value metabolites that their utilization in this manner may help solve pollution problems (Mafakher et al. 2010).

The fermentation of fatty low-value renewable carbon sources like OMWs to production of various added-value metabolites such as lipases, organic acids, microbial biopolymers and lipids, single cell oil , single cell proteins and biosurfactants is very interesting in the sector of industrial microbiology and microbial biotechnology (Darvishi et al. 2009). Thus, more research is needed on the development of new bioremediation technologies and strategies of OMWs, as well as the valorisation by microbial biotechnology (Morillo et al. 2009).

Few investigations dealing with the development of value-added products from these low cost materials, especially OMWs have been conducted. This chapter discusses olive oil microbiology, the most significant recent advances in the various types of biological treatment of OMWs and derived added-value microbial products.

2. Olive oil microbiology

In applied microbiology, specific microorganisms employed to remove environmental pollutants or industrial productions have often been isolated from specific sites. For example, when attempting to isolate an organism that can degrade or detoxify a specific target compound like OMW, sites may be sampled that are known to be contaminated by this material. These environments provide suitable conditions to metabolize this compound by microorganisms.

Recent microbiological research has demonstrated the presence of a rich microflora in the suspended fraction of freshly produced olive oil. The microorganisms found in the oil derive from the olives' carposphere which, during the crushing of the olives, migrate into the oil together with the solid particles of the fruit and micro-drops of vegetation water. Having made their way to the new habitat, some microbic forms succumb in a brief period of time whereas others, depending on the chemical composition of the oil, reproduce in a selective way and the typical microflora of each oil (Zullo et al. 2010).

Newly produced olive oil contains numerous solid particles and micro-drops of olive vegetation water containing, trapped within, a high number of microorganisms that remain during the entire period of olive oil preservation. The microbiological analyses highlighted the presence of yeasts, but not of bacteria and moulds (Ciafardini and Zullo 2002). Several isolated genus of yeasts were identified as *Saccharomyces, Candida* and *Williopsis* (Ciafardini et al. 2006).

Some types of newly produced oil are very bitter since they are rich in the bitter-tasting secoiridoid compound known as oleuropein, whereas after a few months preservation, the bitter taste completely disappears following the hydrolysis of the oleuropein. In fact, the taste and the antioxidant capacity of the oil can be improved by the β-glucosidase-producing yeasts, capable of hydrolysing the oleuropein into simpler and less bitter compounds characterized by a high antioxidant activity. Oleuropein present in olive oil can be hydrolysed by β-glucosidase from the yeasts *Saccharomyces cerevisiae* and *Candida wickerhamii*. The absence of lipases in the isolated *S. cerevisiae* and *C. wickerhamii* examined that the yeasts contribute in a positive way to the improvement of the organoleptic characteristics of the oil without altering the composition of the triglycerides (Ciafardini and Zullo 2002).

On the other hand, the presence of some lipase-producing yeast can worsen oil quality through triglycerides hydrolysis. Two lipase-producing yeast strains *Saccharomyces cerevisiae* 1525 and *Williopsis californica* 1639 were found to be able to hydrolyse olive oil triglycerides. The lipase activity in *S. cerevisiae* 1525 was confined to the whole cells as cell-bound lipase, whereas in *W. californica* 1639, it was detected as extracellular lipase. Furthermore, the free fatty acids of olive oil proved to be good inducers of lipase activity in both yeasts. The microbiological analysis carried out on commercial extra virgin olive oil demonstrated that the presence of lipase-producing yeast varied from zero to 56% of the total yeasts detected (Ciafardini et al. 2006).

Some dimorphic species can also be found among the unwanted yeasts present in the olive oil, considered to be opportunistic pathogens to man as they have often been isolated from immunocompromised hospital patients. Recent studies demonstrate that the presence of dimorphic yeast forms in 26% of the commercial extra virgin olive oil originating from different geographical areas, where the dimorphic yeasts are represented by 3-99.5% of the total yeasts. The classified isolates belonged to the opportunistic pathogen species *Candida parapsilosis* and *Candida guilliermondii*, while among the dimorphic yeasts considered not pathogenic to man, the *Candida diddensiae* species (Koidis et al. 2008; Zullo and Ciafardini 2008; Zullo et al. 2010).

Overall, these findings show that yeasts are able to contribute in a positive or negative way to the organoleptic characteristics of the olive oil. Necessary microbiological research carried out so far on olive oil is still needed. From the available scientific data up to now, it is not possible to establish that other species of microorganisms are useful and harmful in stabilizing the oil quality. In particular, it is not known if the yeasts in the freshly produced olive oil can modify some parameters responsible for the quality of virgin olive oil. Further microbiological studies on olive oil proffer to isolation of new microorganisms with biotechnological potential. The OMWs due to their particular characteristics, in addition to fat and triglycerides, sugars, phosphate, polyphenols, polyalcohols, pectins and metals, could provide microorganisms with biotechnological potential and low-cost fermentation substrates. For example, the exopolysaccharideproducing bacterium *Paenibacillus jamilae* (Aguilera et al. 2001) and the obligate alkaliphilic *Alkalibacterium olivoapovliticus* (Ntougias and Russell 2001) were isolated from olive mill wastes.

3. Olive mill waste as renewable low-cost substrates

According to the last report of Food and Agriculture Organisation of the United Nations (FAOSTAT 2009), 2.9 million tons of olive oil are produced annually worldwide, 75.2% of which are produced in Europe, with Spain (41.2%), Italy (20.1%) and Greece (11.4%) being the highest olive oil producers. Other olive oil producers are Asia (12.4%), Africa (11.2%), America (1.0%) and Oceania (0.2%). Olive oil production is a very important economic activity, particularly for Spain, Italy and Greece; worldwide, there has been an increase in production of about 30% in the last 10 years (FAOSTAT 2009).

Multiple methods are used in the production of olive oil, resulting in different waste products. The environmental impact of olive oil production is considerable, due to the large amounts of wastewater (OMWW) mainly from the three-phase systems and solid waste. The three-phase system, introduced in the 1970s to improve extraction yield, produces three streams: pure olive oil, OMWW and a solid cake-like by-product called olive cake or *orujo*. The olive cake, which is composed of a mixture of olive pulp and olive stones, is transferred to central seed oil extraction plants where the residual olive oil can be extracted. The two-phase centrifugation system was introduced in the 1990s in Spain as an ecological approach for olive oil production since it drastically reduces the water consumption during the process. This system generates olive oil plus a semi-solid waste, known as the two-phase olive-mill waste (TPOMW) or *alpeorujo* (Alburquerque et al. 2004; McNamara et al. 2008; Morillo et al. 2009).

The olive oil industry is characterized by its great environmental impact due to the production of a highly polluted wastewater and/or a solid residue, olive skin and stone (olive husk), depending on the olive oil extraction process (Table 1) (Azbar et al. 2004).

Pressure and three-phase centrifugation systems produce substantially more OMWW than two-phase centrifugation, which significantly reduces liquid waste yet produces large amounts of semi-solid or slurry waste commonly referred to as TPOMW. The resulting solid waste is about 800 kg per ton of processed olives. This "alpeorujo" still contains 2.5–3.5% residual oil and about 60% water in the two-phase decanter system (Giannoutsou et al. 2004).

Production process	Inputs	Outputs
Traditional process (pressing)	Olives (1 ton)	Oil (~200 kg)
	Wash water (0.1-0.12 m3)	Solid waste (~400 kg)
		Wastewater (~600 kg)
	Energy (40-63 kWh)	-
Three-phase process	Olives (1 ton)	Oil (200 kg)
	Wash water (0.1-0.12 m3)	Solid waste (500-600 kg)
	Fresh water for decanter (0.5-1.0 m3)	Wastewater (800-950 kg)
	Water to polish the impure oil (10 kg)	-
	Energy (90-117 kWh)	-
Two-phase process	Olives (1 ton)	Oil (200 kg)
	Wash water (0.1-0.12 m3)	Solid waste (800 kg)
		Wastewater (250 kg)
	Energy (90-117 kWh)	-

Table 1. Inputs and outputs from olive oil industry (Adapted from Azbar et al. 2004)

The average amount of OMWs produced during the milling process is approximately 1000 kg per ton of olives (Azbar et al. 2004). 19.3 million tons of olive are produced annually worldwide, 15% of them used to produce olive oil (FAOSTAT 2009). As an example of the scale of the environmental impact of OMWW, it should be noted that 10 million m³ per year of liquid effluent from three-phase systems corresponds to an equivalent load of the wastewater generated from about 20 million people. Furthermore, the fact that most olive oil is produced in countries that are deficient in water and energy resources makes the need for effective treatment and reuse of OMWW (McNamara et al. 2008). Overall, about 30 million tons of OMWs per year are produced in the world that could be used as renewable negative or low-cost substrates.

4. Microbial biotechnology applications in olive oil industry

Microbial biotechnology applications in olive oil industry, mainly attempts to obtain added-value products from OMWs are summarised in Fig. 1. OMWs could be used as renewable low-cost substrate for industrial and agricultural microbial biotechnology as well as for the production of energy.

The chemical oxygen demand (COD) and biological oxygen demand (BOD) reduction of OMWs with a concomitant production of biotechnologically valuable products such as enzymes (lipases, ligninolytic enzymes), organic acids, biopolymers and biodegradable plastics, biofuels (bioethanol, biodiesel, biogas and biohydrogen), biofertilizers and amendments will be review.

4.1 Olive mill wastes biological treatment

Ironically, while olive oil itself provides health during its consumption, its by-products represent a serious environmental threat, especially in the Mediterranean, region that accounts for approximately 95% of worldwide olive oil production.

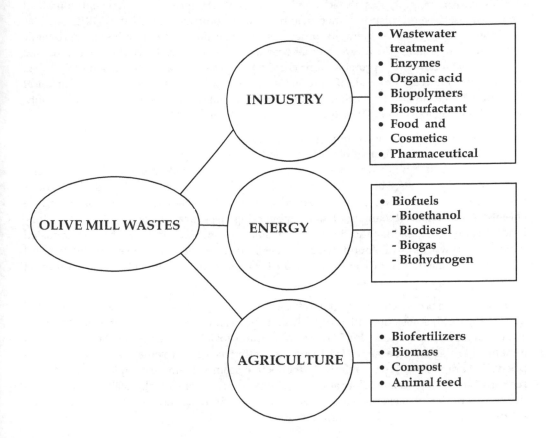

Fig. 1. Potential uses of olive mill wastes in microbial biotechnology.

Moreover, olive oil production is no longer restricted to the Mediterranean basin, and new producers such as Australia, USA and South America will also have to face the environmental problems posed by OMWs. The management of wastes from olive oil extraction is an industrial activity submitted to three main problems: the generation of waste is seasonal, the amount of waste is enormous and there are various types of olive oil waste (Giannoutsou et al. 2004).

OMWs have the following properties: dark brown to black colour, acidic smell, a high organic load and high C/N ratio (chemical oxygen demand or COD) values up to 200 g per litre, a chemical oxygen demand/biological oxygen demand (COD/BOD5) ratio ranging from 2.5 to 5.0, indicating low biodegrability, an acidic pH of between 4 and 6, high concentration of phenolic substances 0.5-25 g per litre with more than 30 different phenolic compounds and high content of solid matter. The organic fraction contains large amounts of proteins (6.7-7.2%), lipids (3.76-18%) and polysaccharides (9.6-19.3%), and also phytotoxic components that inhibit microbial growth as well as the germination and vegetative growth of plants (Roig et al. 2006; McNamara et al. 2008).

OMWs treatment processes tested employ physical, chemical, biological and combined technologies. Several disposal methods have been proposed to treat OMWs, such as traditional decantation, thermal processes (combustion and pyrolysis), agronomic applications (e.g. land spreading), animal-breeding methods (e.g. direct utilisation as animal feed or following protein enrichment), physico-chemical treatments (e.g. precipitation/flocculation, ultrafiltration and reverse osmosis, adsorption, chemical oxidation processes and ion exchange), extraction of valuable compounds (e.g. antioxidants, residual oil, sugars), and biological treatments (Morillo et al. 2009).

Among the different options, biological treatments or bioremediation are considered the most environmentally compatible and the least expensive (Mantzavinos and Kalogerakis 2005). Bioremediation is a treatment process employing naturally microorganisms (bacteria and fungi like yeasts, molds and mushrooms) to break down, or degrade, hazardous substances into less toxic or non-toxic substances. Bioremediation technologies can be classified as *in-situ* (bioaugmentation, bioventing, biosparging) or *ex-situ* (bioreactors, landfarming, composting and biopiles). *In-situ* bioremediation treats the contaminated water or soil where it was found, whereas *ex-situ* bioremediation processes involve removal of the contaminated soil or water to another location prior to treatment (Arvanitoyannis et al. 2008).

Bioremediation occurs either under aerobic or anaerobic conditions. Many aerobic biological processes, technologies and microorganisms have been tested for the treatment of OMWs, aimed at reducing organic load, dark colour and toxicity of the effluents (Table 2). In general, aerobic bacteria appeared to be very effective against some low molecular mass phenolic compounds but are relatively ineffective against the more complex polyphenolics responsible for the dark colouration of OMWs (McNamara et al. 2008). A number of different species of bacteria, yeasts, molds and mushrooms have been tested in aerobic processes to treat OMWs that are listed (Table 2).

A number of studies have utilized bacterial consortia for bioremediation of OMWW. Bioremediation of OMWW using aerobic consortia has been quite successful in these studies, achieving significant reductions in COD (up to 80%) and the concentration of phytotoxic compounds, and complete removal of some simple phenolics (Zouari and Ellouz 1996; Benitez et al. 1997). A combined bacterial–yeast system of *Pseudomonas putida* and *Yarrowia lipolytica* were used to degrade OMWW (De Felice et al. 1997).

Anaerobic bioremediation of OMWs has employed, almost exclusively, uncharacterized microbial consortia derived from municipal and other waste facilities. This technique presents a number of advantages in comparison to the classical aerobic processes: (a) a high degree of purification with high-organic-load feeds can be achieved; (b) low nutrient requirements are necessary; (c) small quantities of excess sludge are usually produced; and (d) a combustible biogas is generated (Dalis et al. 1996; Zouari and Ellouz 1996; Borja et al. 2006). Combined aerobic–anaerobic systems have also been used effectively in the bioremediation of OMWs (Hamdi and Garcia 1991; Borja et al. 1995). Aerobic processes are applied waste streams of OMWs with low organic loads, whereas anaerobic processes are applied waste streams with high organic loads.

Microorganism	Waste type	Method	Results	Reference
Bacteria				
Azotobacter vinelandii	OMWW	Culture in OMWW	70% COD reduction	(Piperidou et al. 2000)
Bacillus pumilus	OMWW	Culture in OMWW	50% phenol reduction	(Ramos-Cormenzana et al. 1996)
Lactobacillus paracasei	OMWW	Culture in OMWW with cheese whey's	47% colour removal 22.7% phenol reduction	(Aouidi et al. 2009)
Lactobacillus plantarum	OMWW	Culture in OMWW	Increase of simple polyphenols content	(Kachouri and Hamdi 2004)
Pseudomonas putida and *Ralstonia* spp.	OMWW	Culture two strains in OMWW	Biodegradation of aromatic compounds	(Di Gioia et al. 2001)
Yeasts				
Candida boidinii	TPOMW	Fed-batch microcosm	57.7% phenol reduction	(Giannoutsou et al. 2004)
Candida cylindracea	OMWW	Culture in OMWW	reduction of phenolic compounds and COD	(Gonçalves et al. 2009)
Candida holstii	OMWW	Culture in OMWW	39% phenol reduction	(Ben Sassi et al. 2008)
Candida oleophila	OMWW	Bioreactor batch culture with OMWW	Tannins content reduction	(Peixoto et al. 2008)
Candida rugosa	OMWW	Culture in OMWW	reduction of phenolic compounds and COD	(Gonçalves et al. 2009)
Candida tropicalis	OMWW	Culture in OMWW	62.8% COD reduction 51.7% phenol reduction	(Fadil et al. 2003)
Geotrichum candidum	OMWW	Culture in bioreactors with OMWW	60% COD reduction	(Asses et al. 2009)
Geotrichum candidum	TPOMW	Fed-batch microcosm	57% phenol reduction	(Giannoutsou et al. 2004)
Saccharomyces spp.	TPOMW	Fed-batch microcosm	61% phenol reduction	(Giannoutsou et al. 2004)
Trichosporon cutaneum	OMWW	Culture in OMWW	removal of mono- and polyphenols	(Chtourou et al. 2004)
Yarrowia lipolytica	OMWW	Culture in OMWW	20-40% COD reduction < 30% phenol reduction	(Lanciotti et al. 2005)
Yarrowia lipolytica W29	OMWW	Culture in OMWW	67-82% COD reduction	(Wu et al. 2009)
Molds				
Aspergillus niger	OMWW	Culture in OMWW	73% COD reduction 76% phenol reduction	(García García et al. 2000)
Aspergillus spp.	OMWW	Culture in OMWW	52.5% COD reduction 44.3% phenol reduction	(Fadil et al. 2003)
Aspergillus terreus	OMWW	Culture in OMWW	63% COD reduction 64% phenol reduction	(García García et al. 2000)
Fusarium oxysporum	DOR	Culture in DOR	16-71% phytotoxicity reduction	(Sampedro et al. 2007a)
Penicillium spp.	OMWW	Culture in OMWW	38% COD reduction 45% phenol reduction	(Robles et al. 2000)

Microorganism	Waste type	Method	Results	Reference
Phanerochaete chrysosporium	OMWW	Culture in bioreactors with OMWW	75% COD reduction 92% phenol reduction	(García García et al. 2000)
Phanerochaete chrysosporium	TPOMW	Culture in TPOMW	9.2% TOC reduction 14.5% phenol reduction	(Sampedro et al. 2007b)
Phanerochaete flavido-alba	OMWW	Culture in bioreactors with OMWW	52% phenol reduction	(Blánquez et al. 2002)
Phanerochaete flavido-alba	TPOMW	Solid-state culture	70% phenol reduction	(Linares et al. 2003)
Mushrooms				
Coriolopsis rigida	TPOMW	Culture in OMWW	9% TOC reduction 89% phenol reduction	(Sampedro et al. 2007b)
Coriolopsis polyzona	OMWW	Culture in OMWW	75% colour removal	(Jaouani et al. 2003)
Coriolus versicolor	OMWW	Culture in OMWW	65% COD reduction 90% phenol reduction	(Yesilada et al. 1997)
Funalia trogii	OMWW	Culture in OMWW	70% COD reduction 93% phenol reduction	(Yesilada et al. 1997)
Lentinula edodes	OMWW	Culture in OMWW	65% COD reduction 88% phenol reduction	(D'Annibale et al. 2004)
Lentinus tigrinus	OMWW	Culture in OMWW	Effective in decolorization	(Jaouani et al. 2003)
Pleurotus eryngii	OMWW	Culture in OMWW	> 90% phenol reduction	(Sanjust et al. 1991)
Pleurotus floridae	OMWW	Culture in OMWW	> 90% phenol reduction	(Sanjust et al. 1991)
Pleurotus ostreatus	OMWW	Culture in OMWW	100% phenol reduction	(Tomati et al. 1991)
Pleurotus ostreatus	OMWW	Culture in bioreactors with OMWW	Phenol reduction nearly complete	(Aggelis et al. 2003)
Pleurotus ostreatus	OMWW	Solid-state culture	67% phenol reduction	(Fountoulakis et al. 2002)
Pleurotus ostreatus	TPOMW	Plastic bag	22% TOC reduction 90% phenol reduction	(Saavedra et al. 2006)
Pleurotus pulmonarius	TPOMW	Culture in TPOMW	9.7% TOC reduction 66.2% phenol reduction	(Sampedro et al. 2007b)
Pleurotus sajor-caju	OMWW	Culture in OMWW	> 90% phenol reduction	(Sanjust et al. 1991)
Pleurotus spp.	OMWW	Culture in OMWW	76% phenol reduction	(Tsioulpas et al. 2002)
Phlebia radiata	TPOMW	Culture in TPOMW	13% TOC reduction 95.7% phenol reduction	(Sampedro et al. 2007b)
Poria subvermispora	TPOMW	Culture in TPOMW	13.2% TOC reduction 72.3% phenol reduction	(Sampedro et al. 2007b)
Pycnoporus cinnabarinus	TPOMW	Culture in TPOMW	7.6% TOC reduction 88.7% phenol reduction	(Sampedro et al. 2007b)
Pycnoporous coccineus	OMWW	Culture in OMWW	Effective in decolorization	(Jaouani et al. 2003)

OMWW: olive oil wastewater, TPOMW: two-phase olive-mill waste, COD: chemical oxygen demand, TOC: Total organic carbon, DOR: olive-mill dry residue.

Table 2. Aerobic treatment of OMWs by microorganisms

In general, available scientific information shows that fungi are more effective than bacteria at degrading both simple phenols and the more complex phenolic compounds present in olive-mill wastes. For example, several species of the genus *Pleurotus* were found to be very effective in the degradation of the phenolic substances present in OMWs (Hattaka 1994). For OMWs biotreatment in large-scale, the use of filamentous fungi have considerable problems because of the formation of fungal pellets and other aggregations. The use of yeast in bioreactors could be a way forward to overcome this limitation.

4.2 Enzymes

In recent years, many researchers have utilized OMWs as growth substrates for microorganisms, obtaining a reduction of the COD level, together with enzyme production. The addition of nutrients can modify the pattern of degrading enzymes production by specific microorganisms from OMWs. (De la Rubia et al. 2008).

Lipases (EC 3.1.1.3) are among the most important classes of industrial enzymes (Darvishi et al. 2009). Many microorganisms are known as potential producers of lipases including bacteria, yeast, and fungi. Several reviews have been published on microbial lipases (Arpigny and Jaeger 1999; Vakhlu and Kour 2006; Treichel et al. 2010).

Lipolytic fungal species, such as *Aspergillus oryzae, Aspergillus niger, Candida cylindracea, Geotrichum candidum, Penicillium citrinum, Rhizopus arrhizus* and *Rhizopus oryzae* were preliminarily screened for their ability to grow on undiluted OMWW and to produce extra-cellular lipase. A promising potential for lipase production was found by *C. cylindracea* NRRL Y-17506 on OMWW (D'Annibale et al. 2006).

Among the different yeasts tested, the *Y. lipolytica* most adapted to grow on OMW. the *Y. lipolytica* strains were produced 16-1041 U/L of lipase on OMWs and also reduced 1.5-97% COD, 80% BOD and 0-72% phenolic compounds of OMWs (Fickers et al. 2011). The yeasts *Saccharomyces cerevisiae* and *Candida wickerhamii* produce β-glucosidase enzyme to hydrolyse oleuropein present in olive oil (Ciafardini and Zullo 2002).

Olive oil cake (OOC) used as a substrate for phytase production in solid-state fermentation using three strains of fungus *Rhizopus* spp. OOC of initial moisture 50% was fermented at 30°C for 72 hours and inoculated with *Rhizopus oligosporus* NRRL 5905, *Rhizopus oryzae* NRRL 1891 and *R. oryzae* NRRL 3562. The results indicated that all three *Rhizopus* strains produced very low titers of enzyme on OOC (Ramachandran et al. 2005).

Tannase could be utilized as an inhibitor of foam in tea production, clarifying agent in beer and fruit juices production, in the pharmaceutical industry and for the treatment of tannery effluents. *Aspergillus niger* strain HA37, isolated from OMW, was incubated on a synthetic medium containing tannic acid and on diluted OMW on a rotary shaker at 30°C. On the medium containing tannic acid, tannase production was 0.6, 0.9 and 1.5 U/ml at 0.2%, 0.5% and 1% initial tannic acid concentration, respectively (Aissam et al. 2005).

Extracellular ligninolytic enzymes such as lignin peroxidase (LiP), manganese peroxidase (MnP) and laccase (Lac) were produced by the white rot fungus *Phanerochaete flavido-alba* with a concomitant decoloration and decrease in phenolic content and toxicity of OMWW. Laccase was the sole ligninolytic enzyme detected in cultures containing monomeric aromatic compounds. Laccase and an acidic manganese-dependent peroxidase (MnPA, pI

62.8) were accumulated in cultures with OMWW or polymeric pigment. Furthermore, modified manganese-dependent peroxidases were observed mainly in OMWW-supplemented cultures. Laccase was more stable to the effect of OMWW toxic components and was accumulated in monomeric aromatic-supplemented cultures, suggesting a more important role than manganese-dependent peroxidases in OMWW detoxification. Alternatively, MnPA accumulated in cultures containing the polymeric pigment seemed to be more essential than laccase for degradation of this recalcitrant macromolecule by *P. flavido-alba*. (Ruiz et al. 2002).

Enzyme laccase, produced by fungus *Pycnoporus coccineus*, is responsible for OMWW decolorization and decrease COD and phenolic compounds. The highest laccase level was 100 000 U/l after 45 incubation-days. The enzyme was stable at pH 7, at room temperature and showed a half-life of 8 and 2 h at 50 and 60°C, respectively (Jaouani et al. 2005). In order to decolourise OMWW efficiently, production and differential induction of ligninolytic enzymes by the white rot *Coriolopsis polyzona*, were studied by varying growth media composition and/or inducer addition (Jaouani et al. 2006). The production of lignin peroxidase (LiP), manganese peroxidase (MnP) and lipases by *Geotrichum candidum* were performed in order to control the decolourisation and biodegradation of OMWW (Asses et al. 2009).

Sequential batch applications starting with adapted *Trametes versicolor* FPRL 28A INI and consecutive treatment with *Funalia trogii*, possible to remove significant amount of total phenolics content and higher decolorization as compared to co-culture applications. Also highest laccase and manganese peroxidase acitivities were obtained with *F. trogii* (Ergul et al. 2010).

4.3 Organic acids

Some *Y. lipolytica* strains are good candidates for the reduction of the pollution potential of OMWW and for the production of enzymes and metabolites such as lipase and citric acid (Lanciotti et al. 2005). *Y. lipolytica* strain ACA-DC 50109 demonstrated efficient growth on media containing mixtures of OMWs and commercial glucose. In nitrogen-limited diluted and enriched with high glucose quantity OMWW, a noticeable amount of total citric acid was produced. The ability of *Y. lipolytica* to grow on relatively high phenolic content OMWs based media and produce in notable quantities citric acid, make this non-conventional yeast worthy for further investigation (Papanikolaou et al. 2008).

The biochemical behavior and simultaneous production of valuable metabolites such as lipase, citric acid (CA), isocitric acid (ICA) and single-cell protein (SCP) were investigate by *Y. lipolytica* DSM 3286 grown on various plant oils as sole carbon source. Among tested plant oils, olive oil proved to be the best medium for lipase and CA production. The *Y. lipolytica* DSM 3286 produced 34.6 ± 0.1 U/ml of lipase and also CA, ICA and SCP as by-product on olive oil medium supplemented with yeast extract. Urea, as organic nitrogen, was the best nitrogen source for CA production. The results of this study suggest that the two biotechnologically valuable products, lipase and CA, could be produced simultaneously by this strain using renewable low-cost substrates such as plant oils in one procedure (Darvishi et al. 2009).

In the other study, a total of 300 yeast isolates were obtained from samples of agro-industrial wastes, and M1 and M2 strains were investigated for their ability to produce lipase and citric acid. Identification tests showed that these isolates belonged to the species *Y. lipolytica*. M1 and M2 strains produced maximum levels of lipase on olive oil, and high levels of citric acid on citric acid fermentation medium (Mafakher et al. 2010).

The highest oxalic acid quantity (5 g/l) was obtained by the strain *Aspergillus* sp. ATHUM 3482 on waste cooking olive oil medium. For strain *Penicillium expansum* NRRL 973 on this medium sole organic acid detected was citric acid with maximum concentration achieved 3.5 g/l (Papanikolaou et al. 2011).

4.4 Biopolymers and biodegradable plastics

Exopolysaccharides (EPSs) often show clearly identified properties that form the basis for a wide range of applications in food, pharmaceuticals, petroleum, and other industries. The production of these microbial polymers using OMWW as a low-cost fermentation substrate has been proposed (Ramos-Cormenzana et al. 1995). This approach could reduce the cost of polymer production because the substrate is often the first limiting factor. Moreover, OMWW contains free sugars, organic acids, proteins and other compounds such as phenolics that could serve as the carbon source for polymer production, if the chosen microorganism is able to metabolize these compounds (Fiorentino et al. 2004).

Xanthan gum, an extracellular heteropolysaccharide produced by the bacterium *Xanthomonas campestris* has been obtained from OMWW. Growth and xanthan production on dilute OMWW as a sole source of nutrients were obtained. Addition of nitrogen and/or salts led to significantly increased xanthan yields with a maximum of 7.7g/l (Lopez and Ramos-Cormenzana 1996).

The fungus *Botryospheria rhodina* has been used for the production of β-glucan from OMWW with yield of 17.2 g/l and a partial dephenolisation of the substrate (Crognale et al. 2003). A metal-binding EPS produced by *Paenibacillus jamilae* from OMWs. Maximum EPS production (5.1 g/l) was reached in batch culture experiments with a concentration of 80% of OMWW as fermentation substrate (Morillo et al. 2007).

Polyhydroxyalkanoates (PHAs) are reserve polyesters that are accumulated as intracellular granules in a variety of bacteria. Of these polymers, poly-β-hydroxybutyrate (PHB) is the most common. Since the physical properties of PHAs are similar to those of some conventional plastics, the commercial production of PHAs is of interest. However, these biodegradable and biocompatible 'plastics' are not priced competitively at the present, mainly because the sugars (i.e. glucose) used as fermentation feed-stocks are expensive. Finding a less expensive substrate is, therefore, a major need for a wide commercialisation of these products. Large amounts of biopolymers containing β-hydroxybutyrate (PHB) and copolymers containing β-hydroxyvalerate (P[HB-co-HV]) are produced by *Azotobacter chroococcum* in culture media amended with alpechin (wastewater from olive oil mills) as the sole carbon source (Pozo et al. 2002).

4.5 Biosurfactants

Rhamnolipids, typical biosurfactants produced by *Pseudomonas aeruginosa*, consist of either one or two rhamnose molecules, linked to one or two fatty acids of saturated or unsaturated alkyl chain between C8 and C12. The *P. aeruginosa* 47 T2 produced two main rhamnolipid homologs, (Rha-C10-C10) and (Rha-Rha-C10-C10), when grown in olive oil waste water or in waste frying oils consisting from olive/sunflower (Pantazaki et al. 2010).

4.6 Food and cosmetics

A few edible fungi, especially species of *Pleurotus*, can also be grown using OMWs as the source of nutrients by the application of different strategies. Recently the cultivation of the oyster mushroom *Pleurotus ostreatus* was suggested on OMWW (KalmIs et al. 2008).

Hydroxy fatty acids (HFAs) are known to have special properties such as higher viscosity and reactivity compared to other normal fatty acids. These special properties used in a wide range of applications including resins, waxes, nylons, plastics, lubricants, cosmetics, and additives in coatings and paintings. Some HFAs are also reported as antimicrobial agents against plant pathogenic fungi and some of food-borne bacteria. Bacterium *Pseudomonas aeruginosa* PR3 produce several hydroxy fatty acids from different unsaturated fatty acids. Of those hydroxy fatty acids, 7,10-dihydroxy-8(E)-octadecenoic acid (DOD) was efficiently produced from oleic acid by strain PR3. DOD production yield from olive oil was 53.5%. Several important environmental factors were also tested. Galactose and glutamine were optimal carbon and nitrogen sources, and magnesium ion was required for DOD production from olive oil (Suh et al. 2011).

4.7 Pharmaceutical

The enhancing effect of various concentrations of 18 oils and a silicon antifoam agent on erythromycin antibiotic production by *Saccharopolyspora erythraea* was evaluated in a complex medium containing soybean flour and dextrin as the main substrates. The highest titer of erythromycin was produced in medium containing 55 g/l black cherry kernel oil (4.5 g/l). The titers of erythromycin in the other media were also recorded, with this result: black cherry kernel > water melon seed > melon seed > walnut > rapeseed > soybean > (corn = sesame) > (olive = pistachio = lard = sunflower) > (hazelnut = cotton seed) > grape seed > (shark = safflower = coconut). In medium supplement with olive oil, concentration of erythromycin was 2.15±0.03 and 2.75±0.02 g/l before and after optimization, respectively (Hamedi et al. 2004).

4.8 Biofuels

It is widely recognised that clean and sustainable technologies, e.g. biofuels, are only part of the solution to the impending energy crisis. Comparing the heating value of biohydrogen (121 MJ/kg), methane (50.2 MJ/kg) and bioethanol (23.4 MJ/kg), the production of hydrogen will be more attractive. Nevertheless, the use of biohydrogen is still not practical and thus there is a higher demand for methane and bioethanol because they can be used directly as biofuels with the existing technology (Duerr et al. 2007).

Ethanol production as a biofuel from OMWs with high content of organic matter is interesting (Li et al. 2007). The two main components of TPOMW (stones and olive pulp) as substrates were used to production of ethanol by a simultaneous saccharification and fermentation process (Ballesteros et al. 2001). In recent study, an enzymatic hydrolysis and subsequent glucose fermentation by baker's yeast were evaluated for ethanol production using dry matter of TPOMW. The results showed that yeasts could effectively ferment TPOMW without nutrient addition, resulting in a maximum ethanol production of 11.2 g/l and revealing the tolerance of yeast to TPOMW toxicity (Georgieva and Ahring 2007).

Anaerobic digestion is a biological process in which organic material is broken down by microorganisms. Unlike composting, the process occurs in the absence of air. Anaerobic digestion is a practical alternative for the treatment of TPOMW, which produces biogas. The TPOMW is biodegradable by anaerobic digestion at mesophilic temperatures in stirred tank reactors, with COD removal efficiencies in the range of 72–89% and an average methane yield coefficient of 0.31 dm^3 CH$_4$ per gramme COD removed. Hydrogen production was coupled with a subsequent step for methane production, giving the potential for production of 1.6 mmol H$_2$ per gramme of TPOMW (Borja et al. 2006).

The OMW used as a sole substrate for the production of hydrogen gas with *Rhodobacter sphaeroides* O.U.001. The bacterium was grown in diluted OMW media, containing OMW concentrations between 20% and 1% in a glass column photobioreactor at 32°C. The released gas was nearly pure hydrogen, which can be utilized in electricity producing systems, such as fuel cells. The maximum hydrogen yield (145 ml) was obtained with 3% and 4% OMW concentrations. However, as well as hydrogen production, COD, BOD and phenol reduction from OMW were recorded (Eroglu et al. 2004).

Biodiesel, a fuel that can be made from renewable biological sources such as vegetable oils or animals fats, has been recognized recently as an environment friendly alternative fuel for diesel engines. Among liquid biofuels, biodiesel derived from vegetable oils is gaining ground and market share as diesel fuel in Europe and the USA. A mixture of frying olive oil and sunflower oil for the production of methyl esters that can be used as biodiesel (Encinar et al. 2005).

4.9 Biofertilizers

As far as agronomic use of the waste is concerned, the idea of re-using microbially treated OMWW as fertiliser has been also proposed. An acidogenic fungus strain *Aspergillus niger* was grown in either free or immobilised form on OMWW with rock phosphate added in order to solubilise it. It was found that at optimized process conditions (moisture 70%; corn steep liquor as a nitrogen source; inoculum size of 3-4 ml; presence of slow release phosphate), the filamentous fungal culture was able to produce 58 U phytase/g dry substrate and 31 mg soluble phosphate per flask (Vassilev et al. 1997; Vassilev et al. 2007).

4.10 Biomass

Already 50 years ago, the production of yeast biomass using OMWW in a chemostat for use in industrial applications was reported. The microbial biomass produced from OMW fermentations either as an additive to animal feed or to improve its agronomic use. For example, an intense degradation of most polluting substances of OMWW and the production of biomass could be used as an animal feed integrator using a chemical–biological method (Morillo et al. 2009).

Seven strains of *Penicillium* isolated from OMWW disposal ponds were tested for biomass production and biodegradation of undiluted OMWW. Best results were obtained by using strain P4, which formed 21.50 g (dry weight) of biomass per litre of undiluted wastewater after 20 days of cultivation. This and other strains also carried out an outstanding reduction of the COD and the phenolic content of OMW, as well as a pH raise (Robles et al. 2000). The *Y. lipolytica* strain ATCC 20255 strain has been effective in the treatment of OMWW that yield of the biomass (single-cell protein) was 22.45 g/l (Scioli and Vollaro 1997).

Microalgal biomass is as a potential source of proteins, carbohydrates, pigments, lipids, and hydrocarbons. In addition, the biomass can be used as a low-release fertilizer. This chemical composition has great variation, depending on the species, culture medium, and the operating conditions. Microalga *Scenedesmus obliquus* was used to biomass production from rinse water (RW) from two-phase centrifugation in the olive-oil extraction industry. Maximum specific growth rate, 0.044 per hour was registered in the culture with 5% RW and reduces 67.4% BOD when operating with 25% RW. The greater specific rate of protein synthesis during the exponential phase was 3.7 mg/g h to 50% RW (Hodaifa et al. 2008).

Microbial lipid (single cell oil or SCO) production has been an object of research and industrial interest for more than 60 years. Microorganisms can store triacylglycerol (TAG) as intracellular oil droplets. *Gordonia* sp. DG accumulated more than 50% lipid with most tested wastes, while only 29, 36 and 41% was accumulated in presence of olive mill waste, hydrolyzed barely seeds and wheat bran, respectively (Gouda et al. 2008).

Carbon-limited cultures were performed on waste cooking olive oil, added in the growth medium at 15 g/l, and high biomass quantities were produced up to 18 g/l. Cellular lipids were accumulated in notable quantities in almost all cultures. *Aspergillus* sp. ATHUM 3482 accumulated lipid up to 64% (w/w) in dry fungal mass. In parallel, extracellular lipase activity was quantified, and it was revealed to be strain and fermentation time dependent, with a maximum quantity of 645 U/ml being obtained by *Aspergillus niger* NRRL 363. Storage lipid content significantly decreased at the stationary growth phase (Papanikolaou et al. 2011).

4.11 Compost

Composting is the aerobic processing of biologically degradable organic waste to produce a reasonably stable, granular material and valuable plant nutrients. Composting removes the phytotoxicity of the residues within a few weeks and allows the subsequent enrichment of croplands with nutrients that were originally taken up by olive tree cultivation. Composting of OMWs requires the proper adjustment of pH, temperature, moisture, oxygenation and nutrients, thereby allowing the adequate development of the microbial populations (Arvanitoyannis and Kassaveti 2007).

Among the possible technologies for recycling the TPOMW, composting is gaining interest as a sustainable strategy to recycle this residue for agricultural purposes. Dry olive cake alone or mixed with municipal biosolids vermicomposted for 9 months in order to examine the behaviour of three specific humic substance-enzyme complexes. During the process, β-glucosidase synthesis and release was observed, whereas no significant change in urease and phosphatase activity was recorded. The vermicomposted olive cake, alone or in blends with biosolids, could be effectively used as amendment due to their ability to reactivate the C, P and N-cycles in degraded soils for regeneration purposes (Benitez et al. 2005).

Olive pomace, a wet solid waste from the three-phase decanters and presses, was composted by using a reactor for a period of 50 days in four bioreactors. Urea was added to adjust C/N ration between 25-30. At the end of 50 days of composting using *Trichoderma harzianum* and *Phanerochaete chrysosporium*, cellulose and lignin were highly degraded. It was found that after 30 days, *P. chrysosporium* and *T. harzianum* degraded approximately 71.9% of the lignin and 59.25% of the cellulose, respectively (Haddadin et al. 2009).

4.12 Animal feed

Treated OMW may find applications as a raw material in various biotechnological processes or as animal food. The appropriate utilization of by-products in animal nutrition can improve the economy and the efficiency of agricultural, industrial and animal production.

The olive pomace was alkali-treated, transferred to culture flasks and inoculated with the above fungi. After inoculation, the fermentation process was carried out at 25°C for 60 days. The results indicated that *Oxysporus* spp. degraded lignin up to 69%, whereas *Phanerochaete chrysosporium* and *Schizophyllum commune* delignified olive pomace 60% and 53%, respectively. However, the potential use of treated olive pomace as a feed for poultry is still under investigation. The fermented olive pomace can be used as a feed for the poultry industry (Haddadin et al. 2002).

5. Conclusion

The olive oil industry generates large amounts of olive mill wastes (OMWs) as by-products that are harmful to the environment. About 30 million tons of OMWs per year are produced in the world. Thus, more research is needed on the development of new bioremediation technologies and strategies of OMWs, as well as the valorisation by microbial biotechnology. The fermentation of fatty low-value renewable carbon sources like OMWs aiming at the production of various added-value metabolites is a noticeable interest in the sector of industrial microbiology and microbial biotechnology.

Microbiological studies show that presence of yeasts, but not of bacteria and moulds in the olive oil. Some of the yeasts are considered useful as they improve the organoleptic characteristics of the oil during preservation, whereas others are considered harmful as they can damage the quality of the oil through the hydrolysis of the triglycerides. Olive oil and its by-products could provide a source of low-cost fermentation substrate and isolation of new microorganisms with biotechnological potentials.

OMWs treatment processes that employ physical, chemical, biological and combined technologies have been tested. Among the different options, biological treatments or bioremediation are considered the most environmentally compatible and the least expensive. Bioremediation occurs either under aerobic or anaerobic conditions. Aerobic processes are applied waste streams of OMWs with low organic loads, whereas anaerobic processes are applied waste streams with high organic loads.

Microbial biotechnology strategies and methods in olive oil industry were used to reduce chemical oxygen demand (COD), biological oxygen demand (BOD) and phenolic compounds of OMWs with a concomitant production of biotechnologically valuable products such as enzymes (lipases, β-glucosidase, phytase, tannase, lignin peroxidase, manganese peroxidise, laccase and pectinases), organic acids (citric, isocitric and oxalic acids), biopolymers and biodegradable plastics (xanthan, β-glucan and polyhydroxyalkanoates), biosurfactants, food and cosmetics, pharmaceutical, biofuels (bioethanol, biogas, biohydrogen and biodiesel), biofertilizers and amendments, biomass (single cell proteins, single cell oil), compost and animal feed.

What has been discussed in this review indicate that microbial biotechnology can be used for the production of value-added products from olive oil by-products and can facilitate a significant reduction in waste treatment costs.

6. References

Aggelis, G., Iconomou, D., Christou, M., Bokas, D., Kotzailias, S., Christou, G., Tsagou, V. & Papanikolaou, S. (2003). Phenolic removal in a model olive oil mill wastewater using *Pleurotus ostreatus* in bioreactor cultures and biological evaluation of the process. *Water Research*, 37:3897-3904

Aguilera, M., Monteoliva-Sánchez, M., Suárez, A., Guerra, V., Lizama, C., Bennasar, A. & Ramos-Cormenzana, A. (2001). *Paenibacillus jamilae* sp. nov., an exopolysaccharide-producing bacterium able to grow in olive-mill wastewater. *International Journal of Systematic Evolution Microbiology*, 51:1687-1692

Aissam, H., Errachidi, F., Penninck, F., Merzouki, M & Benlemlih, M. (2005). Production of tannase by *Aspergillus niger* HA37 growing on tannic acid and olive mill waste waters. *World Journal of Microbiology and Biotechnology*, 21:609-614

Alburquerque, J. A., Gonzalvez, J., Garcia, D. & Cegarra, J. (2004). Agrochemical characterisation of "alperujo", a solid by-product of the two-phase centrifugation method for olive oil extraction. *Bioresource Technology*, 91:195-200

Aouidi, F., Gannoun, H., Ben Othman, N., Ayed, L. & Hamdi, M. (2009). Improvement of fermentative decolorization of olive mill wastewater by *Lactobacillus paracasei* by cheese whey's addition. *Process Biochemistry*, 44:597-601

Arpigny, J. L. & Jaeger, K. E. (1999). Bacterial lipolytic enzymes: Classification and properties. *Biochemical Journal*, 343:177-183

Arvanitoyannis, I. S. & Kassaveti, A. (2007). Current and potential uses of composted olive oil waste. *International Journal of Food Science Technology*, 42:281-295

Arvanitoyannis, I. S., & Kassaveti, A. (2008). Olive Oil Waste Management: Treatment Methods and Potential Uses of Treated Waste. In: *Waste Management for the Food Industries*, Academic Press, Amsterdam, pp 453-568

Asses, N., Ayed, L., Bouallagui, H., Ben Rejeb, I., Gargouri, M. & Hamdi, M. (2009). Use of *Geotrichum candidum* for olive mill wastewater treatment in submerged and static culture. *Bioresource Technology*, 100:2182-2188

Azbar, N., Bayram, A., Filibeli, A., Muezzinoglu, A., Sengul, F. & Ozer, A. (2004). A review of waste management options in olive oil production. *Critical Review of Environmental Science Technology*, 34:209-247

Azbar, N., Keskin, T., & Yuruyen, A. (2008). Enhancement of biogas production from olive mill effluent (OME) by co-digestion. *Biomass and Bioenergy*, 32:1195-1201

Ballesteros, I., Oliva, J. M., Saez, F. & Ballesteros, M. (2001). Ethanol production from lignocellulosic byproducts of olive oil extraction. *Applied Biochemistry and Biotechnology*, 91-93:237-252

Ben Sassi, A., Ouazzani, N., Walker, G. M., Ibnsouda, S., El Mzibri, M. & Boussaid, A. (2008). Detoxification of olive mill wastewaters by Moroccan yeast isolates. *Biodegradation*, 19:337-346

Benitez, E., Sainz, H. & Nogales, R. (2005). Hydrolytic enzyme activities of extracted humic substances during the vermicomposting of a lignocellulosic olive waste. *Bioresource Technology*, 96:785-790

Benitez, J., Beltran-Heredia, J., Torregrosa, J., Acero, J. L. & Cercas, V. (1997). Aerobic degradation of olive mill wastewaters. *Applied Microbiology and Biotechnology*, 47:185-188

Blánquez, P., Caminal, G., Sarrá, M., Vicent, M. T. & Gabarrell, X. (2002). Olive oil mill waste waters decoloration and detoxification in a bioreactor by the white rot fungus *Phanerochaete flavido-alba*. *Biotechnology Progress*, 18:660-662

Borja, R., Alba, J., Garrido, S. E., Martinez, L., Garcia, M. P., Monteoliva, M. & Ramos-Cormenzana, A. (1995). Effect of aerobic pretreatment with *Aspergillus terreus* on the anaerobic digestion of olive-mill wastewater. *Biotechnology and Applied Biochemistry*, 22:233-246

Borja, R., Rincón, B. & Raposo, F. (2006). Anaerobic biodegradation of twophase olive mill solid wastes and liquid effluents: kinetic studies and process performance. *Journal of Chemical Technology and Biotechnology*, 81:1450-1462

Chatjipavlidis, I., Antonakou, M., Demou, D., Flouri, F. & Balis, C. (1996). Bio-fertilization of olive oil mills liquid wastes. The pilot plant in Messinia, Greece. *International Biodeterioration and Biodegradation*, 38:183-187

Chtourou, M., Ammar, E., Nasri, M. & Medhioub, K. (2004). Isolation of a yeast, *Trichosporon cutaneum*, able to use low molecular weight phenolic compounds: application to olive mill waste water treatment. *Journal of Chemical Technology and Biotechnology*, 79:869-878

Ciafardini, G. & Zullo, B. A. (2002). Microbiological activity in stored olive oil. *International Journal of Food Microbiology*, 75:111-118

Ciafardini, G., Zullo, B. A. & Iride, A. (2006). Lipase production by yeasts from extra virgin olive oil. *Food Microbiology*, 23:60-67

Crognale, S., Federici, F. & Petruccioli, M. (2003). β-Glucan production by *Botryosphaeria rhodina* on undiluted olive-mill wastewaters. *Biotechnology Letters*, 25:2013-2015

D'Annibale, A., Casa, R., Pieruccetti, F., Ricci, M. & Marabottini, R. (2004). *Lentinula edodes* removes phenols from olive-mill wastewater: impact on durum wheat (*Triticum durum Desf.*) germinability. *Chemosphere*, 54:887-894

D'Annibale, A., Sermanni, G. G., Federici, F. & Petruccioli, M. (2006). Olive-mill wastewaters: a promising substrate for microbial lipase production. *Bioresource Technology*, 97:1828-1833

Dalis, D., Anagnostidis, K., Lopez, A., Letsiou, I. & Hartmann, L. (1996). Anaerobic digestion of total raw olive-oil wastewater in a two-stage pilot-plant (up-flow and fixed-bed bioreactors). *Bioresource Technology*, 57:237-243

Darvishi, F., Nahvi, I., Zarkesh-Esfahani, H. & Momenbeik, F. (2009). Effect of plant oils upon lipase and citric acid production in *Yarrowia lipolytica* yeast. *Journal of Biomedicine and Biotechnology*, 562943:1-7

De Felice, B., Pontecorvo, G. & Carfagna, M. (1997). Degradation of waste waters from olive oil mills by *Yarrowia lipolytica* ATCC 20255 and *Pseudomonas putida*. *Acta Biotechnologica*, 17:231-239

De la Rubia, T., Lucas, M. & Martínez, J. (2008). Controversial role of fungal laccases in decreasing the antibacterial effect of olive mill waste-waters. *Bioresource Technology*, 99:1018-1025

Dhouib, A., Aloui, F., Hamad, N. & Sayadi, S. (2006). Pilot-plant treatment of olive mill wastewaters by *Phanerochaete chrysosporium* coupled to anaerobic digestion and ultrafiltration. *Process Biochemistry*, 41:159-167

Di Gioia, D., Bertin, L., Fava, F. & Marchetti, L. (2001). Biodegradation of hydroxylated and methoxylated benzoic, phenylacetic and phenylpropenoic acids present in olive mill wastewaters by two bacterial strains. *Research in Microbiology*, 152:83-93

Duerr, M., Gair, S., Cruden, A. & McDonald, J. (2007). Hydrogen and electrical energy from organic waste treatment. *International Journal of Hydrogen Energy*, 32:705-709

Encinar, J. M., Gonzalez, J. F. & Rodriguez-Reinares, A. (2005). Biodiesel from used frying oil; variables affecting the yields and characteristics of the biodiesel. *Industrial Engineering Chemistry Research*, 44:5491-5499

Ergul, F. E., Sargın, S., Ongen, G. & Sukan, F. V. (2010). Dephenolization and decolorization of olive mill wastewater through sequential batch and co-culture applications. *World Journal of Microbiology and Biotechnology*, 27:107-114

Eroglu, E., Eroglu, I., Gündüz, U., Türker, L. & Yücel, M. (2006). Biological hydrogen production from olive mill wastewater with two-stage processes. *International Journal of Hydrogen Energy*, 31:1527-1535

Eroglu, E., Gunduz, U., Yucel, M., Turker, L. & Eroglu, I. (2004). Photobiological hydrogen production by using olive mill wastewater as a sole substrate source. *International Journal of Hydrogen Energy*, 29:163-171

Fadil, K., Chahlaoui, A., Ouahbi, A., Zaid, A. & Borja, R. (2003). Aerobic biodegradation and detoxification of wastewaters from the olive oil industry. *International Biodeterioration and Biodegradation*, 51:37-41

Food and Agriculture Organisation. (2009). *FAOSTAT database*. http://faostat.fao.org/

Fickers. P., Marty, A. & Nicaud, J. M. (2011). The lipases from *Yarrowia lipolytica*: Genetics, production, regulation, biochemical characterization and biotechnological applications. *Biotechnology Advances*, 29:632-644

Fiorentino, A., Gentili, A., Isidori, M., Lavorgna, M., Parrella, A. & Temussi, F. (2004). Olive oil mill wastewater treatment using a chemical and biological approach. *Journal of Agricultural Food and Chemistry*, 52:5151-5154

Fountoulakis, M. S., Dokianakis, S. N., Kornaros, M. E., Aggelis, G. & Lyberatos, G. (2002). Removal of phenolics in olive mill wastewaters using the white-rot fungus *Pleurotus ostreatus*. *Water Research*, 36:4735-4744

García García, I., Jiménez Peña, P. R., Bonilla Venceslada, J. L., Martin Martin, A., Martin Santos, M. A. & Ramos Gomez, E. (2000). Removal of phenol compounds from olive mill wastewater using *Phanerochaete chrysosporium*, *Aspergillus niger*, *Aspergillus terreus* and *Geotrichum candidum*. *Process Biochemistry*, 35:751-758

Gelegenis, J., Georgakakis, D., Angelidaki, I., Christopoulou, N. & Goumenaki, M. (2007). Optimization of biogas production from olive-oil mill wastewater, by codigesting with diluted poultry-manure. *Applied Energy*, 84:646-663

Georgieva, T. I. & Ahring, B. K. (2007). Potential of agroindustrial waste from olive oil industry for fuel ethanol production. *Biotechnology Journal*, 2:1547-1555

Giannoutsou, E. P., Meintanis, C. & Karagouni, A. D. (2004). Identification of yeast strains isolated from a two-phase decanter system olive oil waste and investigation of their ability for its fermentation. *Bioresource Technology*, 93:301-306

Gonçalves, C., Lopes, M., Ferreira, J. P. & Belo, I. (2009). Biological treatment of olive mill wastewater by non-conventional yeasts. *Bioresource Technology*, 100:3759-3763

Gouda, M. K., Omar, S. H. & Aouad, L. M. (2008). Single cell oil production by *Gordonia* sp. DG using agro-industrial wastes. *World Journal of Microbiology and Biotechnology*, 24:1703-1711

Haddadin, M. S., Al-Natour, R., Al-Qsous, S. & Robinson, R. K. (2002). Bio-degradation of lignin in olive pomace by freshly-isolated species of Basidiomycete. *Bioresource Technology*, 82:131-137

Haddadin, M. S. Y., Haddadin, J., Arabiyat, O. I. & Hattar, B. (2009). Biological conversion of olive pomace into compost by using *Trichoderma harzianum* and *Phanerochaete chrysosporium*. *Bioresource Technology*, 100:4773-4782

Hamdi, M. & Garcia, J. L. (1991). Comparison between anaerobic filter and anaerobic contact process for fermented olive mill wastewaters. *Bioresource Technology*, 38:23-29

Hamedi, J., Malekzadeh, F. & Saghafi-nia, A. E. (2004). Enhancing of erythromycin production by *Saccharopolyspora erythraea* with common and uncommon oils. *Journal of Industrial Microbiology and Biotechnology*, 31:447-456

Hattaka, A. (1994). Lignin degrading enzymes from selected white-rot fungi. Production and role in lignin degradation. *FEMS Microbiology Reviews*, 13:125-135

Hodaifa, G., Martinez, M. E. & Sanchez, S. (2008). Use of industrial wastewater from olive-oil extraction for biomass production of *Scenedesmus obliquus*. *Bioresource Technology*, 99:1111-1117

Jaouani, A., Guillén, F., Penninckx, M. J., Martinez, A. T. & Martinez, M. J. (2005). Role of *Pycnoporus coccineus* laccase in the degradation of aromatic compounds in olive oil mill wastewater. *Enzyme and Microbial Technology*, 36:478-486

Jaouani, A., Sayadi, S., Vanthournhout, M. & Penninckx, M. J. (2003). Potent fungi for decolourisation of olive oil mill wastewaters. *Enzyme and Microbial Technology*, 33:802-809

Jaouani, A., Tabka, M. G. & Penninckx, M. J. (2006). Lignin modifying enzymes of *Coriolopsis polyzona* and their role in olive oil mill wastewaters decolourisation. *Chemosphere*, 62:1421-1430

Kachouri, F & Hamdi, M. (2004) Enhancement of polyphenols in olive oil by contact with fermented olive mill wastewater by *Lactobacillus plantarum*. *Process Biochemistry*, 39:841-845

Kalmis, E., Azbar, N., Yildiz, H. & Kalyoncu, F. (2008). Feasibility of using olive mill effluent (OME) as a wetting agent during the cultivation of oyster mushroom, *Pleurotus ostreatus*, on wheat straw. *Bioresource Technology*, 99:164-169

Kestioglu, K., Yonar, T. & Azbar, N. (2005). Feasibility of physico-chemical treatment and Advanced Oxidation Processes (AOPs) as a means of pretreatment of olive mill effluent (OME). *Process Biochemistry*, 40:2409-2416

Koidis, A., Triantafillou, E. & Boskou, D. (2008). Endogenous microflora in turbid virgin olive oils and the physicochemical characteristics of these oils. *European Journal of Lipid Science Technology*, 110:164-171

Lanciotti, R., Gianotti, A., Baldi, D., Angrisani, R., Suzzi, G., Mastrocola, D. & Guerzoni, M. E. (2005). Use of *Yarrowia lipolytica* strains for the treatment of olive mill wastewater. *Bioresource Technology*, 96:317-322

Li, A., Antizar-Ladislao, B. & Khraisheh, M. A. M. (2007). Bioconversion of municipal solid waste to glucose for bio-ethanol production. *Bioprocess and Biosysteme Engineering*, 30:189-196

Linares, A., Manuel Caba, J., Ligero, F., de la Rubia, T. & Martinez, J. (2003). Detoxification of semisolid olive-mill wastes and pine-chip mixtures using Phanerochaete flavido-alba. *Chemosphere*, 51:887-891

Lopez, M. J., Moreno, J. & Ramos-Cormenzana, A. (2001). *Xanthomonas campestris* strain selection for xanthan production from olive mill wastewaters. *Water Research*, 35:1828-1830

Lopez, M. J. & Ramos-Cormenzana A. (1996). Xanthan production from olive-mill wastewaters. *International Biodeterioration and Biodegradation*, 38:263-270

Mafakher, L., Mirbagheri, M., Darvishi, F., Nahvi, I., Zarkesh-Esfahani, H. & Emtiazi, G. (2010). Isolation of lipase and citric acid producing yeasts from agro-industrial wastewater. *New Biotechnology*, 27:337-340

Mantzavinos, D. & Kalogerakis, N. (2005). Treatment of olive mill effluents: Part I. Organic matter degradation by chemical and biological processes--an overview. *Environment International*, 31:289-295

McNamara, C. J., Anastasiou, C. C., O'Flaherty, V. & Mitchell, R. (2008). Bioremediation of olive mill wastewater. *International Biodeterioration and Biodegradation*, 61:127-134

Morillo, J. A., Antizar-Ladislao, B., Monteoliva-Sánchez, M., Ramos-Cormenzana, A. & Russell, N. J. (2009). Bioremediation and biovalorisation of olive-mill wastes. *Applied Microbiology and Biotechnololgy*, 82:25-39

Morillo, J. A., Guerra Del Águila, V., Aguilera, M., Ramos-Cormenzana, A. & Monteoliva-Sánchez, M. (2007). Production and characterization of the exopolysaccharide produced by *Paenibacillus jamilae* grown on olive mill-waste waters. *World Journal of Microbiology and Biotechnology*, 23:1705-1710

Ntougias, S. & Russell, N. J. (2001). *Alkalibacterium olivoapovliticus* gen. nov., sp. nov., a new obligately alkaliphilic bacterium isolated from edible-olive wash-waters. *International Journal of Systematic Evolution Microbiology*, 51:1161-1170

Okafor, N. (2007). *Modern industrial microbiology and bitechnology*. Science Publishers, Enfield, USA

Pantazaki, A. A., Dimopoulou, M. I., Simou, O. M. & Pritsa, A. A. (2010). Sunflower seed oil and oleic acid utilization for the production of rhamnolipids by *Thermus thermophilus* HB8. *Applied Microbiology and Biotechnology*, 88:939-951

Papanikolaou, S., Dimou, A., Fakas, S., Diamantopoulou, P., Philippoussis, A., Galiotou-Panayotou, M. & Aggelis, G. (2011). Biotechnological conversion of waste cooking olive oil into lipid-rich biomass using *Aspergillus* and *Penicillium* strains. *Journal of Applied Microbiology*, 110:1138-1150

Papanikolaou, S., Galiotou-Panayotou, M., Fakas, S., Komaitis, M. & Aggelis, G. (2008). Citric acid production by *Yarrowia lipolytica* cultivated on olive-mill wastewater-based media. *Bioresource Technology*, 99:2419-2428

Peixoto, F., Martins, F., Amaral, C., Gomes-Laranjo, J., Almeida, J. & Palmeira, C. M. (2008). Evaluation of olive oil mill wastewater toxicity on the mitochondrial bioenergetics after treatment with *Candida oleophila*. *Ecotoxicology and Environmental Safety*, 70: 266-275

Piperidou, C., Chaidou, C., Stalikas, C., Soulti, K., Pilidis, G. & Balis, C. (2000). Bioremediation of olive oil mill wastewater: chemical alterations induced by *Azotobacter vinelandii*. *Journal of Agriculture and Food Chemistry*, 48:1941-1948

Pozo, C., Martinez-Toledo, M. V., Rodelas, B. & Gonzalez-Lopez, J. (2002). Effects of culture conditions on the production of polyhydroxyalkanoates by *Azotobacter chroococcum* H23 in media containing a high concentration of alpechin (wastewater from olive oil mills) as primary carbon source. *Journal of Biotechnology*, 97:125-131

Ramachandran, S., Roopesh, K., Nampoothiri, K. M., Szakacs, G. & Pandey, A. (2005). Mixed substrate fermentation for the production of phytase by *Rhizopus* spp. using oilcakes as substrates. *Process Biochemistry*, 40:1749-1754

Ramos-Cormenzana, A., Juarez-Jimenez, B. & Garcia-Pareja, M. P. (1996). Antimicrobial activity of olive mill wastewaters (alpechin) and biotransformed olive oil mill wastewater. *International Biodeterioration and Biodegradation*, 38:283-290

Ramos-Cormenzana, A., Monteoliva-Sanchez, M. & Lopez, M. J. (1995). Bioremediation of alpechin. *International Biodeterioration and Biodegradation*, 35:249-268

Robles, A., Lucas, R., de Cienfuegos, G. A. & Gálvez, A. (2000). Biomass production and detoxification of wastewaters from the olive oil industry by strains of *Penicillium* isolated from wastewater disposal ponds. *Bioresource Technology*, 74:217-221

Roig, A., Cayuela, M. L. & Sánchez-Monedero, M. A. (2006). An overview on olive mill wastes and their valorisation methods. *Waste Management*, 26:960-969

Ruiz, J. C., Teresa, R., Pérez, J. & Lopez, M. J. (2002). Effect of olive oil mill wastewater on extracellular ligninolytic enzymes produced by *Phanerochaete flavido-alba*. *FEMS Microbiology Letters*, 212:41-45

Saavedra, M., Benitez, E., Cifuentes, C. & Nogales, R. (2006). Enzyme activities and chemical changes in wet olive cake after treatment with *Pleurotus ostreatus* or *Eisenia fetida*. *Biodegradation*, 17:93-102

Sampedro, I., D'Annibale, A., Ocampo, J. A., Stazi, S. R. & Garcia-Romera, I. (2007a). Solid-state cultures of *Fusarium oxysporum* transform aromatic components of olive-mill dry residue and reduce its phytotoxicity. *Bioresource Technology*, 98:3547-3554

Sampedro, I., Marinari, S., D'Annibale, A., Grego, S., Ocampo, J. A. & García-Romera, I. (2007b). Organic matter evolution and partial detoxification in two-phase olive mill waste colonized by white-rot fungi. *International Biodeterioration and Biodegradation*, 60:116-125

Sanjust, E., Pompei, R., Rescigno, A., Rinaldi, A. & Ballero, M. (1991). Olive milling wastewater as a medium for growth of four *Pleurotus* species. *Applied Biochemistry and Biotechnology*, 31:223-235

Scioli, C., & Vollaro, L. (1997). The use of *Yarrowia lipolytica* to reduce pollution in olive mill wastewaters. *Water Research*, 31:2520-2524

Suh, M. J., Baek, K. Y., Kim, B. S., Hou, C. T. & Kim, H. R. (2011). Production of 7,10-dihydroxy-8(E)-octadecenoic acid from olive oil by *Pseudomonas aeruginosa* PR3. *Applied Microbiology and Biotechnology*, 89:1721-1727

Tomati, U., Galli, E., Di Lena, G. & Buffone, R. (1991). Induction of laccase in *Pleurotus ostreatus* mycelium grown in olive oil waste waters. *Agrochimica*, 35:275-279

Treichel, H., de Oliveira, D., Mazutti, M. A., Di Luccio, M. & Oliveira, V. J. (2010). A review on microbial lipases production. *Food Bioprocess Technology*, 3:182-196

Tsioulpas, A., Dimou, D., Iconomou, D. & Aggelis, G. (2002). Phenolic removal in olive oil mill wastewater by strains of *Pleurotus* spp. in respect to their phenol oxidase (laccase) activity. *Bioresource Technology*, 84:251-257

Vakhlu, J. & Kour, A. (2006). Yeast lipases: enzyme purification, biochemical properties and gene cloning. *Electronic Journal of Biotechnology*, 9:69-85

Vassilev, N., Fenice, M., Federici, F. & Azcon, R. (1997). Olive mill waster water treatment by immobilized cells of *Aspergillus niger* and its enrichment with soluble phosphate. *Process Biochemistry*, 32:617-620

Vassilev, N., Vassileva, M., Bravo, V., Fernandez-Serrano, M., & Nikolaeva, I. (2007). Simultaneous phytase production and rock phosphate solubilization by *Aspergillus niger* grown on dry olive wastes. *Industrial Crops and Products*, 26:332-336

Wu, L., Ge, G. & Wan, J. (2009). Biodegradation of oil wastewater by free and immobilized *Yarrowia lipolytica* W29. *Journal of Environmental Sciences*, 21:237-242

Yesilada, O., Sik, S. & Sam, M. (1997). Biodegradation of olive oil mill wastewater by *Coriolus versicolor* and *Funalia trogii*: effects of agitation, initial COD concentration, inoculum size and immobilization. *World Journal of Microbiology and Biotechnology*, 14:37-42

Zouari, N. & Ellouz, R. (1996). Microbial consortia for the aerobic degradation of aromatic compounds in olive oil mill effluent. *Journal of Industrial Microbiology*, 16:155-162

Zullo, B. A. & Ciafardini, G. (2008). Lipolytic yeasts distribution in commercial extra virgin olive oil. *Food Microbiology*, 25:970-977

Zullo, B. A., Cioccia, G. & Ciafardini, G. (2010). Distribution of dimorphic yeast species in commercial extra virgin olive oil. *Food Microbiology*, 27:1035-1042.

4

Genetic Improvement of Olives, Enzymatic Extraction and Interesterification of Olive Oil

Fabiano Jares Contesini[1,*], Camilo Barroso Teixeira[1], Paula Speranza[1],
Danielle Branta Lopes[1], Patrícia de Oliveira Carvalho[2],
Hélia Harumi Sato[1] and Gabriela Alves Macedo[1]
*[1]Laboratory of Food Biochemistry, Department of Food Science,
College of Food Engineering, State University of Campinas (UNICAMP), Campinas, SP,
[2]Multidisciplinar Laboratory, University São Francisco, Bragança Paulista, SP,
Brazil*

1. Introduction

Extra virgin olive oil (EVOO) contains a wide range of bioactive compounds which give it its particular aroma and taste. It is a well-known key component in the traditional Mediterranean diet and due to its high levels of phenolics and unsaturated fatty acids, it is believed to be associated with good health and a relatively long life (De Faveri et al., 2008). The phenolic compounds have the ability to reduce the oxidative modification of low-density lipoproteins (Fitó et al., 2005), which play a key role in the development of atherosclerosis and coronary heart disease. Moreover, olive oil is very resistant to peroxidation (Najafian et al., 2009), a fact conferring great oxidative stability to the product (Bendini et al., 2006). The sustainable development of the agriculture and food industry is dependent upon powerful biotechnological tools which meet the demands of the new urbanized population. The improvement in the properties of EVOO is a good example of how useful the application of biotechnology to improve food quality is.

The olive oil extraction process is extremely important for its quality. During this step, the content of some components is significantly altered, depending on the extraction technique employed. A new process for the extraction of olive oil that has been studied is the addition of enzyme preparations during malaxation. This reduces the complexing of hydrophilic phenols with polysaccharides, increasing the concentration of free phenolic compounds in the olive paste and their consequent dissolution in the oil and waste waters during processing (De Faveri et al., 2008). The enzymes most used in the extraction of EVOO are microbial pectinases and cellulases, which hydrolyse the cell wall of the olive fruits, liberating the oil and phenolic compounds. This method has some advantages compared to traditional methods, giving higher oil and phenolic compound extraction yield. It also involves lower energy costs and possibly provides an oil with improved health properties

---

* Corresponding Author

due to the liberation of the phenolic compounds. Microbial lipases can also be used to synthesize structured lipids from olive oil triacylglycerols.

This review discusses mainly the genetic improvement of *Olea europaea* to achieve higher quality EVOO, with lower production costs and greater productivity. Additionally, it reports on the use of enzymes to improve the extraction of virgin olive oil from olives and the enzymatic synthesis of lipids based on olive oil triacylglycerols.

2. Genetic modification of olive cultivars: Crossbreeding

Improvement of the cultivars is one of the major targets of olive biotechnology. The recent diffusion of the olive outside its traditional cultivation area, the Mediterranean basin, together with a continuous trend for industrial modernization, has recently increased the demand for improved cultivars. As a result, clonal selection and crossbreeding programmes have been applied in olive growing countries, aiming at selecting genotypes with improved characteristics. The desirable characteristics are early bearing, resistance to pests and to abiotic stress (such as frost and drought), limited alternate bearing and suitability for intensive cultivation and mechanical harvesting. In relation to the product, the search is for high-quality production with respect to the organoleptic characteristics of both the fruits and oils, and finally a high content of bioactive substances that may favourably affect human health (Fabbri et al., 2009). Olive crossbreeding programmes have provided new genotypes with a wide range of variation for all the characteristics, including the oil composition (Belaj et al., 2010). This technique has been used to generate new cultivars from traditional ones, which are used as the genitors. For instance, in Tunisia, an olive breeding programme started in 1989 with a cross between Tunisian and foreign cultivars. This created new cultivars with a quality of oil superior to that of *Chemlali* (the main olive cultivar in the country, characterized by low levels of oleic acid) and characteristics close to the standards of the international market (Baccouri et al., 2007; León et al., 2011; Manai et al., 2007; Manai et al., 2008; Rjiba et al., 2010). The large variability in all the components of olive oil in these advanced selections suggests that diversity in olive oil composition could be obtained in any crossbreed progeny. Therefore, any breeding programme provides diversity of the oils.

In Spain a breeding programme began in 1992 to obtain new olive cultivars with some of the following advantages: early bearing, high productivity and oil content, resistance to peacock eye (*Spilocaea oleagina*, Cast), suitability for mechanical harvesting and high olive oil quality (León et al., 2004a).

León et al. (2011) selected fifteen genotypes from crosses between the cultivars *Arbequina*, *Frantoio* and *Picual* on the basis of their agronomic characteristics. In this work, the main components of the olive oil were characterized and compared with their genitors. A wide range of variation was observed for all the fatty acids and minor components, and for the related characteristics evaluated, with significant differences between the genotypes, except for the β-tocopherol content. The values obtained in the selections have extended the range of variation of the three genitors for all the characteristics evaluated. The selections showed the highest average values for tocopherols, polyphenols and the C18:1 content, respectively. The breeding procedures used to obtain these selections including crossing, the forced growth protocol and an initial seedling evaluation, are all described (León et al., 2004a; León et al., 2004b; Santos-Antunes et al., 2005).

A wide range of variation was observed for all the fatty acids, minor components and related characteristics evaluated by León et al. (2011). The fatty acid C18:1 was the predominant fatty acid in all the selections, with values ranging from 62 to 81%. Together with C16:0 and C18:2, it accounted for more than 94% of the total fatty acid composition, on average. The genotypes producing olive oils with high oleic acid percentages could be of particular interest for planting in low latitude locations, where the oleic acid content tends to be too low (Ripa et al., 2008). Of the minor components, α-tocopherol represented more than 90% of the total tocopherols, whereas the total polyphenol content varied widely from 67 to 1033 mg/kg. A wide range of variation was also obtained for stability, with values ranging from 16 to 195 h.

The statistical analysis showed that genotypic variance was the main contributor to the total variance for all the fatty acids and ratios evaluated, with significant differences between the genotypes in all cases. In fact, the effect was significant for all the fatty acids, except C18:3, all the minor components and related characteristics evaluated, α-tocopherol and stability, but was lower for the other characteristics. Several studies have demonstrated that the quality of olive oil is greatly determined by genetic (cultivars) factors. For instance, in the Germplasm Banks of Catalonia and Cordoba, Tous et al. (2005) and Uceda et al. (2005), respectively, showed that more than 70% of the variation in the fatty acids (except for C18:3) and several minor components, such as the polyphenol content, bitter index (K225) and oil stability, was due to genetic effects. It should be noted that many other factors including pedoclimatic aspects, olive ripeness, olive harvesting methods and the olive extraction system have also been reported as quality indicators of virgin olive oil (Aguilera et al., 2005; Guerfel et al., 2009; León et al., 2011).

Ayton et al. (2007) found a stronger relationship between the polyphenols content and oil stability when individual cultivars were analyzed separately, which suggests that the relationship between induction time and total polyphenol content is different for each cultivar. In another study (León et al., 2011), the ranking of the cultivars was different for the polyphenols content and oil stability, which could suggest that not only the total polyphenol content, but also different polyphenol profiles in the different cultivars could have distinct antioxidant effects. Similar results have been reported for the analysis of the composition and oxidative stability of virgin olive oil from selected wild olives (Baccouri et al., 2008). The correlation between the different fatty acids also agrees with what was previously reported for olive cultivar collections and breeding progenies (León et al., 2004a).

Significant differences between the genotypes obtained for crosses between *Arbequina*, *Frantoio* and *Picual* were observed for the fatty acid composition, minor components and related characteristics. The multivariate analysis allowed for the classification of the genotypes into four groups according to their olive oil compositions. Further work will be required to determine the best selections to adapt to different environmental conditions, as well as the optimal harvesting periods in terms of optimal oil quality (León et al., 2011).

Ripa et al. (2008) evaluated oil quality, in terms of fatty acid composition and content in phenolic compounds, for many new genotypes previously selected in a breeding programme and cultivated in three different locations of central and southern Italy. The availability of data from many genotypes cultivated in all three locations allowed quantitative analyses of the genetic and environmental effects on the oil quality traits studied. The acidic composition varied greatly both with genotype and with environment, and so did the concentration in phenols, though the effect of genotype on phenols was not significant. The fatty acid composition appeared predominantly under genetic control while the environmental effect explained 0.31 of the total variance. The oil content in phenolic compounds, instead, had lower heritability (0.29) and was more affected by the environment, which explained 0.50 of the total variance. Few genotypes were selected as the best for each location, but none performed best in all locations. This suggests that, in olives, the highest oil quality is difficult to achieve with a single genotype in different environments, due to a strong or even predominant effect of the environment on some quality traits. More likely, combinations of genotypes and territories can produce oils with high and typical quality.

3. The use of enzymes in the extraction of olive oil

The most commonly used method for the extraction of olive oil is the mechanical process, however some of the non-extracted oil remains in the solid residue or cake. The majority of the oil is located in the vacuoles as free oil, but oil dispersed in the cytoplasm is not extractable and is therefore lost in the waste (Najafian et al., 2009). Therefore the cell walls must be destroyed to effectively recover the oil enclosed in the cell. The use of enzymes has been studied for the hydrolysis of the different types of polysaccharides in the cell wall structure (Chiacchierini et al., 2007). The major polysaccharides found in the cell wall of the olive fruit were pectic polysaccharides and the hemicellulosic polysaccharides xyloglucan and xylan (Vierhuis et al., 2003).

Several innovating biotechnological techniques have been studied to obtain high-quality oils and/or improve product outputs. They include the use of microorganisms (Kachouri & Hamdi, 2004, 2006) or enzymes (Vierhuis et al., 2001) during different steps of the oil processing procedure. Several enzyme processing aids have been successfully tested for olives in recent years (De Faveri et al., 2006; García et al., 2001). Different enzymes are naturally present inside the olive fruit, but are strongly deactivated during the pressing phase, most likely due to the formation of oxidized phenols bonding to the enzyme prosthetic group (Vierhuis et al., 2001). In this case, the addition of suitable enzymes to the olive paste during the mixing step was proposed as a tool to replace the deactivated natural ones (Ranalli et al., 2001). Furthermore, the enzyme complexes are water-soluble and after the application, they are found in the olive mill waste waters, indicating that the oil composition is not modified (Chiacchierini et al., 2007).

Ranalli et al. (2003a) estimated the composition of three types of olive oil (Caroleo, Coratina and Leccino) extracted by the application of the Bioliva enzymatic complex. During extraction, the action of the enzymes on the fruit tissues resulted in the release of greater amounts of oil and other constituents, which dissolved in the oily phase (Ranalli et al., 2003b). The enzymatic application resulted in an increase in several key compounds, such as phenols, tocopherols, and flavour compounds, without changing the natural parameters related to product authenticity (waxes, sterols, triterpene alcohols, fatty acids and triacylglycerol composition).

The loss of phenols during processing can be attributed to interactions between the polysaccharides and phenolic compounds present in the olive pastes (Servili et al., 2004). Studies show that the addition of commercial enzyme preparations during the malaxation can reduce the complexation of hydrophilic phenols with polysaccharides. It increases the concentration of free phenols in the olive paste and their consequent release into the oil and waste waters during processing (De Faveri et al., 2008).

A mixture of three enzyme formulations was tested by Aliakbarian et al. (2008) to improve the yield and the quality of the olive oil obtained from the Italian cultivar Coratina. Since no single enzyme is adequate for the efficient maceration and extraction of oil from olives, pectinase, cellulase and hemicellulase were essential for this purpose (Chiacchierini et al., 2007; De Faveri et al., 2008). A homogeneous mixture of the three different enzyme formulations was used at the beginning of the malaxation step in the proportions 33.3:33.3:33.3% (v/v/v). This choice was suggested by the higher efficacy of these enzymes in releasing phenolics into the oil when working as a ternary system (A:B:C), rather than in binary combinations (A:B, A:C, B:C) (De Faveri et al., 2008). In summary, A is a complex formulation containing pectinase plus cellulase and hemicellulase; B shows equilibrated pectinase–hemicellulase activity; C is a pectolytic enzyme. The enzymes selected are naturally present inside the olive fruit, but are strongly deactivated during the critical pressing step, presumably because of the oxidation (Chiacchierini et al., 2007). The highest levels of total polyphenols (874 $\mu g_{CAE}/g_{oil}$), antiradical power (25.1 $\mu g_{DPPH}/\mu L_{extract}$) and o-diphenols ($\mu g_{CAE}/g_{oil}$) were all reached at the highest enzyme concentration (25 mL/kg_{paste}). Moreover, the highest oil extraction yield (17.5 $g_{oil}/100$ g_{paste}) was reached with the longest malaxation time (t = 150 min), always with the highest enzyme concentration.

4. Enzymatic synthesis of structured lipids

The enzymatic synthesis of structured lipids is relatively new in lipid modification. Although enzymes have been used for several years to modify the structure and composition of foods, they have only recently become available for large-scale use, mainly because of the high cost. Within this context lipases are reported for the enzymatic synthesis of structured lipids. They have the ability to carry out hydrolytic reactions, but the manipulation of the reaction at low water levels permits their use also for the synthesis of triacylglycerols. These enzymes can be successfully used in the production of lipids structured for medical purposes (De Araújo, 2011).

Enzymatic modification of olive oil triacylglycerols has been discussed by Boskou (2006, 2009). The development of techniques for the preparation of oils and fats from enzyme-modified olive oil is an attractive prospect for the food industry, given the high oxidative stability of the product at frying temperatures and the health enhancing properties of this material (Criado et al., 2007).

Nunes et al. (2011) produced structured triacylglycerols containing medium chain fatty acids, by the acidolysis of virgin olive oil (VOO) with caprylic or capric acids in a solvent-free media or in n-hexane, catalyzed by immobilized lipases from *Thermomyces lanuginosa*, *Rhizomucor miehei* and *Candida antarctica*. The results indicated that the incorporation was always greater for capric than for caprylic acid, but for both acids, higher incorporation was always attained in solvent-free media. All the biocatalysts presented 1,3-regioselectivity. The lipases from *Rhizomucor miehei* and *Candida antarctica* were the biocatalysts presenting the highest operational stability, together with high incorporation levels and low acyl migration in the batch production of structured lipids by the acidolysis of VOO with caprylic or capric acids. Therefore, these biocatalysts seem to be the most adequate for the implementation of a process aimed at the production of triacylglicerols containing medium and long fatty acids (MLM) rich in caprylic and capric acids. The structured triacylglycerol obtained from VOO has oleic acid at the sn-2 position, indicating a better absorption, whilst medium chain fatty acids will mainly be esterified at the external positions of the TAG molecules.

Criado et al. (2007) have also studied the enzymatic interesterification of extra virgin olive oil with a fully hydrogenated palm oil to produce lipids with desirable chemical, physical and functional properties. The sn-1,3 non-specific immobilized lipases from Candida antarctica and two sn-1,3 specific immobilized lipases from *Thermomyces lanuginosus* and *Rhizomucor miehei* were employed as the biocatalysts. The authors concluded that the oxidative stability increased when the percentage of TAG containing multiple fully saturated residues increased. In all the cases studied, the stability of the physical blend was higher than that of the reaction products. The final products were considered as plastic over wider temperature ranges. The large amount of unsaturated residues present in these samples, primarily oleic acid residues, was the factor leading to the extended plasticity range of these interesterified mixtures.

5. Conclusion

The improvement of the properties of extra virgin olive is crucial, considering the extensive number of functional compounds present in this oil. Genetic improvement techniques are the option showing the most promising results. The genetic modification of olive cultivar crossbreeding is focused on solving agronomic and commercial problems, such as the control of fruit ripening and increase in the oil content and quality. Another improvement suggested by different research groups is the use of enzymes added to the paste to improve extraction of the oil and the bioactive compounds it contains. Enzymes, such as microbial lipases, can also be used to synthesize different high-value products from tryacylglicerol of olive oil. It is possible to obtain structured lipids that have functional and technological properties suitable for applications in the food, pharmaceutical and oleochemical industries. This can be considered as a "green" industrial process, taking into account that these compounds are currently synthesized by chemical processes that use catalysts and generate byproducts.

6. References

Aguilera, M. P., Beltrán, G., Ortega, D., Fernández, A., et al. (2005) Characterization of virgin olive oil of Italian olive cultivars: 'Frantoio' and 'Leccino', grown in Andalusia. *Food Chemistry*. V.89, pp.387–391.

Aliakbarian, B, De Faveri D, Converti A, Perego P. (2008) Optimisation of olive oil extraction by means of enzyme processing aids using response surface methodology. *Biochemical Engineering Journal*.V. 42, pp.34-40.

Ayton, J., Mailer, R. J., Haigh, A., Tronson, D., Conlan, D. (2007) Quality and oxidative stability of Australian olive oil according to harvest date and irrigation. *Journal of Food Lipids*, V.14, pp.138–156.

Baccouri, B., Ben Temime, S., Taamalli, W., Daoud, D. et al. (2007) Analytical characteristics of virgin olive oils from two new varieties obtained by controlled crossing on Meski variety. *Journal of Food Lipids*, V.14, pp.19–34.

Baccouri, B., Zarrouk, W., Baccouri, O., Guerfel, M., et al. (2008) Composition, quality and oxidative stability of virgin olive oils from some selected wild olives (Olea europaea L. subsp. Oleaster). *Grasas y Aceites*, V.59, pp.346–351.

Belaj, A., Munõz-Diez, C., Baldoni, L., Satovic, Z.,Barranco, D. (2010) Genetic diversity and relationships of wild and cultivated olives at regional level in Spain. *Scientific Horticulture*, V.124, pp.323–330.

Bendini A, Cerretani L, Vecchi S, Carrasco-Pancorbo A, Lercker G. (2006) Protective effects of extra virgin olive oil phenolics on oxidative stability in the presence or absence of copper ions. *Journal of Agriculture and Food Chemistry*. V.54, pp.4880–4887.

Boskou D. (Ed.). (2009). *Olive Oil. Minor Constituents and Health*. CRC Press Taylor & Francis Group, ISBN-13:978-1-4200-5993-9, United States.

Boskou D. (Ed.). (2006). *Olive Oil: Chemistry and Technology*. CRC Press Taylor & Francis Group. ISBN-13: 978-1-893997-88-2, ISBN-10: 1-893997-88-X, United States.

Criado M.; Hernandez-Martín E., Lopez-Hernandez A.; Otero C. (2007) Enzymatic Interesterification of Extra Virgin Olive Oil with a Fully Hydrogenated Fat: Characterization of the Reaction and Its Products. *Journal of American Oil Chemical Society*, V.84, pp.717–726.

Chiacchierini E, Mele G, Restuccia D, Vinci G. (2007) Impact evaluation of innovative and sustainable extraction technologies on olive oil quality. *Trends in Food Science and Technology*. V.18, pp.299–305.

De Araujo, M. E. M. B., Campos, P. R. B., Noso, T. M., Oliveira, R. M. A., Cunha, I. B. S., Simas, R. C., Eberlin, M. N., Carvalho, P.O. (2011) Response surface modelling of the production of structured lipids from soybean oil using *Rhizomucor miehei* lipase. *Food Chemistry*, V. 127, pp. 28-33.

De Faveri D, Aliakbarian B, Avogadro M, Perego P, Converti A. (2008) Improvement of olive oil phenolics content by means of enzyme formulations: effect of different enzyme activities and levels. *Biochemical Engineering Journal*, V.41, pp.149-156.

De Faveri D, Torre P, Aliakbarian B, Perego P, Domínguez JM, Rivas Torres B. (2006) Effect of different enzyme formulations on the improvement of phenolic compound content in olive oil, in: Proceedings of the IUFoST 13thWorld Congress of Food Science & Technology, Nantes, France, September, pp. 927-928.

Fabbri A, Lambardi M, Ozden-Tokatli Y. (2009). Olive Breeding. In:*Breeding Plantation Tree Crops: Tropical Species*, Springer Science+Business Media, LLC. Cp.12, p.423-465.

Fitó M, Cladellas M, de la Torre R, Martí J, Alcántara M, Pujadas-Bastardes M, Marrugat J, Bruguera J, López-Sabater MC, Vila J, Covas MI. (2005) Antioxidant effect of virgin olive oil in patients with stable coronary heart disease: a randomized, crossover, controlled, clinical trial. *Atherosclerosis*. V.181, pp.149-158.

García A, Brenes M, Moyano MJ, Alba J, Garcia P, Garrido A. (2001) Improvement of phenolic compound content in virgin olive oils by using enzymes during malaxation. *Journal of Food Engineering*.V.48,pp.189–194.

Guerfel M, Ouni Y, Taamalli A, Boujnah D, Stefanoudaki E, Zarrouk M. (2009) Effect of location on virgin olive oils of the two main Tunisian olive cultivars. *European Journal of Lipid Science and Technology*. V.111, pp.926-932.

Kachouri F, Hamdi M. (2004) Enhancement of polyphenols in olive oil by contact with fermented olive mill wastewater by *Lactobacillus plantarum*. *Process Biochemistry*. V.39, pp.841–845.

Kachouri F, Hamdi M. (2006) Use *Lactobacillus plantarum* in olive oil process and improvement of phenolic compounds content, *Journal of Food Engineering*. V.77, pp.746-752.

León L, Beltrán G, Aguilera MP, Rallo L, Barranco D, De La Rosa R. (2011) Oil composition of advanced selections from an olive breeding program. *European Journal of Lipid Science and Technology*. V.113, pp.870–875.

León L, De la Rosa R, Barranco D, Rallo L. (2007) Breeding for early bearing in olive. *Hort.Science*, V.42, pp.499–502.

León L, Martín LM, Rallo L. (2004b) Phenotypic correlations among agronomic traits in olive progenies. *J. Am. Soc. Hortic. Sci.* 129, 271–276.

León L, Uceda M, Jiménez A, Martín LM, Rallo L. (2004a) Variability of fatty acid composition in olive (Olea europaea, L) progenies. *Spanish J. Agric. Res*, 2, 353–359.

Manai H, Haddada F M, Trigui A, Daoud D, Zarrouk M. (2007) Compositional quality of virgin olive oil from two new Tunisian cultivars obtained through controlled crossings. *Journal of Science and Food Agriculture.*, V.87, pp.600–606.

Manai H, Mahjoub-Haddada F, Oueslati I, Daoud D, Zarrouk M. (2008) Characterization of monovarietal virgin olive oils from six crossing varieties. *Scientia Horticulturae.* V.115, pp.252– 260.

Najafian L, Ghodsvali A, Khodaparast MHH, Diosady LL. (2009) Aqueous extraction of virgin olive oil using industrial enzymes. *Food Research International.* V.42, pp.171-175.

Nunes PA, Pires-Cabral P, Ferreira-Duas, S. (2011) Production of olive oil enriched with medium chain fatty acids catalysed by commercial immobilised lipases *Food Chemistry* V. 127, Pp. 993-998

Ranalli A, Malfatti A, Cabras P. (2001) Composition and quality of pressed virgin olive oils extracted with a new enzyme processing aid. *Journal of Food Science.* V.66, pp.592-603 (a).

Ranalli A, Pollastri L, Contento S, Iannucci E. (2003). The glyceridic and nonglyceridic constituents in virgin olive oil after use of a novel method of enzyme extraction. *International Journal of Food Science and Technology.* V.38, pp.17-27 (a).

Ranalli A, Pollastri L, Contento S, Lucera, Del Re P. (2003) Enhancing the quality of virgin olive oil by use of a new vegetable enzyme extract during processing. *European Food Research Technology.* V.216, pp.109–115 (b).

Ripa V, De Rose F, Caravita MA, Parise MR., et al. (2008) Qualitative evaluation of olive oils from new olive selections and effects of genotype and environment on oil quality. *Advanced Horticulture Science.* V.22, pp.95–103.

Rjiba I, Dabbou S, Gazzah N, Hammami M. (2010) Effect of crossbreeding on the chemical composition and biological characteristics of Tunisian new olive progenies. *Chemistry & Biodiversity.* V.7, pp.649–655.

Santos-Antunes AF, León L, De la Rosa R, Alvarado J, et al. (2005) The length of the juvenile period of olive seedlings as influenced by vigor and the choose of genitors. *Horticulture Science.*V.40, pp.1213–1215.

Servili M, Selvaggini R, Esposto S, Taticchi A, Montedoro G, Morozzi G. (2004) Health and sensory properties of virgin olive oil hydrophilic phenols: agronomic and technological aspects of production that affect their occurrence in the oil. *Journal of Chromatography A*, V. 1054, pp. 113-127.

Tous J, Romero A, Díaz I, Rallo L, Barranco D, Caballero J, Martín A, Del Río C, Tous J, Trujillo I. (2005) *Las Variedades de Olivo Cultivadas en España, Libro II: Variabilidad y Selección*, Junta de Andalucía, MAPA and Ediciones Mundi-Prensa, Madrid (Spain), pp. 357–372.

Uceda M, Beltrán G, Jiménez A, Rallo L, Barranco D, Caballero J, Martín A, Del Río C, Tous J, Trujillo I. (2005) *Las Variedades de Olivo Cultivadas en España, Libro II: Variabilidad y Selección*, Junta de Andalucía, MAPA and Ediciones Mundi-Prensa, Madrid (Spain), pp. 357–372.

Vierhuis E, Korver M, Schols HA, Voragen AGJ. (2003). Structural characteristics of pectic polysacharides from olive fruit (Olea europaea cv moraiolo) in relation to processing for oil extraction. *Carbohydrate Polymers.* V.51, pp.135–148.

Vierhuis E, Servili M, Baldioli M, Schols HÁ, Voragen AGJ, Montedoro GF. (2001) Effect of enzyme treatment during mechanical extraction of olive oil on phenolic compounds and polysaccharides. *Journal of Agriculture and Food Chemistry.* V.49, 1218-1223.

5

Potential Applications of Green Technologies in Olive Oil Industry

Ozan Nazim Ciftci, Deniz Ciftci and Ehsan Jenab
University of Alberta
Canada

1. Introduction

Conventional olive oil production methods create large amounts of waste and by-products. Most production plants do not invest in purification and utilization of those by-products. Purification or conversion methods may add value to those by-products and prevent the environmental pollution.

Global trends show that "green" products and technologies are needed. Increasing environmental concerns, government measures and population drive the search for green processes to replace the conventional ones. This search is essential to achieve sustainable processing and to reduce commercial energy use (Clark, 2011). There are several applications for green technology in the olive oil industry.

This chapter reviews the potential applications of major green processes such as supercritical fluid extraction, membrane technology, bioconversions and molecular distillation in the olive oil industry.

2. Supercritical fluid technology

Supercritical Fluid Technology (SFT) has received growing interest as a green technology, with extraction being the main application in the food industry. Fluids become supercritical by increasing pressure and temperature above the critical point. Supercritical fluids have liquid-like solvent power and gas-like diffusivity. These physical properties make them ideal clean solvents for extraction of lipids.

Carbon dioxide (CO_2) is the most widely used supercritical fluid due to a lack of toxicity and flammability, low cost, wide availability, tunable solvent properties, and moderate critical temperature and pressure (31.1°C and 7.38 MPa) (Black, 1996). Because of the relatively low viscosity, high molecular diffusivity and low surface tension of the system, mass transfer is improved in supercritical CO_2 (SC-CO_2) in comparison to liquid organic solvents (Oliveira & Oliveira, 2000). Moreover, separation of CO_2 from the product can easily be achieved by reduction of pressure, because the products do not dissolve in CO_2 at atmospheric pressure.

Another unique property of supercritical fluids is their selectivity. The density of a supercritical fluid is higher than that of a gas, making them better solvents. Extraction selectivity of supercritical fluids can be changed altering density which is done by adjusting pressure and temperature. Selectivity can also be changed by the addition of a co-solvent such as ethanol, methanol, hexane, acetone, chloroform and water to increase or decrease the polarity. Ethanol is the most preferred co-solvent because it is non-toxic and meets green technology criteria (GRAS status) (Dunford, 2004).

SC-CO$_2$ processing adds value because products obtained may be considered as natural. Although SFT is used for extraction of plants and vegetables of different sources (Table 1), applications in the olive oil industry have been limited. SFT can be used in olive oil processing for extraction and deacidification, as well as separation, purification or concentration of minor components.

Sample	Analyte	Reference
Carrot	Carotenes	Vega et al. (1996)
Tomato skin	Lycopene	Ollanketo et al. (2001)
Mushrooms	Oleoresins	del Valle & Aguilera (1989)
Tea	Caffeine	Calabuig Aracil (1998)
Grape skin	Anthocyanins	Blasco et al. (1999)
Cottonseed	Lipids	Bhattacharjee et al. (2007)
Hops	Humulone, lupulone and essential oils	Langezaal et al. (1990)
Rosemary	Oil	Bensebia et al. (2009)

Table 1. Supercritical fluid extraction of different plants and vegetables.

2.1 Extraction

SC-CO$_2$ has been used to replace hexane in the olive oil industry and meets the growing demand for natural products (Temelli, 2009). The most common applications are extraction of total lipids from olive husk or minor lipid components from olive oil. Extraction of high value minor components without degradation led industry and researchers to focus on SC-CO$_2$ extraction. Fig. 1 represents a typical lab scale SC-CO$_2$ system used for extraction of lipids.

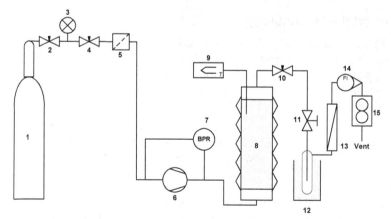

Fig. 1. Schematic diagram of a lab scale SC-CO$_2$ extraction system: 1, CO$_2$ tank; 2,4,10, shut-off valves; 3, pressure gauge; 5, filter; 6, compressor; 7, back pressure regulator; 8, extraction vessel; 9, thermocouple; 11, micrometering valve; 12; sample collector; 13, oil and moisture trap; 14, flowmeter; 15, gas meter.

SC-CO_2 extraction of olive husk oil is superior compared to conventional hexane extraction because the oil is also deacidifed and deodorised during the extraction process, and the resulting extract is free of residual solvent (Esquível & Bernardo-Gil, 1993). Esquivel and Bernardo-Gil (1993) extracted olive husk oil using SC-CO_2 under pressures of 12 to 18 MPa and temperatures of 35 to 45 °C.

2.2 Deacidification

Crude olive oil contains free fatty acids (FFA) and other impurities which must be removed, yielding a triacylglycerol (TAG) rich fraction. A high FFA content decreases the oxidative stability of the oil and leads to rancidity. A reduction in FFA content in virgin olive oil results in an increase in commercial value (Vázquez et al., 2009).

Supercritical fluid extraction has been proposed as an alternative technology for deacidification of oils and has been used for deacidification of olive pomace oil, an important by-product of olive oil industry. Crude olive pomace oil is often very acidic, darkly colored and highly oxidized. Intensive refining is thus required to make it suitable for human consumption. Neutralization is currently applied, but there are drawbacks to this process. Product yield is very low and neutralization increases the cost per unit. Therefore, it is necessary to reduce the FFA content before refining (Fadiloglu et al., 2003). Supercritical deacidification is actually a selective supercritical fluid extraction process. During the process, FFAs preferentially extracted with minimum neutral oil (TAGs, tocopherols, phytosterols) loss (Vázquez et al., 2009). A schematic diagram of a supercritical fluid extraction system for deacidification of oils is shown in Fig. 2. The oil is fed to the extraction column by a pump. The extraction column consists of two sections: an enriching (above of the oil feeding point) section, and a stripping section (below the oil feeding point). Raffinate is first separated from the extract and sent to the stripping section. Then, in the stripping section, the extract is separated from raffinate and transported to the enriching section. Extract rich in minor lipid compounds and CO_2 is separated in the separator. A specified amount of the extract is transferred to the top of the column as reflux (Brunner, 2009). CO_2 can be purified and recycled into the system. Raffinate is collected at the bottom of the column.

Deacidification of different oil sources using supercritical fluids have been performed at laboratory scale by several researchers. Turkay et al. (1996) achieved a selective and quantitative (90%) FFA extraction for deacidification of high acidic black cumin seed oil using SC-CO_2 at relatively low pressure (15 MPa) and relatively high (60 °C) temperature. Ooi et al. (1996) decreased the FFA content of palm oil to 0.1% in a continuous SC-CO_2 extractor.

Brunetti et al. (1989) obtained deacidification of high acidic olive oil with SC-CO_2 at pressures of 20 and 30 MPa, and temperatures of 40 and 60 °C. They reported that the selectivity for FFAs was highest at 20 MPa and 60 °C. Bondioli et al. (1992) studied the supercritical fluid deacidification of olive oil in the pressure range of 9–15 MPa and 40–50 °C. The acidity was reduced from 6.3% to values less than 1% at 40 °C and 13 MPa. In another application, Vázquez et al. (2009) used SC-CO_2 as an extraction solvent to remove FFAs from cold-pressed olive oil in a packed column. The acidity was reduced from 4 to 1.43% at 25 MPa and 40 °C.

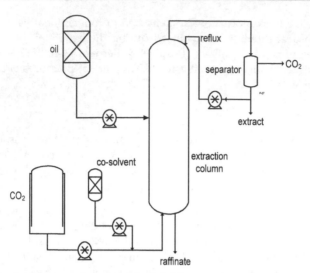

Fig. 2. Schematic diagram of supercritical fluid extraction pilot plant used for deacidification, separation, concentration and purification of oils.

2.3 Separation, concentration and purification of minor lipid compounds

Extraction of high value minor components from natural products is of great interest to food industry. SFT has been applied for purification, separation or concentration of several compounds from vegetable oils, essential oils and deodorizer distillates. These applications include purification of monoacylglycerols (MAGs) and lecithin, removal of cholesterol and limonene, and separation of squalene, tocopherols and fatty acid esters (Brunner, 2009).

Products of the olive oil industry are important sources of high value components such as tocopherols, phytosterols, squalene and fatty acids. The system used for separation of minor lipid compounds is the same as shown in Fig. 1. Fornari et al. (2008) purified squalene from a by-product obtained after distillation, esterification and transesterification of olive oil deodorizer distillates. They obtained 89.4% squalene purity and 64.2% yield at 70 °C and 18 MPa, and obtained a raffinate concentrated in TAGs and sterol compounds.

Dauksas et al. (2002) extracted tocopherols from olive tree leaves using SC-CO$_2$. They obtained a high value extract of 97.1% (w/w) tocopherol at 25 MPa and 40 °C after 1 h of extraction, and 74.48 % at the same pressure and temperature after 2 h. Le Floch et al. (1998) used supercritical fluid extraction for isolation of phenols from olive leave samples using SC-CO$_2$ modified with 10% methanol at 33.4 MPa and 100 °C.

2.4 Use of supercritical fluids as reaction media for enzymatic modification of lipids

Enzymatic interesterification in organic solvents leads to very important modifications of lipids. However, the use of organic solvents in these reactions is a disadvantage. Therefore, biosynthesis in supercritical fluids is attracting much attention. Replacement of organic solvents by supercritical fluids makes the process green and eliminates the need of solvent separation. The lower viscosity and the higher diffusivity of supercritical fluids allow easier transport of substrates to the catalyst and, in the case of enzyme within the pores of enzyme support, this results in an easier access to the enzyme sites leading to higher reaction rates. In addition to the previously mentioned advantages of supercritical fluids, the finding that enzymes can retain their biocatalytic activity at high pressures has also encouraged the use of enzymes under supercritical condition (Rezaei et al., 2007b; Rezaei et al., 2007a).

In general, expansion of the substrates in CO_2 seems to be the main advantage of enzymatic lipid reactions in SC-CO_2. Expanded substrates have better diffusivity, low surface tension and low viscosity. In addition, a lesser amount of substrate available per unit amount of enzyme per unit time will increase the reaction rate (Ciftci & Temelli, 2011). However, at very high pressures, mass transfer properties of the substrates may be affected negatively. High CO_2 densities at high pressures lead to a decrease in enzymatic conversions. It has been reported that diffusion coefficients of fatty acids, fatty acid esters and glycerides in SC-CO_2 may also decrease at high pressures due to increase in the density of CO_2 (Rezaei & Temelli, 2000). Therefore, optimization of the process in terms of pressure and temperatures is crucial. Esmelindro et al. (2008) produced MAGs from olive oil in compressed propane. Their results showed that lipase-catalyzed glycerolysis in compressed propane might be a potential replacement for conventional methods, as high contents of reaction products, MAG and diacylglycerol (DAG), were achieved at mild temperature and pressure conditions (30 °C and 3 MPa) with a low solvent to substrates mass ratio (4:1) in short-reaction times (1 h). Lee et al. (2009) produced biodiesel from various oils, namely, olive, soybean, rapeseed, sunflower and palm oil, using lipase in SC-CO_2. The highest yield (65.18%) was obtained from olive oil at 13 MPa, 45 °C and 20% of lipase concentration (based on weight of oil).

3. Membrane technology

Membrane technology is becoming increasingly important as a green processing and separation method in food processing and waste water treatment Membranes are used as filters in separation processes and have a wide variety of applications Membrane technology is now competitive compared to conventional techniques such as adsorption, ion exchangers and sand filters.

The main advantage of membrane processing is that it avoids the use of any chemicals that have to be discharged. It works with relatively high efficiency and low energy consumption (Mulder, 1996). It also has the advantage of operating at ambient temperature, resulting in preservation of heat-sensitive components and nutritional value of food products (Dewettinck & Le, 2011).

Membrane separation processes differ greatly in the type of membranes and driving forces used for separation, the process design, and the area of application. There are many different membrane processes, including reverse osmosis, micro-, ultra- and nanofiltration, dialysis, electrodialysis, Donnan dialysis, pervaporation, gas seperation, membrane contactors, membrane distillation, membrane based solvent extraction, membrane reactors, etc. Among them, the innovative methods preferred by the food industry are pressure driven separation processes such as reverse osmosis, nanofiltration, ultrafiltration and microfiltration. These preferred methods facilitate the separation of components with a large range of particle sizes. The obtained products are generally of high quality and less post-treatment procedures are required (Baker, 2004).

3.1 Applications of membrane technology in the olive oil industry

Membrane technology has been used in the edible oil industry for degumming, deacidification, waste water treatment, recovery of solvent from micelles, condensate return, catalyst recovery and hydrolysis or synthesis of structured lipids with two-phase membrane reactors, involving pigment removal, separation and concentration of minor compounds in the oil. Despite its use in other sectors of the edible oil industry, this technology has not been broadly extended to olive oil processing.

3.1.1 Deacidification

Conventional chemical and physical deacidification methods have some drawbacks such as use of large amount of water and chemicals, and loss of neutral oil (Kale et al., 1999). Membrane technology may be proposed as a new alternative deacidification process for edible oils (Bhosle & Subramanian, 2005).

A membrane-based process for deacidification of lampante olive oil was undertaken by Hafidi et al. (2005a). Their objective was to deacidify, while also preserving the sensitive and bioactive components in the oil by operating at ambient temperature. The results showed that oils were obtained almost FFA- and soap-free in a single step. In another study, the impact of this process on some minor components and on the organoleptic characteristics of the purified olive oils was investigated (Hafidi et al., 2005b). It was reported that, while a complete deacidification was achieved, some desirable components, mainly phenolics, were eliminated during the filtering process. Thus, it was suggested to focus on reducing the elimination of phenolic compounds and the improvement of the organoleptic characteristics of the filtered oils.

3.1.2 Wastewater treatment

Olive mill wastewater (OMW), a by-product of olive oil extraction, is one of the most contaminated effluents. The polluting load is due to organic substances such as sugars, tannins, polyphenols, polyalcohols, pectins, lipids, proteins and organic acids, (Cassano et al., 2011). Phenolic compounds can act as phytotoxic components, inhibiting microbial growth as well as plant germination and vegetative growth (Morillo et al., 2009).

Biochemical oxygen demand (BOD5) and chemical oxygen demand (COD) of OMW may be as high as 100 and 200 g L^{-1}, respectively (de Morais Coutinho et al., 2009). Besides, OMWs are considered as a potential source for the recovery of antioxidant, antiatherogenic and anti-inflammatory biophenols (Obied et al., 2005). Detoxification and recovery of valuable components from wastewater are among the most useful treatments based on membrane technology.

In the study of Paraskeva et al. (2007), combinations of different membrane processes were used for the fractionation of OMW. Ultrafiltration in combination with nanofiltration and/or reverse osmosis were found to be very efficient for this process. It was shown that better efficiency of the OMW treatment was achieved by applying reverse osmosis after ultrafiltration. The ultrafiltration concentrate was found to contain the largest portion of fats, lipids, solids, etc. Further processing with nanofiltrationmay be employed for the separation of a greater part of phenols.

In another study, OMW was used to investigate the variation of COD and total organic carbon (TOC) removal efficiencies together with permeate fluxes for ultrafiltration process (Akdemir & Ozer, 2009). Two types of ultrafiltration membranes which are JW (polyvinylidine-difluoride) and MW (ultrafilic) gave close removal efficiencies. Ultrafiltration membranes with bigger molecular weight cut-offs for OMW were suggested to increase flux value and decrease efficiency loss. In their previous work, observed COD removal efficiency by ultrafiltration without pretreatment was found higher than 80% by promising value for OMW (Akdemir & Ozer, 2008). El-Abbassi et al. (2009) studied the treatment of OMW to obtain high value-added compounds such as sugar and polyphenols, by membrane distillation. Two types of commercial membranes, polytetrafluoroethylene (TF200) and polyvinilydene fluoride (GVHP), were compared and the effects of membrane parameters on direct contact membrane distillation (DCMD) performance (i.e. permeate flux and polyphenols retention) were investigated. Their results demonstrated that TF200 had a better separation coefficient (99%) after 9 h of DCMD operation than that of GVHP (89%). OMW concentration factor was found to be 1.72 for TF200, whereas it was only 1.4 for GVHP after 9 h.

Another OWM treatment was tested by Dhaouadi and Marrot (2008). Diluted solutions of OMW were treated in a ceramic membrane bioreactor with biomass specially acclimated to phenol. It gave stabilized permeate flux with zero suspended solid and no phenolic compounds. No fouling problems occurred during the experiments. OMW treatment in a membrane bioreactor can be used as a pre-treatment stage for the removal of phenolic compounds before a conventional biological process.

Recently, Coskun et al. (2010) studied the treatment of OWM using nanofiltration and reverse osmosis membranes. They reported that overall COD removal efficiencies were 97.5%. It was shown that reverse osmosis membranes are capable of producing a higher quality effluent from OMW than nanofiltration membranes. NF270 membranes were found to be most applicable among nanofiltration membranes due to their higher fluxes and higher removal efficiencies. In addition, it was found that centrifugation alone can be used as a promising option for primary treatment of OMWs with nanofiltration process.

In summary, there appears to be a potential for the use of membrane technology in the olive oil industry. Membranes can provide an opportunity to develop alternative environmentally friendly processes for the refining of olive oils and treatment of OWM. Despite promising results, further studies must be done on this new approach, namely, to evaluate the effect of the process on the oil composition, to improve flow rate, to reduce fouling inclusions and to assess economic viability.

4. Bioconversions

Lipids require modification in order to be used for special purposes or production of value added products. Lipids can be modified by hydrogenation, blending, fractionation and chemical or enzymatic reactions such as hydrolysis, direct esterification and interesterification.

4.1 Enzymatic conversions

Interesterification reactions are widely studied to produce margarines and shortenings with zero-*trans* fatty acids, cocoa butter equivalents, structured lipids with specific nutritional properties, partial glycerides and biodiesel. Chemical interesterification uses metal alcoholate catalysts to incorporate fatty acids randomly. This reaction produces a complete positional randomization of acyl groups in TAGs. In enzymatic interesterification the final structure of TAGs is controlled and a desired acyl group can be guided into TAGs using nonspecific, regiospecific (sn-1,3- or 2- specific) and fatty acid specific lipases as catalysts. This results in products with predictable composition. Enzymatic interesterification is becoming a more attractive method to convert cheap oils such as olive pomace oil, soya bean oil, rape seed oil, lard, tallow, etc. to high-value-added products and modified fats (An et al., 2007; Liua et al., 1997; Macrae, 1983; Miller et al., 1991; Pomier et al., 2007; Xu, 2003). Furthermore, enzymatic interesterification has milder reaction conditions and produces less waste than the chemical alternative. In addition, the same immobilized enzyme can be used many times (Akoh et al., 1998; Marangoni & Rousseau, 1995; Willis et al., 1998; Willis & Marangoni, 2002). Therefore, intensive research has aimed at replacing chemical interesterification with enzymatic interesterification.

There are three types of interesterification reactions: acidolysis, which is the reaction between an ester and a fatty acid, alcoholysis, the reaction between an ester and an alcohol, and transesterification, the reaction of an ester with another ester, also called ester-ester exchange (Macrae, 1983; Xu, 2003). Production of structured lipids and biodiesel has been the major topics of enzymatic interesterification studies.

4.1.1 Structured lipids

Structured lipids are novel modified TAGs produced by the incorporation of desirable fatty acids at specific positions or by changing the position of the fatty acids on the glycerol backbone. These processes allow for specific characteristics to be obtained such as melting behavior, functionality, and metabolism. Lipases, especially those which are sn-1,3 specific, are used for this purpose because these enzymes can make changes at sn-1 and sn-3 positions by keeping sn-2 ester group position unchanged.

Cocoa butter (CB) has a narrow melting range due to its unique TAG composition. This melting behavior is critical. The steepness of the melting profile (% solid fat as a function of time) has an impact on flavor release and crystallization. The high price of cocoa butter has prompted the industry search for CB alternatives. CB equivalents (CBEs) can be produced from palm oil and exotic fats by means of fractionation. Enzymatic synthesis of CBEs from cheap oils and fats using sn-1,3 specific lipases is also an alternative method. CB-like fats could be produced which have even more desirable properties than natural CB. Ciftci et al. (2010) used olive pomace oil for the production of CB-like fat using sn-1,3 specific lipase. They interesterified refined olive pomace oil, palmitic acid and stearic acid at a molar ratio of 1:2:6, respectively, at 45°C using a pack bed reactor filled with sn-1,3 specific lipase. They reported that the CB-like fat could replace CB up to 30% without significantly changing the physical and chemical properties of the product. Chang et al. (1990) also produced CB-like fat by enzymatic interesterification of fully hydrogenated cotton seed and olive oils. The melting point of their CB-like fat was 39°C; close to 36°C, the melting point of CB.

Any lipid containing medium-chain and long-chain unsaturated fatty acids might be useful for certain applications and functionalities. Nunes et al. (2011) produced structured lipids containing medium-chain fatty acids at sn-1,3 position and long-chain unsaturated fatty acids at the sn-2 position by acidolysis of virgin olive oil and caprylic or capric acids using

1,3-selective *Rhizopus oryzae* heterologous lipase (rROL) immobilized in Eupergit C and modified sepiolite. These structured lipids are low caloric and and have dietetic properties for controlling obesity and malabsorption. They showed that rROL immobilized in Eupergit C was able to catalyze the incorporation of 21.6 and 34.8 mol% of caprylic or capric acid into virgin olive oil, after 24 h at 40 °C in solvent-free media. Fumoso and Akoh (2002) also used lipase-catalyzed acidolysis of olive oil and caprylic acid to produce structured lipids. They used a *sn*-1,3-specific lipase from *Rhizomucor miehei* in a bench-scale packed bed bioreactor. They studied the effect of solvent, temperature, substrate mol ratio, and flow rate/residence time. The optimal solvent-free production of structured lipid was obtained at a substrate flow rate of 1 ml/min, a residence time 2.7 h, 60 °C, and a mol ratio 1:5 (olive oil/caprylic acid). The structured lipid produced at optimal conditions had 7.2% caprylic acid, 69.6% oleic acid, 21.7% linoleic acid and 1.5% palmitic acid at the *sn*-2 position. Another structured lipid used as a constituent of infant formulas, consisting mainly of UPU triglycerides (U=unsaturated acyl chains, P=palmitic acyl group), can be prepared by lipase catalyzed reactions of fractionated palm oil, rich in tripalmitin, and oleic acid from olive oil (Schmid et al., 1998).

4.1.2 Biodiesel

Biodiesel can be obtained from vegetable oils, animal fats, recycled grease, or algae and can be produced by the reaction of TAGs with methanol (methanolysis). Lipase-catalyzed methanolysis is more attractive than conventional base-catalyzed method since the glycerol produced as a by-product can easily be recovered and the purification process for fatty acid methyl esters (FAMEs) is relatively simple. In the oil and fat industry, conversion of waste edible oil and soapstock (a by-product generated in alkali refining of vegetable oils) to biodiesel has attracted a great deal of attention (Azócar et al., 2010; Safieddin Ardebili et al., 2011; Singaram, 2009). Unlike the conventional chemical routes for synthesis of diesel fuels, biocatalytic routes permit one to carry out the interesterification of a wide variety of oil feedstocks in the presence of excess FFAs.

Olive pomace oil was used by Yucel (2011) for enzymatic production of biodiesel. Yucel (2011) immobilized microbial lipase from *Thermomyces lanuginosus* on olive pomace by covalent binding, and then used this immobilized lipase for the methanolysis of olive pomace oil. Under the optimized conditions for solvent-free reaction, the maximum yield was reported to be 93% at 25 °C after 24 h. Sanchez and Vasudevan (2006) produced biodiesel by transesterification of olive oil triolein with methanol using lipase. They studied the effects of the molar ratio of methanol to triolein, semibatch (stepwise addition of methanol) vs batch operation, enzyme activity, and reaction temperature on overall conversion. Because of the inactivation of the enzyme by insoluble methanol, stepwise methanolysis with a 3:1 methanol to triolein molar ratio and an overall ratio of 8:1 gave the best results.

4.2 Enzymatic deacidification

One method to reduce the FFA content in fats and oils is to convert the FFAs to TAGs. This is carried out by direct esterification of fatty acids with glycerol.

A reported application of enzymatic deacidification of olive pomace oil is the enzymatic glycerolysis of highly acidic (32%) olive pomace oil (Fadiloglu et al., 2003). FFAs of olive pomace oil were esterified with glycerol using a nonspecific immobilized lipase, reducing the acidity of the oil to 2.36%. In another study, the FFA content of high acidic (31.6%) degummed and dewaxed olive oil was reduced to 3.7%.

4.3 Bioremediation

Bioremediation, generally classified as *in situ* or *ex situ*, is the use of microorganism metabolism to remove pollutants. *In situ* bioremediation involves treating the contaminated material at the site, while *ex situ* involves the removal of the contaminated material to be treated elsewhere. Some examples of bioremediation technologies are phytoremediation, bioventing, bioleaching, landfarming, composting, bioaugmentation, rhizofiltration, and biostimulation (Shukla et al., 2010). Besides being cost effective, bioremediation can result in the complete mineralization of the pollutant, considered a permanent solution of the pollution problem. Furthermore, it is a non-invasive technique, leaving the ecosystem intact. Bioremediation can deal with lower concentrations of contaminants where cleanup by physical or chemical methods would not be feasible. Unfortunately, it presents some major drawbacks which still limit the application of these techniques, including long processing times and less predictable results compared to conventional methods (Perelo, 2010).

The disposal of OMW is predominantly carried out via land spreading or by means of evaporation ponds, although a wide number of chemical and biological decontamination and valorisation technologies have been reported. The two-phase centrifugation system, as an alternative ecological approach for olive oil production, drastically reduces the water consumption during the process. This system generates olive oil plus a semi-solid waste, known as the two-phase olive-mill waste (Morillo et al., 2009).

Ramos-Cormenzana et al. (1996) performed aerobic biodegradation on OMW by using bacterium *Bacillus pumilus* to reduce the phenol content. They reported 50% reduction in phenol content using *Bacillus pumilus*. The detoxification of OMW following inoculation with *Azotobacter vinelandii* (strain A) was performed for two successive 5-day-period cycles in an aerobic, biowheel-type reactor, under non-sterile conditions by Ehaliotis et al. (1999). The authors indicated that the phytotoxicity of the processed product was reduced by over 90% at the end of both cycles. However, aerobic bacteria cannot generally biodegrade complex phenolic compounds which are responsible for the dark color of OMW. Fungi, compared to bacteria, are more effective at degrading both simple and complex phenolic compounds presenting in olive mill wastes. This is due to the presence of compounds analogous to lignin monomers, which are more easily degraded by wood-rotting fungi (Garcia Garcia et al., 2000).

Demirer et al. (2000) generated biogas containing about 77% methane by anaerobic bioconversion of OMW (57.5 L methane per liter of wastewater). Ammary (2005) treated OMW using a lab scale anaerobic sequencing batch reactor, achieving more than 80% COD removal at 3 d hydraulic retention time. Anaerobic bioconversion has some advantages compared to aerobic processes: (a) high organic load feeds are used, (b) low nutrient requirements are necessary, (c) small quantities of excess sludge are usually produced, and (d) a combustible biogas is generated. However, the nutrient imbalance of OMW, mainly due to its high C/N ratios, low pH and the presence of biostatic and inhibitory substances, cause a problem. Not quite clear Rephrase An additional problem of two-phase olive-mill waste is its high consistency making its transport, storage and handling difficult (Morillo et al., 2009).

Olive mill wastes can be treated with other methods such as composting to produce fertilizers (Ntoulas et al., 2011); using as a culture medium to produce useful microbial biomass (de la Fuente et al., 2011); using as a low-cost fermentation substrate for producing microbial biopolymers for production of polysaccharides and biodegradable plastics (Ntaikou et al., 2009); and as a base-stock for production of biofuels (Rincon et al., 2010).

5. Molecular distillation

Molecular distillation, also called short path distillation, has become an important alternative for separation of heat sensitive compounds or substances with very high boiling points. Molecular distillation is characterized by a short time exposure of the distilled liquid to elevated temperature and high vacuum, with a small distance between the evaporator and the condenser (Lutišan et al., 2002). The small distance between the evaporator and the condenser and a high vacuum in the distillation gap results in a specific mass transfer mechanism with evaporation outputs as high as 20–40 $gm^{-2} s^{-1}$ (Cvengroš et al., 2000). Due to short residence time and low temperature, distillation of heat-sensitive materials is accomplished without thermal decomposition. Another advantage of the process is the absence of solvents. Therefore, molecular distillation is considered as a promising method in the separation, purification and concentration of natural products (Martins et al., 2006).

Vegetable oil deodorization process produces a distillate rich in high value components such as phytosterols, tocopherols, and fatty acids, depending on the oil or fat. Martins et al. (2006) separated FFAs from soybean oil deodorizer distillate to obtain a tocopherol concentrate, which contained only 6.4% of FFA and 18.3% of tocopherols (from a raw material containing 57.8% of FFA and 8.97% of tocopherols.) The specific processing conditions were an evaporator temperature of 160 °C and a feed flow rate of 10.4 $gmin^{-1}$. Under these conditions, they achieved 96% FFA elimination and 81% tocopherol recovery.

Although molecular distillation is a promising separation and purification method, it is not commonly applied in the olive oil industry. One relevant application is the purification of the structured lipids enzymatically produced from olive oil and caprylic acid (Fomuso & Akoh, 2002). If the advantages and efficiency of the system are further considered, it may be used in the olive oil industry for deacidification and separation of nutraceuticals. The cost of the system and possible alterations in the structure of the oil during the process seem to be serious disadvantages. Therefore, optimization of each particular system is necessary for a successful industrialization.

6. Use of by-products of olive oil industry for waste treatment

The use of by-products of the olive oil industry for waste treatment is another green approach. Solid olive wastes were used for water purification by El-Hamouz et al. (2007). The solid olive residue was processed to yield relatively high-surface area active carbon after extraction of the oil from the residue. The resulting carbon was used to reversibly adsorb chromate ions from water, aiming at a purification process with reusable active carbon. In another study, olive pomace was used as reactive dye biosorbent material for the removal of RR198 textile dye from aqueous solutions (Akar et al., 2009).

Vlyssides et al. (2004) developed an integrated pollution prevention method which decreased wasterwater production 50% from the 3-phase olive oil extraction process. The process included mechanical separation, crushing, mixing, composting, malaxation, 3-phase centrifugation, coagulation flocculation, chemical oxidation, biological treatment, and reed beds steps. Furthermore, a Fenton oxidation process was used to detoxify the wastewater, with the possibility of extracting commercially valuable antioxidant products. They also produced high-quality compost from the solid residues.

7. Conclusions

Current trends show that future oil processing technologies will be based on green processes. Laboratory and pilot scale applications of such processes in the olive oil industry show that they can be used as alternatives to conventional processes. Further optimization studies are necessary for more successful applications. In spite of the high first capital investment, these processes are advantageous considering the market value of the natural products obtained and remediation of environmental pollution.

8. References

Akar, T., Tosun, I., Kaynak, Z., Ozkara, E., Yeni, O., Sahin, E.N. & Akar, S.T. (2009). An Attractive Agro-Industrial by-Product in Environmental Cleanup: Dye Biosorption Potential of Untreated Olive Pomace. *Journal of Hazardous Materials*, 166, 2-3, 1217-1225

Akdemir, E.O. & Ozer, A. (2008). Application of a Statistical Technique for Olive Oil Mill Wastewater Treatment Using Ultrafiltration Process. *Separation and Purification Technology*, 62, 1, 222-227

Akdemir, E.O. & Ozer, A. (2009). Investigation of Two Ultrafiltration Membranes for Treatment of Olive Oil Mill Wastewater. *Desalination*, 249, 2, 660-666

Akoh, C.C., Lee, K.T. & Fomuso, L.B. (1998). 3, In: *Structural Modified Food Fats: Synthesis, Biochemistry, and Use*, Christophe, A.B., pp. (46-72), The American Oil Chemists Society, Illinois

Ammary, B.Y. (2005). Treatment of Olive Mill Wastewater Using an Anaerobic Sequencing Batch Reactor. *Desalination*, 177, 1-3, 157-165

An, G., Ma, W., Sun, Z., Liu, Z., Han, B., Miao, S., Miao, Z. & Ding, K. (2007). Preparation of Titania/Carbon Nanotube Composites Using Supercritical Ethanol and Their Photocatalytic Activity for Phenol Degradation under Visible Light Irradiation. *Carbon*, 45, 9, 1795-1801

Azócar, L., Ciudad, G., Heipieper, H.J. & Navia, R. (2010). Biotechnological Processes for Biodiesel Production Using Alternative Oils. *Applied Microbiology and Biotechnology*, 88, 3, 621-636

Batistella, C.B., Moraes, E.B., Maciel Filho, R. & Wolf Maciel, M.R. (2002). Molecular Distillation Process for Recovering Biodiesel and Carotenoids from Palm Oil. *Applied Biochemistry and Biotechnology - Part A Enzyme Engineering and Biotechnology*, 98-100, 1149-1159

Bensebia, O., Barth, D., Bensebia, B. & Dahmani, A. (2009). Supercritical Co2 Extraction of Rosemary: Effect of Extraction Parameters. *The Journal of Supercritical Fluids*, 49, 161-166

Bhosle, B.M. & Subramanian, R. (2005). New Approaches in Deacidification of Edible Oils - a Review. *Journal of Food Engineering*, 69, 4, 481-494

Black, H. (1996). Supercritical Carbon Dioxide: The "Greener" Solvent: It's Not Toxic and It Doesn't Harm the Ozone Layer, but Can Supercritical Co2 Do the Job in Major Industrial Applications? *Environmental Science and Technology*, 30, 3,

Bondioli, P., Mariani, C., Lanzani, A., Fedeli, E., Mossa, A. & Muller, A. (1992). Lampante Olive Oil Refining with Supercritical Carbon Dioxide. *Journal of the American Oil Chemists' Society*, 69, 5, 477-480

Brunetti, L., Daghetta, A., Fedell, E., Kikic, I. & Zanderighi, L. (1989). Deacidification of Olive Oils by Supercritical Carbon Dioxide. *Journal of the American Oil Chemists Society*, 66, 2, 209-217

Brunner, G. (2009). Counter-Current Separations. *Journal of Supercritical Fluids*, 47, 3, 574-582

Cassano, A., Conidi, C. & Drioli, E. (2011). Comparison of the Performance of Uf Membranes in Olive Mill Wastewaters Treatment. *Water Research*, 45, 10, 3197-3204

Cermak, S.C., John, A.L. & Evangelista, R.L. (2007). Enrichment of Decanoic Acid in Cuphea Fatty Acids by Molecular Distillation. *Industrial Crops and Products*, 26, 1, 93-99

Chang, M.K., Abraham, G. & John, V.T. (1990). Production of Cocoa Butter-Like Fat from Interesterification of Vegetable Oils. *Journal of the American Oil Chemists' Society*, 67, 11, 832-834

Ciftci, O.N. & Temelli, F. (2011). Continuous Production of Fatty Acid Methyl Esters from Corn Oil in a Supercritical Carbon Dioxide Bioreactor. *Journal of Supercritical Fluids*, 58, 1, 79-87

Çiftçi, O.N., Göğüş, F. & Fadlloğlu, S. (2010). Performance of a Cocoa Butter-Like Fat Enzymatically Produced from Olive Pomace Oil as a Partial Cocoa Butter Replacer. *Journal of the American Oil Chemists' Society*, 87, 9, 1013-1018

Clark, J.H. (2011). In: *Alternatives to Conventional Food Processing*, Proctor, A., pp. (1-10), RCS, Cambridge

Coskun, T., Debik, E. & Demir, N.M. (2010). Treatment of Olive Mill Wastewaters by Nanofiltration and Reverse Osmosis Membranes. *Desalination*, 259, 1-3, 65-70

Cvengroš, J., Lutišan, J. & Micov, M. (2000). Feed Temperature Influence on the Efficiency of a Molecular Evaporator. *Chemical Engineering Journal*, 78, 1, 61-67

Daukšas, E., Venskutonis, P.R. & Sivik, B. (2002). Supercritical Fluid Extraction of Tocopherol Concentrates from Olive Tree Leaves. *Journal of Supercritical Fluids*, 22, 3, 221-228

de la Fuente, C., Clemente, R., Martinez-Alcala, I., Tortosa, G. & Bernal, M.P. (2011). Impact of Fresh and Composted Solid Olive Husk and Their Water-Soluble Fractions on Soil Heavy Metal Fractionation; Microbial Biomass and Plant Uptake. *Journal of Hazardous Materials*, 186, 2-3, 1283-1289

de Morais Coutinho, C., Chiu, M.C., Basso, R.C., Ribeiro, A.P.B., Gonçalves, L.A.G. & Viotto, L.A. (2009). State of Art of the Application of Membrane Technology to Vegetable Oils: A Review. *Food Research International*, 42, 5-6, 536-550

Demirer, G.N., Duran, M., Güven, E., Ugurlu, O., Tezel, U. & Ergüder, T.H. (2000). Anaerobic Treatability and Biogas Production Potential Studies of Different Agro-Industrial Wastewaters in Turkey. *Biodegradation*, 11, 6, 401-405

Dewettinck, K. & Le, T.T. (2011). In: *Alternatives to Conventional Food Processing*, Proctor, A., pp. (184-253), RSC, Cambridge

Dhaouadi, H. & Marrot, B. (2008). Olive Mill Wastewater Treatment in a Membrane Bioreactor: Process Feasibility and Performances. *Chemical Engineering Journal*, 145, 2, 225-231

Dunford, N.T. (2004). In: *Nutritionally Enhanced Edible Oil and Oilseed Processing*, Dunford, N.T.&Dunford, H.B., pp. AOCS Press, Illinois

Ehaliotis, C., Papadopoulou, K., Kotsou, M., Mari, I. & Balis, C. (1999). Adaptation and Population Dynamics of Azotobacter Vinelandii During Aerobic Biological Treatment of Olive-Mill Wastewater. *FEMS Microbiology Ecology*, 30, 4, 301-311

El-Abbassi, A., Hafidi, A., García-Payo, M.C. & Khayet, M. (2009). Concentration of Olive Mill Wastewater by Membrane Distillation for Polyphenols Recovery. *Desalination*, 245, 1-3, 670-674

El-Hamouz, A., Hilal, H.S., Nassar, N. & Mardawi, Z. (2007). Solid Olive Waste in Environmental Cleanup: Oil Recovery and Carbon Production for Water Purification. *Journal of Environmental Management*, 84, 1, 83-92

Esmelindro, A.F.A., Fiametti, K.G., Ceni, G., Corazza, M.L., Treichel, H., de Oliveira, D. & Oliveira, J.V. (2008). Lipase-Catalyzed Production of Monoglycerides in Compressed Propane and Aot Surfactant. *Journal of Supercritical Fluids*, 47, 1, 64-69

Esquível, M.M. & Bernardo-Gil, G. (1993). Extraction of Olive Husk Oil with Compressed Carbon Dioxide. *The Journal of Supercritical Fluids*, 6, 2, 91-94

Fadiloglu, S., Çiftçi, O.N. & Göğüş, F. (2003). Reduction of Free Fatty Acid Content of Olive-Pomace Oil by Enzymatic Glycerolysis. *Food Science and Technology International*, 9, 1, 11-15

Feuge, R.O., Lovegren, N.V. & Cosler, H.B. (1958). Cocoa Butter-Like Fats from Domestic Oils. *Journal of the American Oil Chemists' Society*, 35, 5, 194-199

Fomuso, L.B. & Akoh, C.C. (2002). Lipase-Catalyzed Acidolysis of Olive Oil and Caprylic Acid in a Bench-Scale Packed Bed Bioreactor. *Food Research International*, 35, 1, 15-21

Fornari, T., Va?zquez, L., Torres, C.F., Iba?n?ez, E., Sen?ora?ns, F.J. & Reglero, G. (2008). Countercurrent Supercritical Fluid Extraction of Different Lipid-Type Materials: Experimental and Thermodynamic Modeling. *Journal of Supercritical Fluids*, 45, 2, 206-212

García García, I., Jiménez Peña, P.R., Bonilla Venceslada, J.L., Martín Martín, A., Martín Santos, M.A. & Ramos Gómez, E. (2000). Removal of Phenol Compounds from Olive Mill Wastewater Using Phanerochaete Chrysosporium, Aspergillus Niger, Aspergillus Terreus and Geotrichum Candidum. *Process Biochemistry*, 35, 8, 751-758

Hafidi, A., Pioch, D. & Ajana, H. (2005a). Soft Purification of Lampante Olive Oil by Microfiltration. *Food Chemistry*, 92, 1, 17-22

Hafidi, A., Pioch, D. & Ajana, H. (2005b). Effects of a Membrane-Based Soft Purification Process on Olive Oil Quality. *Food Chemistry*, 92, 4, 607-613

Kale, V., Katikaneni, S.P.R. & Cheryan, M. (1999). Deacidifying Rice Bran Oil by Solvent Extraction and Membrane Technology. *Journal of the American Oil Chemists' Society*, 76, 6, 723-727

Le Floch, F., Tena, M.T., Ríos, A. & Valcárcel, M. (1998). Supercritical Fluid Extraction of Phenol Compounds from Olive Leaves. *Talanta*, 46, 5, 1123-1130

Lee, J.H., Kwon, C.H., Kang, J.W., Park, C., Tae, B. & Kim, S.W. (2009). Biodiesel Production from Various Oils under Supercritical Fluid Conditions by Candida Antartica

Lipase B Using a Stepwise Reaction Method. *Applied Biochemistry and Biotechnology,* 156, 1-3, 24-34

Liua, K.J., Chengb, H.M., Changc, R.C. & Shawa, J.F. (1997). Synthesis of Cocoa Butter Equivalent by Lipase-Catalyzed Interesterification in Supercritical Carbon Dioxide. *Journal of the American Oil Chemists' Society,* 74, 11, 1477-1482

Lutišan, J., Cvengroš, J. & Micov, M. (2002). Heat and Mass Transfer in the Evaporating Film of a Molecular Evaporator. *Chemical Engineering Journal,* 85, 2-3, 225-234

Macrae, A.R. (1983). Lipase-Catalyzed Interesterification of Oils and Fats. *Journal of the American Oil Chemists' Society,* 60, 2, 243A-246A

Marangoni, A.G. & Rousseau, D. (1995). Engineering Triacylglycerols: The Role of Interesterification. *Trends Food Sci. Technol.,* 6, 329-335

Martins, P.F., Ito, V.M., Batistella, C.B. & MacIel, M.R.W. (2006). Free Fatty Acid Separation from Vegetable Oil Deodorizer Distillate Using Molecular Distillation Process. *Separation and Purification Technology,* 48, 1, 78-84

Miller, D.A., Blanch, H.W. & Prausnitz, J.M. (1991). Enzyme-Catalyzed Interesterification of Triglycerides in Supercritical Carbon Dioxide. *Industrial and Engineering Chemistry Research,* 30, 5, 939-946

Morillo, J.A., Antizar-Ladislao, B., Monteoliva-Sánchez, M., Ramos-Cormenzana, A. & Russell, N.J. (2009). Bioremediation and Biovalorisation of Olive-Mill Wastes. *Applied Microbiology and Biotechnology,* 82, 1, 25-39

Mulder, M. (1996). *Basic Principles of Membrane Technology* (2nd), Kluwer Academic Publishers, Dordrecht

Ntaikou, I., Kourmentza, C., Koutrouli, E.C., Stamatelatou, K., Zampraka, A., Kornaros, M. & Lyberatos, G. (2009). Exploitation of Olive Oil Mill Wastewater for Combined Biohydrogen and Biopolymers Production. *Bioresource Technology,* 100, 15, 3724-3730

Ntoulas, N., Nektarios, P.A. & Gogoula, G. (2011). Evaluation of Olive Mill Waste Compost as a Soil Amendment for Cynodon Dactylon Turf Establishment, Growth, and Anchorage. *HortScience,* 46, 6, 937-945

Nunes, P.A., Pires-Cabral, P. & Ferreira-Dias, S. (2011). Production of Olive Oil Enriched with Medium Chain Fatty Acids Catalysed by Commercial Immobilised Lipases. *Food Chemistry,* 127, 3, 993-998

Obied, H.K., Allen, M.S., Bedgood, D.R., Prenzler, P.D., Robards, K. & Stockmann, R. (2005). Bioactivity and Analysis of Biophenols Recovered from Olive Mill Waste. *Journal of Agricultural and Food Chemistry,* 53, 4, 823-837

Oliveira, J.V. & Oliveira, D. (2000). Kinetics of the Enzymatic Alcoholysis of Palm Kernel Oil in Supercritical CO_2. *Industrial and Engineering Chemistry Research,* 39, 12, 4450-4454

Ooi, C.K., Bhaskar, A., Yener, M.S., Tuan, D.Q., Hsu, J. & Rizvi, S.S.H. (1996). Continuous Supercritical Carbon Dioxide Processing of Palm Oil. *Journal of the American Oil Chemists' Society,* 73, 2, 233-237

Paraskeva, C.A., Papadakis, V.G., Tsarouchi, E., Kanellopoulou, D.G. & Koutsoukos, P.G. (2007). Membrane Processing for Olive Mill Wastewater Fractionation. *Desalination,* 213, 1-3, 218-229

Perelo, L.W. (2010). Review: In Situ and Bioremediation of Organic Pollutants in Aquatic Sediments. *Journal of Hazardous Materials,* 177, 1-3, 81-89

Pomier, E., Delebecque, N., Paolucci-Jeanjean, D., Pina, M., Sarrade, S. & Rios, G.M. (2007). Effect of Working Conditions on Vegetable Oil Transformation in an Enzymatic

Reactor Combining Membrane and Supercritical Co2. *Journal of Supercritical Fluids*, 41, 3, 380-385

Ramos-Cormenzana, A., Juárez-Jiménez, B. & Garcia-Pareja, M.P. (1996). Antimicrobial Activity of Olive Mill Wastewaters (Alpechin) and Biotransformed Olive Oil Mill Wastewater. *International Biodeterioration & Biodegradation*, 38, 3-4, 283-290

Rezaei, K. & Temelli, F. (2000). Using Supercritical Fluid Chromatography to Determine Diffusion Coefficients of Lipids in Supercritical Co2. *Journal of Supercritical Fluids*, 17, 1, 35-44

Rezaei, K., Temelli, F. & Jenab, E. (2007a). Effects of Pressure and Temperature on Enzymatic Reactions in Supercritical Fluids. *Biotechnology Advances*, 25, 3, 272-280

Rezaei, K., Jenab, E. & Temelli, F. (2007b). Effects of Water on Enzyme Performance with an Emphasis on the Reactions in Supercritical Fluids. *Critical Reviews in Biotechnology*, 27, 4, 183-195

Rincon, B., Borja, R., Martin, M.A. & Martin, A. (2010). Kinetic Study of the Methanogenic Step of a Two-Stage Anaerobic Digestion Process Treating Olive Mill Solid Residue. *Chemical Engineering Journal*, 160, 1, 215-219

Safieddin Ardebili, M., Ghobadian, B., Najafi, G. & Chegeni, A. (2011). Biodiesel Production Potential from Edible Oil Seeds in Iran. *Renewable and Sustainable Energy Reviews*, 15, 6, 3041-3044

Sanchez, F. & Vasudevan, P.T. (2006). Enzyme Catalyzed Production of Biodiesel from Olive Oil. *Applied Biochemistry and Biotechnology*, 135, 1, 1-14

Schmid, U., Bornscheuer, U.J., Soumanou, M.M., McNeill, G.P. & Schmid, R.D. (1998). Optimization of the Reaction Conditions in the Lipase-Catalyzed Synthesis of Structured Triglycerides. *Journal of the American Oil Chemists' Society*, 75, 11, 1527-1531

Shukla, K.P., Singh, N.K. & Sharma, S. (2010). Bioremediation: Developments, Current Practices and Perspectives. *Genetic Engineering and Biotechnology Journal*,

Singaram, L. (2009). Biodiesel: An Eco-Friendly Alternate Fuel for the Future - a Review. *Thermal Science*, 13, 3, 185-199

Temelli, F. (2009). Perspectives on Supercritical Fluid Processing of Fats and Oils. *Journal of Supercritical Fluids*, 47, 3, 583-590

Vázquez, L., Hurtado-Benavides, A.M., Reglero, G., Fornari, T., Ibáñez, E. & Señoráns, F.J. (2009). Deacidification of Olive Oil by Countercurrent Supercritical Carbon Dioxide Extraction: Experimental and Thermodynamic Modeling. *Journal of Food Engineering*, 90, 4, 463-470

Vlyssides, A.G., Loizides, M. & Karlis, P.K. (2004). Integrated Strategic Approach for Reusing Olive Oil Extraction by-Products. *Journal of Cleaner Production*, 12, 6, 603-611

Willis, W.M. & Marangoni, A.G. (2002). 30, In: *Food Lipids: Chemistry, Nutrition, and Biotechnology* Akoh, C.C.&Min, D.B., pp. (807-840), CRC Press,

Willis, W.M., Lencki, R.W. & Marangoni, A.G. (1998). Lipid Modification Strategies in the Production of Nutritionally Functional Fats and Oils. *Critical Reviews in Food Science and Nutrition*, 38, 8, 639-674

Xu, X. (2003). Engineering of Enzymatic Reactions and Reactors for Lipid Modification and Synthesis. *Eur. J. Lipid Sci. Technol.*, 105, 289-304

Yücel, Y. (2011). Biodiesel Production from Pomace Oil by Using Lipase Immobilized onto Olive Pomace. *Bioresource Technology*, 102, 4, 3977-3980

Part 2

Bioavailability and Biological Properties of Olive Oil Constituents

6

Oleocanthal: A Naturally Occurring Anti-Inflammatory Agent in Virgin Olive Oil

S. Cicerale, L. J. Lucas and R. S. J. Keast
School of Exercise and Nutrition Sciences,
Centre for Physical Activity and Nutrition (CPAN),
Sensory Science Group, Deakin University, Melbourne,
Australia

1. Introduction

Research on the non-steroidal anti-inflammatory olive oil phenolic, (-)- decarboxymethyl ligstroside aglycone (more commonly known as oleocanthal) has supported speculation that this compound may confer some of the health benefits associated with the traditional Mediterranean diet. Oleocanthal elicits a peppery, stinging sensation at the back of the throat similar to that of the non-steroidal anti-inflammatory drug (NSAID), ibuprofen (Beauchamp et al., 2005) and this localized irritation is due to stimulation of the transient receptor potential cation channel A1 (TRPA1) (Peyrot des Gachons et al., 2011). The perceptual similarity between oleocanthal and ibuprofen spurred the hypothesis that these two compounds may possess similar pharmacological properties. Further investigation demonstrated that oleocanthal inhibits inflammation in the same way as ibuprofen, and moreover, is substantially more potent on a equimolar basis (Beauchamp et al., 2005). Subsequent studies have shown that oleocanthal exhibits various modes of action in reducing inflammatory-related disease, including neuro-degenerative disease (Pitt et al., 2009, Li et al., 2009), joint-degenerative disease (Iacono et al., 2010) and specific cancers (Elnagar et al., 2011). Therefore, long term consumption of extra virgin olive oil (EVOO) containing oleocanthal may contribute to the health benefits associated with the Mediterranean dietary pattern. This chapter summarizes the current knowledge on oleocanthal, in terms of its sensory and physiological properties, its extraction from the oil matrix and subsequent identification and quantification, and finally the factors that may influence the concentration of oleocanthal in EVOO.

2. Olive oil, a hallmark of the Mediterranean diet

The health benefits of following a Mediterranean eating pattern were first acknowledged in the Seven Countries Study (Keys, 1970). Thereafter over a period of 30 years, a number of investigators have reported that the Mediterranean diet is associated with low rates of degenerative diseases such as cardiovascular disease (CVD) (Estruch et al., 2006, Pitsavos et al., 2005), coronary heart disease (CHD) (Fung et al., 2009), stroke (Fung et al., 2009), certain types of cancers (La Vecchia, 2004, Dixon et al., 2007), diabetes (Martinez-Gonzalez et al., 2008), Alzheimer's disease (Scarmeas et al., 2009) and non-alcoholic fatty liver disease (Fraser et al., 2008). Research has also demonstrated that Mediterranean populations have increased life expectancy (Hu, 2003, Visioli et al., 2005, Trichopoulou et al., 2005), reduced risk of developing disorders such as metabolic syndrome (Tortosa et al., 2007, Babio et al., 2008) and have decreased levels of systematic inflammation (Dai et al., 2009, Fragopoulou et al., 2010, Panagiotakos et al., 2009).

The traditional Mediterranean diet is defined as the pattern of eating observed in the olive growing areas of the Mediterranean region, namely Greece and Southern Italy in the 1960s. An integral component of this dietary pattern is the consumption of EVOO (Stark and Madar, 2002, Kok and Kromhout, 2004). EVOO, the pillar of Mediterranean recipes, is commonly incorporated into cooked dishes as well as in salads. Typically, the intake of EVOO ranges from 25 to 50 ml per day in the Mediterranean diet (Corona et al., 2009). Therefore, the apparent health benefits have been partially attributed to the dietary consumption of EVOO by Mediterranean populations. Figure 1 displays the Mediterranean diet pyramid featuring EVOO as a core component of this dietary pattern.

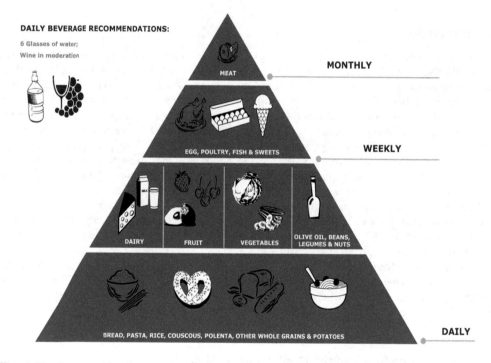

Fig. 1. Food pyramid reflecting the traditional Mediterranean diet. As depicted, EVOO is an integral food component of this diet residing in the consume 'daily' food group.

3. Olive oil phenolic compounds

Historically, the health promoting properties of EVOO were attributed to the high concentration of monounsaturated fatty acids (MUFAs), in particular oleic acid, contained in EVOO. However other seed oils (i.e. sunflower, soybean, and rapeseed), which also contain high concentrations of oleic acid, do not exhibit the same health benefits as EVOO (López-

Miranda et al., 2010, Harper et al., 2006, Aguilera et al., 2004). In addition to oleic acid, EVOO contains a minor, yet significant phenolic fraction that other seed oils lack and this fraction has generated much interest regarding its health promoting properties.

Currently, 36 phenolic compounds have been identified in EVOO and studies (human, animal, *in vivo* and *in vitro*) have demonstrated that olive oil phenolics have positive effects on certain physiological parameters such as plasma lipoproteins, oxidative damage, inflammatory markers, platelet and cellular function, antimicrobial activity, and bone health (for review see (Cicerale et al., 2010)), possibly reducing the risk of chronic disease development.

4. Discovery of oleocanthal

The phenolic compound (-)- decarboxymethyl ligstroside aglycone was first reported in EVOO by Montedoro *et al.* (Montedoro and Servili, 1993) in 1993 (Figure 2). A decade after its discovery, Andrewes and colleagues (Andrewes et al., 2003) reported that decarboxymethyl ligstroside aglycone was responsible for the throat irritation and pungency elicited by some EVOOs. In 2005, Beauchamp *et al.* (Beauchamp et al., 2005) confirmed that the phenolic compound, decarboxymethyl ligstroside aglycone was indeed responsible for the throat irritation elicited by EVOOs post-ingestion. This confirmation was carried out by isolating the compound from various EVOOs and measuring the throat irritation elicited. However, at that stage, there was a possibility that co-elution of a minor component or a mixture of components along with decarboxymethyl ligstroside aglycone may collectively cause the localized throat irritation. Therefore, the authors chemically synthesized decarboxymethyl ligstroside aglycone and dissolved it in non-irritating corn oil. Throat irritation elicited by the synthesized decarboxymethyl ligstroside aglycone was found to be dose-dependent on the concentration of this phenolic in corn oil and mimicked that of EVOO containing this compound naturally. Decarboxymethyl ligstroside aglycone was thus deemed the sole throat irritant in EVOO and was named oleocanthal (*oleo* for olive, *canth* for sting, and *al* for aldehyde) (Beauchamp et al., 2005).

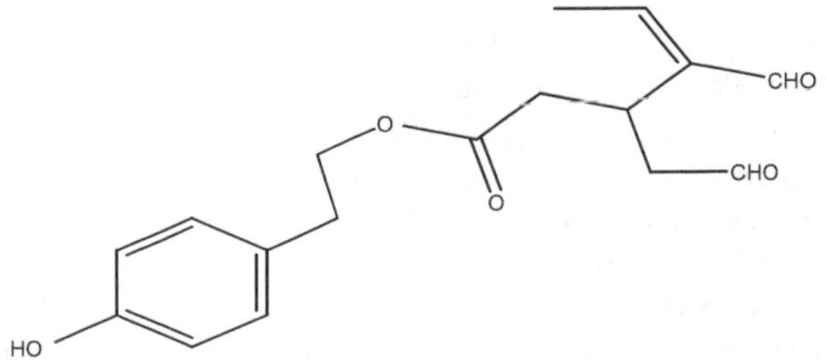

Fig. 2. Oleocanthal structure

5. Sensory properties of oleocanthal

The intake of EVOO is often associated with a peppery sting that is localized to the oropharyngeal region in the oral cavity (Figure 3). There is wide inter-individual variation in sensitivity to oleocanthal which can range from a slight irritation in the throat, to an irritation that is strong enough to produce a cough in those highly sensitive. Of particular interest is the spatial location of irritation produced by oleocanthal. Irritating, pungent compounds often aggravate all regions in the oral cavity rather than acting on one localized area (Peyrot des Gachons et al., 2011), which implies a sensory receptor specific to oleocanthal exists in the oropharyngeal region of the oral cavity. Recent investigations have indeed verified that the TRPA1 is the receptor linked to oleocanthal and the anatomical location of this receptor has been found in the oropharyngeal region of the oral cavity (Peyrot des Gachons et al., 2011).

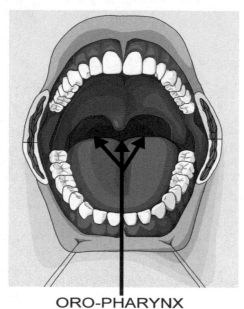

ORO-PHARYNX

Fig. 3. Oleocanthal irritation occurs solely in the oropharyngeal region (area shaded in black) of the oral cavity.

A large variability among subjects in the perceived irritation from oleocanthal has been noted (Cicerale et al., 2009a). Such individual variation in perception of oleocanthal may be related directly to the quantity of the TRPA1 receptor, as has been reported for other oral stimuli such as the bitter compounds: 6-n-propylthiouracil (PROP) and phenylthiocarbamide (PTC) (Hansen et al., 2006), which activate the TAS2R38 bitter receptor (Hayes et al., 2011). The important link between the perceptual aspects of oleocanthal and health benefits is the notion that variation in sensitivity to oleocanthal irritation may relate to potential differences in sensitivity to the anti-inflammatory action of this compound. However, further research is required to investigate this.

6. Physiological properties of oleocanthal and its putative health benefits

Research conducted by Beauchamp and colleagues (Beauchamp et al., 2005) demonstrated that oleocanthal inhibits cyclooxygenase (COX) enzymes in a dose-dependent manner, mimicking the anti-inflammatory action exerted by ibuprofen. Cyclooxygenase 1 and 2 (COX 1 and COX 2) enzymes are responsible for the conversion of arachidonic acid to prostaglandins and thromboxane which are produced in response to inflammatory or toxic stimuli. COX 1 and COX 2 can be harmful to the body. In particular, COX 2 has been implicated in the pathogenesis of several cancers in both human and animal studies (Harris et al., 2003, Boland et al., 2004, Subbaramaiah et al., 2002, Ristimäki et al., 2002), and may also play a role in atherosclerosis (Chenevard et al., 2003). The novel findings presented by Beauchamp and colleagues (Beauchamp et al., 2005) demonstrate that oleocanthal not only mimics the mode of ibuprofen action, it exhibits increased potency (compared with ibuprofen) in inhibiting COX 1 and COX 2 enzymes at equimolar concentrations. For instance, oleocanthal (25 μM) inhibited 41-57% of COX activity in comparison to ibuprofen (25 μM) which inhibited only 13-18% of COX activity.

Moreover, Beauchamp and colleagues (Beauchamp et al., 2005) put forth the suggestion that chronic ingestion of small quantities of oleocanthal via EVOO consumption, may be responsible, in part, for the lowered prevalence of disease associated with the Mediterranean diet. Thus, if an olive oil consumer ingests around 50 g of EVOO a day containing approximately 200 μg/kg of oleocanthal, the person would ingest approximately 10 mg/day of oleocanthal. This would equate to a relatively low (10%) equivalent dose of ibuprofen (recommended for adult pain relief). Chronic low doses of ibuprofen and other COX inhibitors such as aspirin are known to have important health benefits in the prevention of cancer development (e.g. colon and breast) (Garcia-Rodriguez and Huerta-Alvarez, 2001, Harris et al., 2006) and CVD (Hennekens, 2002). Therefore, long term ingestion of oleocanthal via EVOO consumption may contribute to a reduction in chronic disease development and certainly emerging evidence supports this notion.

Finally, it is important to note oleocanthal's bioavailability within the body. Only one study has investigated this to date. A study by Garcia-Villalba et al. (Garcia-Villalba et al., 2010) noted that oleocanthal was readily metabolized however further studies are required to gain a more thorough understanding of the metabolism and bioavailability of this compound.

6.1 Oleocanthal and neuro-degenerative disease

Ibuprofen is known to exert beneficial effects on markers of neuro-degenerative disease (Van Dam et al., 2008) and based on the similar oral irritant properties and shared anti-inflammatory mode of action via COX inhibition, oleocanthal was investigated for potential neuro-protective properties. Li and collegues (Li et al., 2009) presented significant findings demonstrating that oleocanthal inhibits tau fibrillization *in vitro* by forming an adduct with PHF6 peptide, which is a VQIXXK motif that resides in the microtubule binding region, and is crucial in the formation of tau fibrils (Li and Virginia, 2006) Hyperphosphorylated tangles of tau are lesions that are observed in neuro-degenerative disease (i.e. Alzheimer's disease) and PHF6 enables the phosphorylation of tau, thus the convalent modification of PHF6 peptide disrupts tau-tau interaction and subsequent fibril formation (Li et al., 2009) (Figure 4). B-amyloid peptides (Aβ) are another type of lesion that are characteristic of Alzheimer's disease (Guela et al., 1998), as Aβ derived diffusible ligands (ADDLs) are neurotoxic factors proposed to instigate the onset of Alzheimer's disease (Pitt et al., 2009). Pitt and colleagues (2009) demonstrated that *in vitro*, oleocanthal alters the structure of ADDLs and augments antibody clearance of ADDLs, therefore protecting hippocampal neurons from ADDL toxicity. This data supports research showing a 40% decrease in Alzheimer's disease in populations consuming a Mediterranean style diet (Scarmeas et al., 2009).

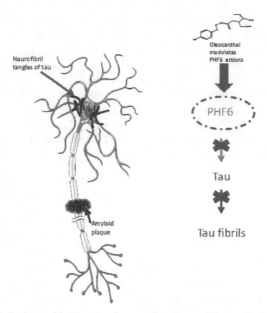

Fig. 4. Oleocanthal inhibits tau fibrilisation by covalently modifying the PHF-6 peptide which is crucial for the formation of tau fibrils. The fibrilisation of tau leads to neurofibrillary tangles which are inherently associated with neurodegenerative diseases such as Alzheimer's disease.

6.2 Oleocanthal and joint-degenerative disease

In vitro research draws attention to oleocanthal, as a potential therapeutic compound that may be of interest in the quest to find suitable natural NSAIDs for the treatment of joint degenerative disease. Pro-inflammatory cytokines up-regulate the synthesis of cartilage to degrading enzymes, and stimulate nitric oxide (NO) production (Scher et al., 2007), as well as increase prostaglandin PGE_2 production, which have all been implicated in the development of arthritic pain and thus joint-degenerative disease. COX enzymes are a catalyst for the formation of prostaglandins and have also been reported to be highly expressed in arthritic spine in an animal model (Procházková et al., 2009). Therefore, oleocanthal may influence arthritic pain through inhibition of PGE_2 synthesis accompanying COX inhibition.

NO plays an integral role in joint-degenerative disease and the stable end product of NO, nitrite (NO_2), is significantly expressed in arthritic synovial fluid (Iacono et al., 2010). In osteoarthritis arthritis (OA) pathogenesis, diseased cartilage synthesizes NO spontaneously from diseased chrondocytes (Tung et al., 2002). NO is biosynthesized by nitric oxide synthase (NOS). Another form of NOS is inducible NOS (iNOS) which is largely responsible for the inflammatory actions of NOS (Espey et al., 2000) (Figure 5). Iacano and collegues (2010) have shown that oleocanthal and synthesized derivatives, decrease production of iNOS protein expression in LPS challenged murine chondrocytes, dose dependently, further highlighting the anti-inflammatory actions of oleocanthal and the pharmacological potential. Also, as oleocanthal mediates prostaglandin synthesis via inhibitory actions on COX enzymes, it is possible that oleocanthal may exert pharmacological actions in the treatment of both rheumatoid arthritis and osteoarthritis through COX inhibition.

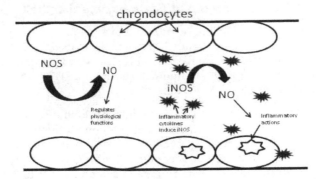

Fig. 5. Nitric oxide (NO) derived from nitric oxide synthase (NOS), functions as a neurotransmitter and vasodilator, and is important in normal physiological responses. Inducible nitric oxide synthase (iNOS) is a third form of NOS and is not present in resting cells, but rather is induced by inflammatory cytokines. NO produced from iNOS promotes inflammation in chrondocytes and is associated with cartilage degenerative diseases such as osteoarthritis.

6.3 Oleocanthal and cancer

Recent evidence demonstrates that oleocanthal may exert therapeutic properties against the pathogenesis of c-Met kinase induced malignancies. Elnagar and colleagues (Elnagar et al., 2011) have reported that oleocanthal possesses anti-proliferative effects in human breast and prostate cancer lines (Elnagar et al., 2011). Also Khanal and collegues (Khanal et al., 2011) recently showed that oleocanthal has an anti-proliferative effect and prevents tumour induced cell transformation in mouse epidermal JB6 Cl41cells. The mechanism of action in which oleocanthal achieved this was via the inhibition of extracellaullar signal-regulated kinases 1/2 and p90RSK phosphorylation. Furthermore, oleocanthal has also been shown to promote cell apoptosis by activating caspase-3 and poly-adenosine diphosphate-ribose polymerase, the phosphorylation of p53 (Ser15), and also induced fragmentation of DNA in HT-29 cells derived from human colon adenocarcinoma (Khanal et al., 2011). These findings suggest that oleocanthal may have potential as a therapeutic agent in the inhibition of carcinoma progression and supports substantial evidence that populations residing in the Mediterranean region have a reduced incidence of prostate, breast, lung and gastrointestinal cancer (Trichopoulou et al., 2000, La Vecchia, 2004, Fortes et al., 2003, Dixon et al., 2007). It is important to note that while there is strong evidence that oleocanthal is an effective anti-inflammatory agent and demonstrates pharmacological characteristics *in vitro,* future *in vivo* studies are required to fully elucidate the efficacy of this natural NSAID. Caution is required when extrapolating results from a single compound out of the matrix in which it normally exists. Oleocanthal is one of many phenolic compounds contained in EVOO, and it is probable that the synergistic and interactive actions of these phenolics combined are responsible for the low incidence of chronic inflammation associated with EVOO intake. Furthermore, the bioavailability of oleocanthal needs to be firmly established in future research to consolidate the pharmacological potential of this compound.

7. Extraction, identification and quantification of oleocanthal

The method used for the extraction, identification and quantification of oleocanthal described herein, is selective for oleocanthal and was developed by Beauchamp and co-workers (Beauchamp et al., 2005). More recently, this method was adapted and involves the quantification of oleocanthal using an internal standard (ISTD), 3,5 dimethoxyphenol (Beauchamp et al., 2005, Cicerale et al., 2009b).

The extraction of oleocanthal from the oil matrix involves liquid-liquid partitioning using both hexane and acetonitrile, whereby the phenolic fraction partitions into the acetonitrile phase. Acetonitrile is then removed and the dried down extract is dissolved in methanol:water and analyzed by HPLC. Separation of the oleocanthal phenolic compound is carried out using a HPLC system with a diode array detector set to 278 nm. A reverse phase-C18 column (250 mm × 4.6 mm ID, 5 µm) is used for the separation at a constant temperature of 25°C using the gradients listed in Table 1. A flow rate of 1 ml/min is used, and the injection volume is 20 µl. See Figure 6 for HPLC chromatogram.

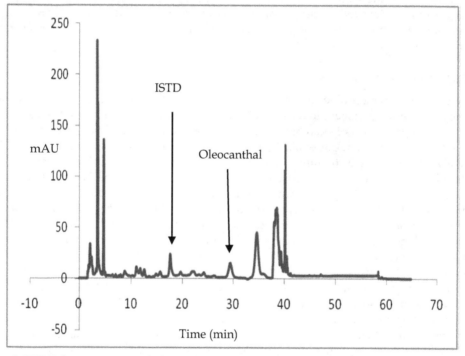

Fig. 6. HPLC chromatogram of olive oil phenolic extract containing oleocanthal and 3,5 dimethoxyphenol (ISTD) (Cicerale et al., 2009b).

Via mass spectrometry (6210 MSDTOF), oleocanthal is further identified under the following conditions: drying gas, nitrogen (N_2) (7 mL^{-1}, 350°C); nebulizer gas, N_2 (15 psi); capillary voltage 4.0 kV; vaporizer temperature, 350°C; and cone voltage, 60 V. Figure 7 displays the negative ion mass spectrum of oleocanthal with the characteristic [M-H]$^-$ ion at m/z 303.12 highlighted.

Time	Gradient
0-35 min	75% water, 25% acetonitrile
36-55 min	90% acetonitrile, 10% methanol
56-65 min	75% water, 25% acetonitrile

Table 1. Mobile phase gradient for oleocanthal separation.

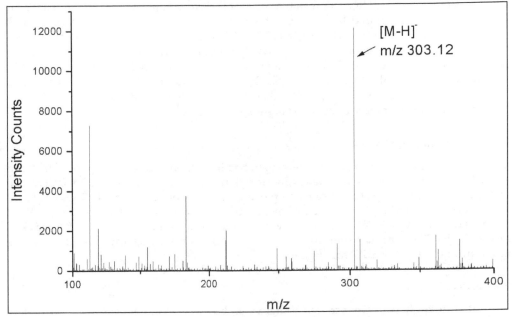

Fig. 7. Negative ion mass spectrum of oleocanthal with the characteristic [M-H]- ion at *m/z* 303.12 (Cicerale et al., 2009b).

8. Oleocanthal concentration in olive oil and factors which may affect its concentration

Oleocanthal concentration in EVOO is highly variable, ranging from as little as 0.2 mg/kg to as high as 498 mg/kg (Gomez-Rico et al., 2006). Such variation in oleocanthal concentration amongst differing EVOOs may be due to multiple factors that have the capacity to modify the concentration of this compound (Carrasco-Pancorbo et al., 2005). These factors include: method of phenolic extraction and quantification, geographic region of olive growth, olive tree cultivar, agricultural techniques applied to cultivate olives, olive maturity, processing of the olives to oil, storage of oil, and domestic heat application to oil.

8.1 Extraction and quantification

The analytical method used to quantify phenolics present in EVOO has an influence on the reported concentration and therefore is an important consideration when interpreting and comparing the data of such investigations (Carrasco-Pancorbo et al., 2005). Regarding the analysis of oleocanthal, Beauchamp and colleagues (Beauchamp et al., 2005) in collaboration with Impellizzeri et al. (Impellizzeri and Lin, 2006), developed an extraction and quantification method specific for this compound. Utilizing this method, Impellizeri et al. (Impellizzeri and Lin, 2006) and Franconi et al. (Franconi et al., 2006) found the concentration of oleocanthal ranged between 8.3 ± 4.0 and 189.9 ± 2.7 mg/kg. An adaption of the Beauchamp and co-workers' (Beauchamp et al., 2005) methodology, was also utilized in studies by Cicerale and colleagues (Cicerale et al., 2009a, Cicerale et al., 2009b, Cicerale et al., 2011b, Cicerale et al., 2011a). The resultant oleocanthal concentrations in these investigations were similar to those found by Franconi et al. (Franconi et al., 2006) and Impellizzeri et al. (Impellizzeri and Lin, 2006) (53.9 ± 7.7 to 152.2 ± 10.5 mg/kg). However, a number of methods not specific for oleocanthal have also been used to quantify this compound and may account partially for the large variation in oleocanthal concentration observed (5.0 ± 0.3 – 498.0 ± 47.0 mg/kg) (Vierhuis et al., 2001, Servili et al., 2007b, Servili et al., 2007a, Romero et al., 2002, Morello et al., 2004, Tovar et al., 2001, De Stefano et al., 1999, Montedoro et al., 1992, Gomez-Alonso et al., 2003, Gomez-Rico et al., 2006, Allouche et al., 2007)

8.2 Geographic region

Geographic region in which olives are grown has been shown to be an important factor in regards to phenolic composition and concentration in general (Vinha et al., 2005, Cerretani et al., 2005). Beauchamp et al. (Beauchamp et al., 2005) demonstrated that EVOOs produced in differing countries had variable oleocanthal concentrations. For instance, EVOO produced in the U.S.A., contained a low concentration of oleocanthal (22.6 ± 0.6 mg/kg), however EVOOs produced in Italy contained some of the highest quantities of this compound (up to 191.8 ± 2.7 mg/kg).

8.3 Cultivar

Several studies have demonstrated differences between olive cultivar and oleocanthal concentration in the oil produced. In one study, the Coratina cultivar EVOO contained 78.2 ± 0.5 mg/kg oleocanthal, whereas the Oliarola cultivar EVOO contained 21.0 ± 0.8 mg/kg oleocanthal, a 3-fold difference (De Stefano et al., 1999). In another study, EVOO produced from the Frantoio cultivar had an oleocanthal concentration of 43.8 ± 3.1 mg/kg, whilst EVOO obtained from the Coratina cultivar, contained a 2-fold higher oleocanthal content at 92.8 ± 7.8 mg/kg (Servili et al., 2007b). A study by Franconi and colleagues (Franconi et al., 2006) also showed significant differences in oleocanthal concentration amongst differing olive cultivars. For instance, an oleocanthal concentration of 8.3 ± 4.0 mg/kg and 53.0 ± 12.0 mg/kg in EVOOs produced from the Taggiasca and Seggianese cultivars respectively, was noted.

8.4 Agricultural methods

The concentration of phenolic compounds in EVOO is greatly affected by agricultural techniques used in the cultivation of olive fruit (Gomez-Rico et al., 2006, Ayton et al., 2007). Tovar and co-workers (Tovar et al., 2001) demonstrated that with increased irrigation applied to the olive tree, oleocanthal concentration decreased. For instance, in the EVOO obtained from the least irrigated olive trees (46 mm water per year) oleocanthal concentration was determined to be 50.9 ± 6.5 mg/kg. For the EVOO produced from highly irrigated olive trees (259 mm water per year), oleocanthal concentration was 23.1 ± 1.3 mg/kg. Gomez-Rico et al.

(Gomez-Rico et al., 2006) also demonstrated the negative effect of irrigation on oleocanthal concentration. Rain-fed olive trees produced EVOO containing higher concentrations of oleocanthal (229.0 ± 48.0 to 498 ± 47.0 mg/kg) compared to those that underwent the highest amount of irrigation (206 mm water per year), (119.0 ± 36.0 to 336.0 ± 81.0 mg/kg). Two additional studies (Romero et al., 2002, Servili et al., 2007a) are also in agreement with this data in that, they both observed a 37-38% decrease in oleocanthal concentration amongst the EVOOs from the highly irrigated olive trees compared to those least irrigated.

8.5 Olive maturation

Maturation of the olive fruit at harvest is an important predictor of the phenolic composition and concentration in EVOO. In regards to oleocanthal, one study found that with extended picking date and increased olive fruit ripeness, the concentration of oleocanthal in EVOO decreased by 43% (148.0 mg/kg to 84.0 mg/kg) over a short two month period (Morello et al., 2004). The researchers, Gomez-Rico et al. (Gomez-Rico et al., 2006) also observed a similar decrease of 20% and 54% in oleocanthal with increasing maturity index using two olive cohorts.

8.6 Processing

In general, the processing of olive fruit to oil has a substantial effect on the concentrations of phenolic compounds in EVOO (Kalua et al., 2006b, Cerretani et al., 2005, Vierhuis et al., 2001, Romero et al., 2004, Gimeno et al., 2002). EVOO produced by the traditional processing method (whereby the entire olive fruit is crushed, including the stone), was found to contain lower quantities of oleocanthal (43.8 ± 3.1 mg/kg) compared to EVOO produced by the stoning method in which the olive stone is removed before crushing (54.8 ± 3.1 mg/kg). The researchers from this study hypothesized that the differences may be due to the increased peroxidase (POD) activity that tends to accompany the crushed olive stone, which has an oxidizing effect on oleocanthal concentration (Servili et al., 2007b).

EVOO produced under nitrogen (N_2) flushing and with use of enzymatic treatment (which aids cell wall degradation and thus improves phenolic extraction) (Vierhuis et al., 2001), was found to contain oleocanthal concentrations of 31.4 ± 1.0 mg/kg. EVOO produced with no added enzymes and without nitrogen (N_2) flushing (therefore allowing oxygen (O_2) to be present) was found to contain a lower amount of oleocanthal (24.8 ± 1.9 mg/kg). EVOOs produced with use of N_2 flushing alone and enzymatic treatment alone, contained 28.4 ± 1.4 mg/kg and 29.4 ± 0.8 mg/kg oleocanthal respectively (Vierhuis et al., 2001).

EVOOs produced using two-phase centrifugation which uses no added water in the processing method, was found to contain a higher phenolic concentration compared to EVOOs obtained from three-phase centrifugation which utilizes a considerable amount of water (approximately 400 L/h) (De Stefano et al., 1999). De Stefano and et al. (De Stefano et al., 1999) found oleocanthal concentration in EVOO obtained from the two-phase centrifuge to be higher (78.2 ± 0.5 mg/kg) than that produced from the three-phase method (67.3 ± 2.6 mg/kg). The addition of water in the three-phase centrifugation method, may have a reducing effect on the more water-soluble phenolics from the oil phase during processing, thus reducing the concentration of oleocanthal in the resultant EVOO (Cicerale et al., 2009c).

8.7 Storage of EVOO

Immediately following oil extraction from the olive fruit, there is potential for the phenolic quality of the oil to decline, via oxidation catalysed by oxygen (O_2) and light (Kalua et al., 2006a, Morello, 2004). One study to date, has investigated the effect O_2, light, and storage time have on oleocanthal concentration. In this study, oleocanthal concentration decreased somewhat (15 - 37%) over a 10-month storage period, depending on the storage conditions. The largest decrease was seen in EVOO stored under exposure to O_2 and light (37%) and the smallest loss was found in the EVOO stored under O_2 and light limiting conditions (15%). Oils stored under sole exposure to O_2 or light were found to have a similar rate of oleocanthal degradation over the 10-month period (28% and 25% respectively) (Cicerale et al., 2011b).

8.8 Domestic heat application

In general, research has shown that olive oil phenolic compounds are subject to degradation upon the application of heat during cooking (Brenes et al., 2002, Gomez-Alonso et al., 2003). However, oleocanthal has demonstrated to be relatively stable upon heating when the EVOO contains a considerable quantity of this compound initially.

One study found a 20% decrease in oleocanthal (96.7 ± 8.5 to 77.5 ± 2.4 mg/kg) upon 12 frying operations (each frying operation 10 min in length, at a temperature of 180°C) (Gomez-Alonso et al., 2003). Similarly, another study observed a 24% decrease (41.5 ± 0.3 to 31.4 ± 0.1 mg/kg) in oleocanthal after heating at 180°C for 36 hr (Allouche et al., 2007). However, for EVOO which naturally contained a lower quantity of oleocanthal to begin with (7.9 ± 0.5 mg/kg), oleocanthal degradation was substantially higher at 71%. It appears that oleocanthal possesses an antioxidative effect, in that oleocanthal is able to withstand heating and therefore protect itself to a greater degree when there is a higher concentration of it in EVOO.

Cicerale and co-workers (Cicerale et al., 2009b) also demonstrated oleocanthal to be stable upon heating at high temperatures (240°C) for extended periods of time (up to 90 min). The authors postulated that the minimal degradation of oleocanthal may be partially due to the chemical structure of this compound and subsequent antioxidant activity. The antioxidant capacity of phenolic compounds is dependent upon the number of hydroxyl groups bonded to the aromatic ring (Sroka and Cisowski, 2003). When free radicals are produced through oxidation, phenolic compounds with a higher number of hydroxyls and therefore increased antioxidant capacity, quickly diminish because they react rapidly with lipid radicals and are therefore consumed (Gomez-Alonso et al., 2003). Oleocanthal possesses one hydroxyl group. Moreover, the site of bonding and mutual position of hydroxyls in the aromatic ring was also postulated to play a role in the antioxidant potential of phenolic compounds (Sroka and Cisowski, 2003, Rice-Evans et al., 1996). Rice-Evans and co-workers (Rice-Evans et al., 1996) reported that a hydroxyl group in the *ortho* position in the aromatic ring results in increased antioxidant capacity compared to compounds with hydroxyl groups in the *meta* and *para* positions. Oleocanthal contains its one hydroxyl group in the *para* position. These structural features may help in explaining why a minimal degradation of oleocanthal was observed (Cicerale et al., 2009c).

9. Perspectives and future directions

In summary, EVOO a key component of the Mediterranean diet contains a number of phenolics, one being oleocanthal. The unique sensory qualities and anti-inflammatory actions of oleocanthal, have prompted research to further verify its therapeutic potential. Oleocanthal has been shown, *in vitro*, to exert beneficial physiological effects in terms of neuro-degenerative disease, joint-degenerative disease and cancer. Therefore, it has been postulated that the long term ingestion of this compound via EVOO consumption may have significant health promoting action over time, thereby reducing the development of chronic disease.

However, the studies conducted on the health promoting potential of oleocanthal have involved *in vitro* investigations and it is difficult to extrapolate data from *in vitro* studies to actual physiological benefits. *In vivo* research is therefore required to substantiate the *in vitro* findings. Furthermore, the bioavailability of oleocanthal has not been adequately investigated. To date, only one study has reported on the post-ingestive fate of oleocanthal, noting that it was readily metabolized however the mechanism was not investigated. The degree to which oleocanthal is metabolized is an important consideration when reviewing the health benefits associated with ingestion, and further research on this is warranted. The link between the perceptual aspects of oleocanthal and health benefits is the notion that variation in sensitivity to oleocanthal oro-pharyngeal irritation may relate to potential differences in sensitivity to the anti-inflammatory action of this compound and this is also worthy of future investigations.

Finally, it was not the purpose of this overview to attribute the health benefits of the Mediterranean diet solely to one component and we did not aim to credit oleocanthal as being the lone therapeutic agent contained in EVOO. Other constituents of the Mediterranean diet and EVOO will contribute, either with independent actions, or in a synergistic and complementary manner to impart beneficial health effects (Lucas et al., 2011, Fogliano and Sacchi, 2006). However, the studies conducted to date investigating the pharmacological actions of oleocanthal are encouraging and show that this compound possesses substantial health benefiting properties. Further research will no doubt provide new insight into the pharmacological potential of oleocanthal and assess the role that oleocanthal has in the clinical treatment of chronic disease.

10. References

Aguilera, C. M., Mesa, M. D., Ramirez-Tortosa, M. C., Nestares, M. T., Ros, E. & Gil, A. 2004. Sunflower oil does not protect against LDL oxidation as virgin olive oil does in patients with peripheral vascular disease. *Clinical Nutrition*, 23, 673-681.

Allouche, Y., Jimenez, A., Gaforio, J. J., Uceda, M. & Beltran, G. 2007. How heating affects extra virgin olive oil quality indexes and chemical composition. *Journal of Agricultural and Food Chemistry*, 55, 9646-9654.

Andrewes, P., Busch, J. L., De Joode, T., Groenewegen, A. & Alexandre, H. 2003. Sensory properties of virgin olive oil polyphenols: identification of deacetoxy-ligstroside aglycon as a key contributor to pungency. *Journal of Agricultural and Food Chemistry*.

Ayton, J., Mailer, R. J., Haigh, A., Tronson, D. & Conlan, D. 2007. Quality and oxidative stability of Australian olive oil according to harvest date and irrigation. *Journal of Food Lipids*, 14, 138-156.

Babio, N., Bullo, M., Basora, J., Martinez-Gonzalez, M. A., Fernandez Ballart, J., Marquez-Sandoval, F., et al. 2009. Adherence to the Mediterranean diet and risk of metabolic syndrome and its components. *Nutrition, Metabolism and Cardiovascular Diseases*, 19, 563-570

Beauchamp, G. K., Keast, R. S., Morel, D., Lin, J., Pika, J., Han, Q., et al. 2005. Phytochemistry: ibuprofen-like activity in extra-virgin olive oil. *Nature,* 437, 45-46.

Boland, G., Butt, I., Prasad, R., Knox, W. & Bundred, N. 2004. COX-2 expression is associated with an aggressive phenotype in ductal carcinoma in situ. *British Journal of Cancer,* 90, 423-429.

Brenes, M., Garcia, A., Dobarganes, M. C., Velasco, J. & Romero, C. 2002. Influence of thermal treatments simulating cooking processes on the polyphenol content in virgin olive oil. *Journal of Agricultural and Food Chemistry,* 50, 5962-5967.

Carrasco-Pancorbo, A., Cerretani, L., Bendini, A., Segura-Carretero, A., Gallina-Toschi, T. & Fernandez-Gutierez, A. 2005. Analytical determination of polyphenols in olive oils. *Journal of Separation Science,* 28, 837-858.

Cerretani, L., Bendini, A., Rotondi, A., Lercker, G. & Toschi, T. G. 2005. Analytical comparison of monovarietal virgin olive oils obtained by both a continuous industrial plant and low-scale mill. *European Journal of Lipid Science and Technology,* 107, 93-100.

Chenevard, R., Hurlimann, D., Bechir, M., Enseleit, F., Spieker, L., Hermann, M., et al. 2003. Selective COX-2 inhibition improves endothelial function in coronary artery disease. *Circulation,* 10000051r.

Cicerale, S., Breslin, P. A. S., Beauchamp, G. K. & Keast, R. S. J. 2009a. Sensory characterization of the irritant properties of oleocanthal, a natural anti-inflammatory agent in extra virgin olive oils. *Chemical Senses,* 34 333-339.

Cicerale, S., Conlan, X. A., Barnett, N. W. & Keast, R. S. 2011a. The concentration of oleocanthal in olive oil waste. *Natural Product Research,* 25, 542-8.

Cicerale, S., Conlan, X. A., Barnett, N. W. & Keast, R. S. J. 2011b. Storage of extra virgin olive oil and its effect on the biological activity and concentration of oleocanthal *Food Research International,* doi:10.1016/j.foodres.2011.03.046

Cicerale, S., Conlan, X. A., Barnett, N. W., Sinclair, A. J. & Keast, R. S. 2009b. Influence of heat on biological activity and concentration of oleocanthal- a natural anti-inflammatory agent in virgin olive oil. *Journal of Agricultural and Food Chemistry,* 57, 1326-1330.

Cicerale, S., Conlan, X. A., Sinclair, A. J. & Keast, R. S. J. 2009c. Chemistry and health of olive oil phenolics. *Critical Reviews in Food Science and Nutrition,* 49, 218-236.

Cicerale, S., Lucas, L. & Keast, R. 2010. Biological activities of phenolic compounds present in virgin olive oil. *International Journal of Molecular Sciences,* 11, 458-79.

Corona, G., Spencer, J. P. E. & Dessi, M. A. 2009. Extra virgin olive oil phenolics: absorption, metabolism, and biological activities in the GI tract. *Toxicology and Industrial Health,* 25, 285-293.

Dai, J., Miller, A. H., Bremner, D., Goldberg, J., Jones, L., Shallenberger, L., et al. 2009. Adherence to the Mediterranean diet is inversely associated with circulating interleukin-6 among middle-aged men. A twin study. *Circulation,* 117, 169-175.

De Stefano, G., Piacquadio, P., Servili, M., Di Giovacchino, L. & Sciancalepore, V. 1999. Effect of extraction systems on the phenolic composition of virgin olive oils. *Lipid-Fett,* 101, 328-332.

Dixon, B. L., Subar, A. F., Peters, U., Weissfeld, J. L., Bresalier, R. S., Risch, A., et al. 2007. Adherence to the USDA food guide, DASH eating plan, and Mediterranean dietary pattern reduces risk of colorectal adenoma. *Journal of Nutrition,* 137, 2443-2450.

Elnagar, A., Sylvester, P. & El Sayed, K. 2011. -)-Oleocanthal as a c-Met Inhibitor for the Control of Metastatic Breast and Prostate Cancers. *Planta medica.* Vol- 77, pages-1013-1019.

Espey, M. G., Miranda, K. M., Feelisch, M., Fukuto, J., Grisham, M. B., Vitek, M. P., et al. 2000. Mechanisms of cell death governed by the balance between nitrosative and oxidative stress. *Annals of the New York Academy of Science*, 899, 209-21.

Estruch, R., Martinez-Gonzalez, M. A., Corella, D., Salas-Salvado, J., Ruiz-Gutierrez, V., Covas, M. I., et al. 2006. Effects of a Mediterranean-style diet on cardiovascular risk factors: a randomized trial. *Annals of Internal Medicine*, 145, 1-11.

Fogliano, V. & Sacchi, R. 2006. Oleocanthal in olive oil: between myth and reality. *Molecular Nutrition & Food Research*, 50, 5-6.

Fortes, C., Forastiere, F., Farchi, S., Mallone, S., Trequattrinni, T., Anatra, F., et al. 2003. The protective effect of the Mediterranean diet on lung cancer. *Nutrition and Cancer*, 46, 30-37.

Fragopoulou, E., Panagiotakos, D. B., Pitsavos, C., Tampourlou, M., Chrysohoou, C., Nomikos, T., et al. 2010. The association between adherence to the Mediterranean diet and adiponectin levels among healthy adults: the ATTICA study. *The Journal of Nutritional Biochemistry*, 21, 285-289.

Franconi, F., Coinu, R., Carta, S., Urgeghe, P. P., Ieri, F., Mulinacci, N., et al. 2006. Antioxidant effect of two virgin olive oils depends on the concentration and composition of minor polar compounds. *Journal of Agricultural and Food Chemistry*, 54, 3121-3125.

Fraser, A., Abel, R., Lawlor, D. A., Fraser, D. & Elhayany, A. 2008. A modified Medieterranean diet is associated with the greatest reduction in alanine aminotransferase levels in obese type 2 diabetes patients: results of a quasi-randomised controlled trial. *Diabetologia*, 51, 1616-1622.

Fung, T. T., Rexrode, K. M., Mantzoros, C. S., Manson, J. E., Willett, W. C. & Hu, F. B. 2009. Mediterranean diet and incidence of and mortality from coronary heart disease and stroke in women. *Circulation*, 119, 1093-1100.

Garcia-Rodriguez, L. A. & Huerta-Alvarez, C. 2001. Reduced risk of colorectal cancer among long-term users of aspirin and nonaspirin nonsteroidal anti-inflammatory drugs. *Epidemiology*, 12, 88-93.

Garcia-Villalba, R., Carrasco-Pancorbo, A., Nevedomskaya, E., Mayboroda, O. A., Deelder, A. M., Segura-Carretero, A., et al. 2010. Exploratory analysis of human urine by Lc-Esi-Tof MS after high intake of olive oil: understanding the metabolism of polyphenols. *Analytical and Bioanalytical Chemistry*, 398, 463-75.

Gimeno, E., Castellote, A. I., Lamuela-Raventos, R. M., De La Torre, M. C. & Lopez-Sabater, M. C. 2002. The effects of harvest and extraction methods on the antioxidant content (phenolics, alpha-tocopherol, and beta-carotene) in virgin olive oil. *Food Chemistry*, 78, 207-211.

Gomez-Alonso, S., Fregapane, G., Salvador, M. D. & Gordon, M. H. 2003. Changes in phenolic composition and antioxidant activity of virgin olive oil during frying. *Journal of Agricultural and Food Chemistry*, 51, 667-672.

GOMEZ-RICO, A., SALVADOR, M. D., LA GRECA, M. & FREGAPANE, G. 2006. Phenolic and volatile compounds of extra virgin olive oil (Olea europaea L. Cv. Cornicabra) with regard to fruit ripening and irrigation management. *Journal of Agricultural and Food Chemistry*, 54, 7130-7136.

Guela, C., Wu, C., Saroff, D., Lorenzo, A., Yuan, M. & Yankner, B. 1998. Aging renders the brain vulnerable to amyloid -protein neurotoxicity. *Nature medicine*, 4, 827-831.

Hansen, J. L., Reed, D. R., Wright, M. J., Martin, N. G. & Breslin, P. A. 2006. Heritability and genetic covariation of sensitivity to PROP, SOA, quinine HCl, and caffeine. *Chemical Senses*, 31, 403-413.

Harper, C. R., Edwards, M. C. & Jacobson, T. A. 2006. Flaxseed oil supplementation does not affect plasma lipoprotein concentration or particle size in human subjects. *Journal of Nutrition,* 136, 2844-2848.

Harris, R. E., Beebe-Donk, J. & Alshafie, G. A. 2006. Reduction in the risk of human breast cancer by selective cyclooxygenase-2 (COX-2) inhibitors. *BMC Cancer,* 6, 27.

Harris, R. E., Chlebowski, R. T., Jackson, R. D., Frid, D. J., Ascenseo, J. L., Anderson, G., et al. 2003. Breast cancer and nonsteroidal anti-inflammatory drugs: prospective results from the Women's Health Initiative. *Cancer Research,* 63, 6096-6101.

Hayes, J. E., Wallace, M. R., Knopik, V. S., Herbstman, D. M., Bartoshuk, L. M. & Duffy, V. B. 2011. Allelic Variation in TAS2R Bitter Receptor Genes Associates with Variation in Sensations from and Ingestive Behaviors toward Common Bitter Beverages in Adults. *Chemical Senses,* 36, 311.

Hennekens, C. H. 2002. Update on aspirin in the treatment and prevention of cardiovascular disease. *American Journal of Managed Care,* 8, 691S-700S.

Hu, F. B. 2003. The Mediterranean diet and mortality--olive oil and beyond. *New England Journal of Medicine,* 348, 2595-2596.

Iacono, A., Gómez, R., Sperry, J., Conde, J., Bianco, G., Meli, R., et al. 2010. Effect of oleocanthal and its derivatives on inflammatory response induced by LPS in chondrocyte cell line. *Arthritis & Rheumatism.* Vol- 62, pages- 1675-1682.

Impellizzeri, J. & Lin, J. 2006. A simple high-performance liquid chromatography method for the determination of throat-burning oleocanthal with probated antiinflammatory activity in extra virgin olive oils. *Journal of Agricultural and Food Chemistry,* 54, 3204-3208.

Kalua, C. M., Bedgood, D. R., Bishop, A. G. & Prenzler, P. 2006a. Discrimination of storage conditions and freshness in virgin olive oil. *Journal of Agricultural and Food Chemistry,* 54, 7144-7151.

Kalua, C. M., Bedgood, D. R., Jr., Bishop, A. G. & Prenzler, P. D. 2006b. Changes in volatile and phenolic compounds with malaxation time and temperature during virgin olive oil production. *Journal of Agricultural and Food Chemistry,* 54, 7641-7651.

Keys, A. 1970. Coronary heart disease in seven countries. *Circulation,* 42, 1-211. Vol- 240, pages- 189-197.

Khanal, P., Oh, W. K., Yun, H. J., Namgoong, G. M., Ahn, S. G., Kwon, S. M., et al. 2011. p-HPEA-EDA, a phenolic compound of virgin olive oil, activates AMP-activated protein kinase to inhibit carcinogenesis. *Carcinogenesis,* 32, 545.

Kok, F. J. & Kromhout, D. 2004. Atherosclerosis--epidemiological studies on the health effects of a Mediterranean diet. *European Journal of Nutrition,* 43 Suppl 1, 1-5.

La Vecchia, C. 2004. Mediterranean diet and cancer. *Public Health Nutrition,* 7, 965-968.

Li, W., Sperry, J. B., Crowe, A., Trojanowki, J. Q., Smith, A. B. & Lee, V. M. Y. 2009. Inhibition of tau fibrillization by oleocanthal via reaction with amino groups of tau. *Journal of Neurochemistry,* 110, 1339-1351.

Li, W. & Virginia, M. Y. L. 2006. Characterization of two VQIXXK motifs for tau fibrillization in vitro. *Biochemistry,* 45, 15692-15701.

López-Miranda, J., Pérez-Jiménez, F., Ros, E., De Caterina, R., Badimón, L., Covas, M., et al. 2010. Olive oil and health: summary of the II international conference on olive oil and health consensus report, Jaén and Córdoba (Spain) 2008. *Nutrition, Metabolism and Cardiovascular Diseases,* 20, 284-294.

Lucas, L., Russell, A. & Keast, R. 2011. Molecular Mechanisms of Inflammation. Anti-Inflammatory Benefits of Virgin Olive Oil and the Phenolic Compound Oleocanthal. *Current Pharmaceutical Design,* 17, 754-768.

Martinez-Gonzalez, M. A., De La Fuente-Arrillaga, C., Nunez-Cordoba, J. M., Basterra-Gortari, F. J., Beunza, J. J., Vazquez, Z., et al. 2008. Adherence to Mediterranean diet and risk of developing diabetes: prospective cohort study. *British Medical Journal*, 336, 1348-1351.

Montedoro, G. & Servili, M. 1993. Simple and hydrolyzable compounds in virgin olive oil. Spectroscopic characterizations of the secoiridoid derivatives. *Journal of Agricultural and Food Chemistry*, 41, 2228-2234.

Montedoro, G., Servili, M., Baldioli, M. & Miniati, E. 1992. Simple and Hydrolyzable phenolic compounds in virgin olive oil. 1. Their extraction, separation, and quantitative and semiquantitative evaluation by HPLC. *Journal of Agricultural and Food Chemistry*, 40, 1571-1576.

Morello, J. R. 2004. Changes in commercial virgin olive oil (cv Arbequina) during storage, with special emphasis on the phenolic fraction. *Food Chemistry*, 85, 357-364.

Morello, J. R., Romero, M. P. & Motilva, M. J. 2004. Effect of the maturation process of the olive fruit on the phenolic fraction of drupes and oils from Arbequina, Farga, and Morrut cultivars. *Journal of Agricultural and Food Chemistry*, 52, 6002-6009.

Panagiotakos, D. B., Dimakopoulou, K., Katsouyanni, K., Bellander, T., Grau, M., Koenig, W., et al. 2009. Mediterranean diet and inflammatory response in myocardial infarction survivors. *International Journal of Epidemiology*, 38, 856.

Peyrot Des Gachons, C., Uchida, K., Bryant, B., Shima, A., Sperry, J. B., Dankulich-Nagrudny, L., et al. 2011. Unusual Pungency from Extra-Virgin Olive Oil Is Attributable to Restricted Spatial Expression of the Receptor of Oleocanthal. *The Journal of Neuroscience*, 31, 999.

Pitsavos, C., Panagiotakos, D. B., Tzima, N., Chrysohoou, C., Economou, M., Zampelas, A., et al. 2005. Adherence to the Mediterranean diet is associated with total antioxidant capacity in healthy adults: the ATTICA study. *American Journal of Clinical Nutrition*, 82, 694-699.

Pitt, J., Roth, W., Lacor, P., Blankenship, M., Velasco, P., De Felice, F., et al. 2009. Alzheimer's-associated A-beta oligomers show altered structure, immunoreactivity and synaptotoxicity with low doses of oleocanthal. *Toxicology and Applied Pharmacology*, doi:10.1016/j.taap.2009.07.018.

Procházková, M., Zanvit, P., Doležal, T., Prokešová, L. & Kršiak, M. 2009. Increased gene expression and production of spinal cyclooxygenase 1 and 2 during experimental osteoarthritis pain. *Physiological research*, 58, 419-425.

Rice-Evans, C. A., Miller, N. J. & Paganga, G. 1996. Structure-antioxidant activity relationships of flavonoids and phenolic acids. *Free Radical Biology and Medicine*, 20, 933-956.

Ristimäki, A., Sivula, A., Lundin, J., Lundin, M., Salminen, T., Haglund, C., et al. 2002. Prognostic significance of elevated cyclooxygenase-2 expression in breast cancer. *Cancer Research*, 62, 632.

Romero, C., Brenes, M., Yousfi, K., Garcia, P., Garcia, A. & Garrido, A. 2004. Effect of cultivar and processing method on the contents of polyphenols in table olives. *Journal of Agricultural and Food Chemistry*, 52, 479-484.

Romero, M. P., Tovar, M. J., Girona, J. & Motilva, M. J. 2002. Changes in the HPLC phenolic profile of virgin olive oil from young trees (Olea europaea L. Cv. Arbequina) grown under different deficit irrigation strategies. *Journal of Agricultural and Food Chemistry*, 50, 5349-5354.

Scarmeas, N., Luchsinger, J. A., Schupf, N., Brickman, A. M., Cosentino, S., Tang, M. X., et al. 2009. Physical activity, diet, and risk of Alzheimer disease. *Journal of American Medical Association*, 302, 627-637.

Scher, J., Pillinger, M. & Abramson, S. 2007. Nitric oxide synthases and osteoarthritis. *Current Rheumatology Reports*, 9, 9-15.

Servili, M., Esposto, S., Londolini, E., Selvaggini, R., Taticchi, A., Urbani, S., et al. 2007a. Irrigation effects on quality, phenolic composition, and selected volatiles of virgin olive oils Cv. Leccino. *Journal of Agricultural and Food Chemistry*, 55, 6609-6618.

Servili, M., Taticchi, A., Esposto, S., Urbani, S., Selvaggini, R. & Montedoro, G. 2007b. Effect of olive stoning on the volatile and phenolic composition of virgin olive oil. *Journal of Agricultural and Food Chemistry*, 55, 7028-7035.

Sroka, Z. & Cisowski, W. 2003. Hydrogen peroxide scavenging, antioxidant and anti-radical activity of some phenolic acids. *Food and Chemical Toxicology*, 41, 753-758.

Stark, A. H. & Madar, Z. 2002. Olive oil as a functional food: epidemiology and nutritional approaches. *Nutrition Reviews*, 60, 170-176.

Subbaramaiah, K., Norton, L., Gerald, W. & Dannenberg, A. J. 2002. Cyclooxygenase-2 is overexpressed in HER-2/neu-positive breast cancer. *Journal of Biological Chemistry*, 277, 18649 - 18657.

Tortosa, A., Bes-Rastrollo, M., Sanchez-Villegas, A., Basterra-Gortari, F. J., Nunez-Cordoba, J. M. & Martinez-Gonzalez, M. A. 2007. Mediterranean diet inversely associated with the incidence of metabolic syndrome. *Diabetes Care*, 30, 2957-2959.

Tovar, M. J., Motilva, M. J. & Romero, M. P. 2001. Changes in the phenolic composition of virgin olive oil from young trees (Olea europaea L. cv. Arbequina) grown under linear irrigation strategies. *Journal of Agricultural and Food Chemistry*, 49, 5502-5508.

Trichopoulou, A., Lagiou, P., Kuper, H. & Trichopoulos, D. 2000. Cancer and Mediterranean dietary traditions. *Cancer Epidemiology, Biomarkers and Prevention*, 9, 869-873.

Trichopoulou, A., Orfanos, P., Norat, T., Bueno-De-Mesquita, B., Ocke, M. C., Peeters, P. H., et al. 2005. Modified Mediterranean diet and survival: EPIC-elderly prospective cohort study. *British Medical Journal*, 330, 991-998.

Tung, J., Venta, P. & Caron, J. 2002. Inducible nitric oxide expression in equine articular chondrocytes: effects of antiinflammatory compounds. *Osteoarthritis and Cartilage*, 10, 5-12.

Van Dam, D., Coen, K. & De Deyn, P. 2008. Ibuprofen modifies cognitive disease progression in an Alzheimer's mouse model. *Journal of Psychopharmacology*. Vol- 24, pages- 383-388.

Vierhuis, E., Servili, M., Baldioli, M., Schols, H. A., Voragen, A. G. & Montedoro, G. F. 2001. Effect of enzyme treatment during mechanical extraction of olive oil on phenolic compounds and polysaccharides. *Journal of Agricultural and Food Chemistry*, 49, 1218-1223.

Vinha, A. F., Ferreres, F., Silva, B. M., Valentao, P., Goncalves, A., Pereira, J. A., et al. 2005. Phenolic profiles of Portuguese olive fruits (Olea europaea L.): Influences of cultivar and geographical origin. *Food Chemistry*, 89, 561-568.

Visioli, F., Bogani, P., Grande, S. & Galli, C. 2005. Mediterranean food and health: building human evidence. *Journal of Physiology and Pharmacology*, 56 Suppl 1, 37-49.

7

Metabolism and Bioavailability of Olive Oil Polyphenols

María Gómez-Romero[1], Rocío García-Villalba[2],
Alegría Carrasco-Pancorbo[1] and Alberto Fernández-Gutiérrez[1]
[1]University of Granada
[2] CEBAS CSIC of Murcia
Spain

1. Introduction

The significance of virgin olive oil (VOO), hinged to its many virtues in both gastronomy and health, is nowadays undeniable. Their protective effects are attributed to its high content of monounsaturated fatty acids and to the presence of some minor components, which add up to 2% of the weight. Among its several minor constituents, polar phenolic compounds, usually characterized as polyphenols, have become the subject of intensive research because of their biological activities, their influence on the organoleptic properties of VOO and their contribution to its oxidative stability (Bendini et al., 2007).

The phenolic fraction of VOO consists of a heterogeneous mixture of compounds belonging to several families with varying chemical structures. A brief description of the main classes of phenolic compounds contained in VOO is given below:

- *Phenolic acids*. There are two main series of these acids, depending on the carbon skeleton: benzoic acids (C6-C1: 3-hydroxybenzoic, *p*-hydroxybenzoic, protocatechuic, gentisic, vanillic, syringic and gallic acids) and cinnamic acids (C6-C3: *o*-coumaric, *p*-coumaric, caffeic, ferulic and sinapic acids).
- *Phenolic alcohols*. The two most important in VOO are hydroxytyrosol (Hyty) and tyrosol (Ty), although two Hyty derivatives, its acetate and its glucoside, can be also found. Hyty and Ty only differ in a hydroxyl group in the *meta* position.
- *Secoiridoids*. They are present exclusively in plants of the Olearaceae family. The olives mainly contain the polar oleuropein (Ol) and ligstroside (Lig) glycosides. Ol is the ester of elenolic acid (EA) with Hyty, and Lig is the ester of EA with Ty. Ol and Lig aglycones (Ol Agl and Lig Agl, respectively) are formed by removal of the glucose moiety from glycosides by endogenous β-glucosidases during ripening, oil extraction and storage.
- *Lignans*. (+)-1-Pinoresinol, (+)-1-hydroxypinoresinol and (+)-1-acetoxypinoresinol are the most reported compounds in olive oil.
- *Flavonoids*. The main flavonoids present in VOO are apigenin and luteolin, which are originated from their corresponding glucosides present in the drupe.

The qualitative and quantitative composition of VOO hydrophilic phenols is strongly affected by the agronomic and technological conditions of production (Servili et al., 2004). Among agronomic parameters, the cultivar, the fruit ripening degree, the agronomic techniques used and the pedoclimatic conditions are the aspects more extensively studied (Tovar et al., 2001; Uceda et al., 1999). Moreover, by modulating technology, it is possible to some extent to optimize the transfer of some polar minor constituents into the oil or reduce their level (Boskou, 2009). The influence of variety, extraction system, ripening degree and storage in the polyphenolic content of a VOO has been extensively discussed in the literature (Boskou, 2009; Uceda et al., 1999).

Wide ranges of total polar phenols concentration have been reported in olive oils (50-1000 mg/kg), although the most usual value is found between 100-350 mg/kg (Boskou et al., 2006). In general, the most abundant phenolic compounds in VOO are aglycones deriving from secoiridoids. Trying to establish levels of individual phenols, Servili & Montedoro (2002) calculated average values of 7 phenolic compounds from a considerable number of samples of industrial olive oils. They concluded that Hyty and Ty were found only in trace amounts (less than 10 mg/kg oil) and the most abundant phenols were decarboxylated Ol Agl (63-840 mg/kg), Ol Agl (85–310 mg/kg), and decarboxylated Lig Agl (15-33 mg/kg). Brenes et al. (2002) published values ranging from 3-67 mg/kg for 1-acetoxypinoresinol, and from 19-41 mg/kg for pinoresinol in 5 Spanish olive oils, data that can be completed with the researches carried out by Romero et al. (2002) and Tovar et al. (2001). Levels of luteolin have been found to be around 10 mg/kg in some Spanish olive oils (Brenes et al., 1999) or ranging between 0.2-7 mg/kg for Greek oils (Murkovic et al., 2004). Carrasco-Pancorbo et al. (2006) developed a method to quantify 14 individual phenols belonging to different families in 7 Spanish extra-virgin olive oils (EVOOs). They also quantified them, finding the following contents (mg/kg): simple phenols: 6.8-11.5; complex phenols: 70.5-799.5; lignans: 0.81-20.6; and flavonoids: 1.4-8.6.

Intake of olive oil in the Mediterranean countries is estimated to be 30–50 g/day, based on the per capita olive oil consumption of 10–20 kg/year in Greece, Italy and Spain (Boskou, 2000; Food and Agricultural Organization, 2000). A daily consumption of 50 g olive oil with a concentration of 180 mg/kg of phenols would result in an estimated intake of about 9 mg of olive oil phenols per day (de la Torre, 2008; Vissers et al., 2004), of which at least 1 mg is derived from free Hyty and Ty, and 8 mg are related to their elenolic esters and also to Ol Agl and Lig Agl (de la Torre, 2008). Some other estimations have been made. For the Greek population (Dilis & Trichopolou, 2009), the daily per-capita intake is about 17 mg. Vissers et al. (2004) estimated that about 1 mg of the phenol intake per day (6 mmol) is derived from Hyty and Ty, about 8 mg (23 mmol) from the aglycones, and so the total phenol intake would be about 29 mmol.

2. Bioavailability of olive oil polyphenols

Accumulating evidence suggests that VOO may have health benefits; it can be considered as an example of a functional food containing a variety of components that may contribute to its overall therapeutic characteristics (Stark & Madar, 2002; Visioli & Bernardini, 2011). To explore and determine the mechanisms of action of olive oil polyphenols and their role in disease prevention, understanding the factors that constrain their release from the olive oil, their extent of absorption, and their fate in the organism is crucial. These issues can be described under the term *bioavailability*, borrowed from the field of pharmacology, redefined as "that fraction of an oral dose, either parent compound or active metabolite, from a particular preparation that reaches the systemic circulation" (Stahl et al., 2002). To simplify this definition, D'Archivio et al. (2010) explained that it simply means how much of the ingested amount of polyphenols is able to exert its beneficial effects in the target tissues. It is important to realize that the most abundant phenolic compounds in our diet are not necessarily those that have the best bioavailability profile, either because they have a lower intrinsic activity or because they are poorly absorbed from the intestine, highly metabolized, or rapidly eliminated. In addition, the metabolites that are found in blood and target organs, resulting from digestive or hepatic activity, may differ from the native compounds in terms of biological activity (Manach et al., 2004).

Although the information concerning the bioavailability of most olive oil polyphenols is limited, intensive research has been carried out in the past decade. This fact is reflected in the number of reviews published since 2002 (Corona et al., 2009; Covas et al., 2009; de la Torre, 2008; Fitó et al., 2007; Tuck & Hayball, 2002; Visioli et al., 2002; Vissers et al., 2004). To address the bioavailability of olive oil phenolic compounds, we have reviewed *in vitro* and *in vivo* (both animal and human) studies on the absorption, transport, metabolism and excretion of olive oil phenolic compounds.

2.1 Absorption and disposition

Direct evidence on bioavailability of olive oil phenolic compounds has been obtained by measuring the concentration of the polyphenols and their metabolites in biological fluids, mostly plasma and urine, after ingestion of pure compounds or of olive oil, either pure or enriched with the phenolics under study. The majority of research regarding the bioavailability of olive oil polyphenols has been focused on three major phenolics: Hyty, Ty and Ol, as can be seen in **Tables 1** and **2**.

After ingestion, olive oil polyphenols can be partially modified in the acidic environment of the stomach. The effect of such environment on aglycone secoiridoids has been examined *in vitro* by incubating the compounds at 37 °C in simulated gastric pH conditions and during normal physiological time frames (Corona et al., 2006; Pinto et al., 2011). Although hydrolysis takes place releasing free phenolic alcohols, a significant amount remains intact and thus, enters the small intestine unmodified. Ol Agl and its dialdehydic form, however, are likely not absorbed as such in the small intestine; the major metabolites detected using the perfused rat intestine model were the glucuronide conjugates of the reduced form of both compounds (Pinto et al., 2011).

Manna et al. (2000) carried out studies on the transport kinetics of radiolabeled Hyty using differentiated Caco-2 cells. The only metabolite found in the culture medium was the methylated derivative (i.e. homovanillic alcohol - HVAlc). They also demonstrated that Hyty was transported across the membrane of the human enterocytes by a bidirectional passive diffusion mechanism. Caco-2/TC7 cell monolayers have been used to study the metabolism of other olive oil polyphenols, such as Ty, *p*-coumaric acid, pinoresinol, luteolin (Soler et al., 2010) and Hyty acetate (Mateos et al., 2011). Results showed that the methylated conjugates are the main metabolites and that the acetylation of Hyty significantly increases its transport across the small intestinal epithelial cell barrier, enhancing the delivery of Hyty to the enterocytes.

To study the potential hepatic metabolism of olive oil phenols, human hepatoma HepG2 cells were incubated for 2 and 18 h with Ty, Hyty and Hyty acetate (Mateos et al., 2005). Extensive uptake and metabolism of Hyty and Hyty acetate were observed, with scarce metabolism of Ty. Hyty acetate was converted into free Hyty and then metabolized;

Tested Phenol	Model system[a]	Methods	Metabolites Detected	Study Outcome	Ref.
[14C] Hyty	Caco-2 cell monolayers	Transport kinetics: incubation with increasing concentrations (50-500 µM) at 37 and 4 °C for 2 min. Transepithelial transport: incubation with 100 µM Hyty, glucose and mannitol	HVAlc	Hyty transport occurs via a passive diffusion mechanism, bidirectionally and in a dose-dependent manner. Hyty is quantitatively absorbed in the intestine	Manna et al., 2000
Ol glycoside	Isolated rat intestine	In situ intestinal perfusion technique: infusion of aqueous solution (1 mM, 50 µl/min) at 37 °C during 40 min in both iso-osmotic and hypotonic luminal conditions		Ol in aqueous solution can be absorbed, albeit poorly, from isolated perfused rat intestine. The P_{app} of Ol in hypotonic conditions is significantly higher	Edge-combe et al., 2000
Hyty, Ty, Hyty-Ac	Hepatoma HepG2 cells	Cell uptake and metabolism of phenols: incubation with 100 µM at 37 °C for 2 and 18 h	Hyty mono-gluc and methyl-gluc, HVA, Ty gluc, Hyty-Ac mono-gluc	Extensive uptake and hepatic metabolism of Hyty and Hyty-Ac with scarce metabolism of Ty; main derivatives formed: glucuronidated and methylated conjugates	Mateos et al., 2005
Hyty, Ty, Ol	Caco-2 cell monolayers and rat segments of jejunum and ileum		Hyty and Ty gluc, HVAlc, Hyty glutathionylated	Hyty and Ty were transferred across the cell monolayers and rat segments of intestine and were subjected to classic phase I/II biotransformation. No absorption of Ol	Corona et al., 2006
Hyty, Ty, p-coumaric acid, pinoresinol, luteolin	Caco-2/TC7 cell monolayers	Phenols metabolism: incubation with 40, 50 and 100 µM at 37 °C for 1, 6 and 24 h. Transport experiments in the AP, cellular and BL compartments: AP loading of phenol at 100 µM	*Hyty:* methyl, sulfate, methyl-sulfate. *Ty:* methyl, sulfate. *p-Coumaric acid:* disulfate, methyl. *Pinoresinol:* gluc, sulfate. *Luteolin:* gluc, methyl, methyl-gluc,	Limited intestinal metabolism. Major metabolites: methylated conjugates. Time-dependent efflux of various free and conjugated forms, showing preferential AP to BL transport after 24 h of incubation	Soler et al., 2010
Hyty, Hyty-Ac	Caco-2/TC7 cell monolayers	Metabolism experiments and transport experiments in the AP and BL compartments: incubation with 50 µM at 37 °C for 1, 2 and 4 h	*Hyty:* HVAlc. *Hyty-Ac:* Hyty, HVAlc, mono-gluc.	Hyty-Ac is better absorbed than free Hyty and serves to enhance delivery of Hyty to the enterocytes for subsequent metabolism and BL efflux)	Mateos et al., 2011
Ol Agl, dialdehydic form of Ol Agl	Human Caco-2 cell monolayers and isolated lumen of rat intestine (jejunum and ileum)	Transport experiments using Caco-2 cells: incubation with 50, 100 and 200 µM at 37 °C for 2 h; AP loading. Transport experiments using rat intestine: perfusion of methanol solution (100 µM) at 37 °C during 80 min	Hyty, HVAlc, Hyty and HVAlc gluc conjugates of the reduced forms of tested compounds	Caco-2 cells expressed limited metabolic activity. Major metabolites using the perfused rat intestine model: gluc of the reduce forms. Secoiridoids in the parental form were little absorbed in the small intestine	Pinto et al., 2011

[a] Caco-2 cells: model system of the human intestinal epithelium; HepG2 cells: model system of the human liver; TC7 cells: spontaneously differentiating clone derived from the original Caco-2 cell population.

Abbreviations: AP: apical; BL: basolateral; gluc: glucuronide; Hyty: hydroxytyrosol; Hyty-Ac: hydroxytyrosol acetate; HVA: homovanillic acid; HVAlc homovanillic alcohol; Ol: oleuropein; Ol Agl: oleuropein aglycone; P_{app}: apparent permeability coefficient; Ty: tyrosol.

Table 1. *In vitro* studies carried out with olive oil polyphenols.

glucurono- and methyl-, but no sulfo-conjugates, were found. Olive oil phenols are metabolized by the liver as well, as suggested by these results.

The colonic metabolism of olive oil polyphenols is scarcely reported. Corona et al. (2006) demonstrated that secoiridoids, which appear not to be absorbed in the small intestine, suffer bacterial catabolism in the large intestine with Ol undergoing rapid degradation by the colonic microflora producing Hyty as the major end product.

It is essential to establish whether olive oil phenolics are absorbed in the intestine *in vivo* and how they are distributed in the organism. **Table 1** shows the *in vivo* bioavailability studies of olive oil polyphenols carried out so far. For practical reasons, rats are used as the model of choice for *in vivo* studies. Bai et al. (1998) studied the absorption and pharmacokinetics of Hyty in rats, finding that the absorption of Hyty after the ingestion of a single dose is very fast. The metabolic fate of Hyty and Ty in rats has been also evaluated by administration of the radiolabeled polyphenols. Hyty appeared in plasma at maximum levels 5 min after oral administration, although the proportion of free aglycones in some tissues differed to that observed in plasma (D'Angelo et al., 2001). In all of the investigated tissues, Hyty was enzymatically converted in oxidized and/or methylated derivatives, whereas the major urinary products were sulfo-conjugates. Tuck et al. (2001) compared the elimination of Hyty and Ty in rat urine within 24 h after administration, both orally (in oil- and water-based solutions) and intravenously (in saline). When orally administrated, polyphenols will be subjected to first-pass metabolism, so that the contribution of intestinal metabolism will be quite relevant. If the administration is intravenous, only hepatic contribution to its disposition will be seen. Results showed that Hyty and Ty can be absorbed into the systemic circulatory system after oral dosing and that their bioavailability when administered as an olive oil solution is almost complete. Later, urine samples were re-examined and Hyty and five of its metabolites were detected (Tuck et al., 2002). Three were conclusively identified as monosulfate and 3-O-glucuronide conjugates of Hyty, and homovanillic acid (HVA), and one was tentatively identified as O-glucuronide conjugate of HVA. Although there is no disagreement between studies, a major limitation is that they were done with rats and some researches suggest that comparisons between the model species might not be adequate. Visioli et al. (2003) observed a 25 fold higher basal excretion of Hyty and of its main metabolites in rats than humans.

In a well-designed approach, Vissers et al. (2002) measured the absorption and urinary excretion of olive oil polyphenols in healthy ileostomy subjects and subjects with a colon after the ingestion of increasing doses of extracted phenols. Only a small amount of the ingested compounds was recovered in the urine, supporting the hypothesis that humans absorb a major fraction of the olive oil phenols consumed. Furthermore, the comparison between the absorbed polyphenols in normal and ileostomy subject showed similar results, which implies that the small intestine is the major site for the absorption of those compounds. Free Hyty and Ty and their glucurono-conjugates were the only metabolites detected in the urine samples. Another study carried out in human subjects assessed quantitatively the uptake of phenolics from olive oils containing different amounts of Ty and Hyty (Visioli et al., 2000). It was observed that these compounds were absorbed in a dose-dependent manner, that they were excreted in urine as glucuronide conjugates and that, as the concentration of phenols administered increased, the proportion of conjugation with glucuronic acid also increased. Upon re-examination of samples two more metabolites of Hyty were identified: HVA and HVAlc (Caruso et al., 2001).

Administered Polyphenol	Administration and Dose	Biological Sample	Concentration in Plasma	Excretion in Urine[a]	Metabolites Detected	Other Measurements	Analysis Methods	Ref.
Synthetic Hyty in 0.5% tragacanth solution	Oral, 1 ml single dose: Hyty 10 mg/ml	Rat plasma	Hyty 0.89-3.26 µg/ml (after 10 min)				GC-MS	Bai et al., 1998
Olive oil enriched with increasing concentrations of phenols	Oral, 50 ml single dose. Phenolic content (mg/l): total phenols 487.5-1950; Hyty 20-84; Ty 36-140	Human urine		Total Hyty 30-60%; Total Ty 20-22%	Hyty and Ty gluc		GC-MS	Visioli et al., 2000
Synthetic Hyty in aqueous solution	Oral, single dose: Hyty: 20 mg/kg	Rat plasma	Hyty 1.91 µg/ml (after 10 min)				HPLC-UV (280 nm)	Ruiz-Gutiérrez et al., 2000
EVOO	Oral (a) Sustained doses for 1 month of 50 g EVOO/day; (b) 100 g single dose	Human plasma				Plasma antioxidant capacity. Hyty, Ty and vitamin E content in LDL		Bonanome et al., 2000
Olive mill waste water extracts with increasing concentrations of Hyty	Oral, single doses (a) 1 mg/kg of extract: 41.4 µg/kg of Hyty (b) 5 mg/kg of extract: 207 µg/kg of Hyty (c) 10 mg/kg of extract: 414 µg/kg of Hyty	Rat plasma and urine			Hyty gluc	Plasma antioxidant capacity		Visioli et al., 2001
Radiolabeled synthetic Hyty and Ty in different solutions	(a) Oral, single dose; 225 mg oil-based solution (23.5 mg Hyty or 14.7 mg Ty in 1300 mg EVOO) or water-based solution (25.5 mg Hyty or 14.4 mg Ty in 1300 mg water) (b) IV, 950 mg saline solution (6.5 mg Hyty or 9.8 mg Ty added to 5 ml of 9 g/l NaCl)	Rat urine and feces		(a) Oral in oil (%): Hyty 94.1, Ty 72.9; oral in water, Hyty 70.9, Ty 53.2; (b) IV (%): Hyty 94.9, Ty 74.4	Gluc and sulfate conjugates	Feces: Hyty: < 3% and Ty: 25-30% of administered amount (after 24 h)	HPLC-radiometric detection	Tuck et al., 2001
Radiolabeled synthetic Hyty	IV, 0.3 mg single dose: 1.5 mg Hyty/kg body weight	Rat blood, urine, feces, tissues and GI content	Hyty: 8% of administered radioactivity (after 5 min), 6% associated with plasma and 1.9% with cell fraction	Hyty: 90% of administered radioactivity (after 5 h)	Sulfo-conjugated, HVAlc, HVA, DOPAC, DOPAL	Hyty (% of administered radioactivity): brain 0.89, heart 0.39, kidney 0.8, liver 3.19, lung 0.53, skeletal muscle 61, GI content 9 (after 5 min). Feces: 3.2% (after 5 h). Measurements of detected metabolites in urine and tissues as well	Radioactivity measures, HPLC-UV for metabolite identification	D'Angelo et al., 2001
Olive oil enriched with increasing concentrations of phenols	Oral, 50 ml single dose (mg): total Ol Agl 12.6-39.5; free Hyty 1.9-7.1; total Hyty 7-23.2	Human urine		% of total metabolites: Hyty 16.8-23.7%; HVA 53.9-61.8%; HVAlc 22.0-22.4 %	HVAlc, HVA		GC/LC-MS	Caruso et al., 2001

Administrated Polyphenol	Administration and Dose	Biological Sample	Concentration in Plasma	Excretion in Urine[1]	Metabolites Detected	Other Measurements	Analysis Methods	Ref.
EVOO	Oral, 50 ml single dose: 1650 µg of Ty	Human urine		Ty 17-43%	Ty conjugates		GC-MS	Miró-Casas et al., 2001a
VOO	Ora; 50 ml single dose: 1055 µg of Hyty, and 655 µg of Ty	Human urine		Hyty 32-98.8%; Ty 12.1-52%; total free Hyty and Ty ~15%	Hyty and Ty conjugates		GC-MS	Miró-Casas et al., 2001b
Aqueous and oil solutions of radiolabelled synthetic Hyty	(a) Oral, oil solution (b) IV, aqueous solution (For a detailed description see Tuck et al., 2001)	Rat urine		(a) Free Hyty 4.10 %; Hyty sulfate 48.42%; Hyty gluc 9.53%; HVA 10.26%; other metabolites: 20.27%; (b) Free Hyty 2.35 %; Hyty sulfate 34.24%; Hyty gluc 3.58%; HVA 18.69%; other metabolites 30.87%	Hyty monosulfate, Hyty 3-O-gluc; HVA	Determination of the radical scavenging ability of authentic HVA and HVAlc and of each metabolite using DPPH radical scavenging test.	HPLC-radiometric detection; HPLC-MS/MS, 1H NMR for metabolite identification	Tuck et al., 2002
Supplements containing nonpolar and polar phenols extracted from EVOO, and Ol glycoside (commercially available capsules)	Oral, single doses: 100 mg of phenols (a) Ileostomy subjects. Phenolic content (µmol): nonpolar 371; polar 498; Ol glycoside 190; (b) Subjects with a colon. Phenolic content (µmol): nonpolar 382; polar 526	Human urine and ileostomy effluent		(a) Nonpolar 12%; polar 6%; Ol glycoside 16% (b) Nonpolar 6%; polar 5%	Hyty and Ty in free form or gluc conjugated	Total excretion in ileostomy effluent over 24 h (µmol): nonpolar < 127; polar < 153; Ol glycoside < 51	HPLC-MS/MS, GC-MS, HPLC-DAD	Vissers et al., 2002
VOO from Arbequina cultivar	Oral (a) 50 ml single dose (µg): Ty 1720; Hyty 1370 (b) 25 ml /day sustained doses for 1 week (µg): Ty 860; Hyty 685	Human urine		(a) Ty 16.9%; Hyty 78.5% (b) Ty 19.4%; Hyty 121.5% (at the end of the sustained period)			GC-MS	Miró-Casas et al, 2003a
VOO	Oral, 25 ml single dose (mg/l): free Hyty 6.2; Hyty after acidic treatment 49.3	Human urine and plasma	Hyty conjugate 25.83 µg/l (after 32 min); HVAlc 3.94 µg/l (after 53 min)	Different results according to hydrolytic treatment (µg): acidic conditions: Hyty 714.7, HVAlc 188.0; enzymatic hydrolysis: Hyty 479.6, HVAlc 122.9 (after 12 h)	HVAlc, Hyty gluc		GC-MS	Miró-Casas et al., 2003b
EVOO and synthetic Hyty in ROO and low-fat yogurt	Oral, single dose (a) Rats: 50.3 µg total Hyty 0.5 ml in EVOO (201.2 µg/kg) (b) Humans: 3.2 mg total Hyty in 30 ml EVOO (45.7 µg/kg); 7 mg Hyty in 30 ml ROO; 20 mg synthetic Hyty in 125 g yogurt.	Rat and human urine		% of total Hyty administered: (a) Hyty + HVAlc 7.6 (b) EVOO: Hyty + HVAlc 44.2. ROO: Hyty + HVAlc 23.0; Yogurt: Hyty + HVAlc 6.7	HVAlc		GC/LC-MS	Visioli et al., 2003

Administrated Polyphenol	Administration and Dose	Biological Sample	Concentration in Plasma	Excretion in Urine[a]	Metabolites Detected	Other Measurements	Analysis Methods	Ref.
Ol in soya oil and distilled water	Oral, 350g single dose (Ol 100 mg/kg)	Rat plasma and urine	Ol 200 ng/ml (after 2 h)	Ol gluc 91%; Hyty gluc 97%	Ol and Hyty gluc		LC-MS/MS	Del Boccio et al., 2003
Ol in saline solution	IV, 100 µl single dose: 25 mg/kg of Ol in NaCl (0.9%, w/v)	Rat plasma	Approximate values: 3.5 µg/ml Ol; 20 ng/ml Hyty (after 10 min)				HPLC-fluorescence detection	Tan et al., 2003
EVOO and pure Ol (isolated from olive tree leaves)	Oral, sustained doses for 80 days (g/kg) (a) EVOO 50 (b) Ol 0.15	Rat urine		ng/ml: (a) free Ty 321; free Hyty 253.2; total Ty 1855.6; total Hyty 404.3 (b) free Ty 183.6; free Hyty 154.4; total Ty 814.5; total Hyty 1036.7	Gluc conjugates		GC-MS/MS (for urine samples) and HPLC-DAD (for EVOO extracts)	Bazoti et al.,2005
EVOO with increasing concentrations of polyphenols	Oral, sustained doses for 4 days. 25 ml EVOO: phenol content (mg/kg) (a) high 486 (b) moderate 133 (c) low 10 (%.: Hyty 6.3, Ty 5.3, Ol Agl 40.0; Lig Agl 26.2; luteolin 11.7; apigenin 2.6)	Human urine		Approximate values (mmol): (a) Ty 1.2; Hyty 0.5; HVAlc 0.2 (b) Ty 2.8; Hyty 2.2; HVAlc 0.6; (c) Ty 3.5; Hyty 4.8; HVAlc 1.0			HPLC	Weinbrenner et al., 2004
VOO, COO or ROO from Picual cultivar	Oral, 25 ml sustained doses for 3 weeks. Total phenols in olive oil (mg/kg): VOO 366; COO 164; ROO 2.7	Human plasma and urine				Detection of etheno-DNA adducts in plasma. Measurement in 24 h urine of Ty and Hyty as biomarkers of the type of olive oil ingested	LC-MS/MS	Hillestrom et al., 2006
VOO from Picual cultivar	Oral, 50 ml VOO single dose	Human blood			Hyty mono-gluc, Hyty monosulfate, Ty gluc, Ty and HVA sulfate	Metabolites in LDL (ng/mg apo-B): Hyty mono-gluc 2.11; Hyty monosulfate 24.27; Ty gluc 1.16; Ty sulfate 14.87; HVA sulfate 27.16 (after 1 h)	HPLC-DAD-MS/MS	de la Tore-Carbot et al., 2006
EVOO from Picual cultivar	Oral, 50 ml single dose. Phenolic content (µg/ml): total 648; Hyty 70.6; Ty 27.01	Human blood			Hyty mono-gluc isomers, Hyty monosulfate, Ty and HVA sulfate, Ty gluc	Metabolites in LDL (ng/mg apo-B): total phenolics 105.43; Hyty mono-gluc I 2.45; Hyty mono-gluc II 2.55; Hyty monosulfate 34.22; Ty gluc 0.96; Ty monosulfate: 17.23; HVA sulfate 48.02 (after 1 h)	HPLC-DAD-MS/MS	de la Tore-Carbot et al., 2007
VOO	Oral, 30 ml single dose: total phenols 400 mg/kg	Human plasma	11 free phenolicss and 9 metabolites (after 1 and 2 h)		Hyty and apigenin gluc; Hyty, Ty, HVA, vanillin, vanillic acid, dihydroferulic acid and coumaric acid sulfates		UPLC-ESI-MS/MS	Suárez et al., 2009

Administrated Polyphenol	Administration and Dose	Biological Sample	Concentration in Plasma	Excretion in Urine[a]	Metabolites Detected	Other Measurements	Analysis Methods	Ref.
EVOO enriched with OI and pure OI (isolated from olive tree leaves)	Oral, sustained doses for 80 days (a) EVOO: 1.1 g/kg (b) OI supplement: 0.33 mg/kg	Rat plasma	After enzymatic treatment (ng/ml): (a) HVAlc 61.9 (b) HVAlc 53.5; (Hyty not detected)		HVAlc		LC-MS/MS	Bazoti et al., 2010
Spanish VOO (Picual cultivar) and ROO	Oral, 25 ml/day sustained doses for 3 weeks (a) VOO, phenol concentration (mg/l): total 629; Hyty 63.5; Ty 24.4; OI derivatives 327.2; Lig derivatives 208; other 6 phenolics quantified (b) ROO	Human blood			Hyty, Ty and HVA sulfates	LDL composition before and after consumption of ROO and VOO	HPLC-DAD-MS/MS	de la Torre-Carbot et al., 2010
EVOO (50% v/v Arbequina and Picual cultivars)	Oral, 50 ml single dose. Phenolic content (mg/kg): Hyty 8.31; Ty 5.33; pinoresinol 3.25; luteolin 2.65; apigenin 0.64; EA 34.91; Lig Agl 40.58	Human urine		28 phase I and 32 and II metabolites	Phase I reactions: hydrogenation, hydroxylation, hydration, methylation Phase II reactions: glucuronidation, sulfoconjugation	Excretion kinetics for 6 h of main metabolites identified as biomarkers in human urine after EVOO intake	RRLC-ESI-TOF-MS	García-Villalba et al., 2010
Spanish VOO	Oral, 50 ml single dose. Phenolic content in μmol: Hyty 21.96; Ty 15.20; HVAlc 0.27	Human plasma and urine		% of ingested Hyty: Total Hyty 9.2; free Hyty 1.8; Hyty-4-O-gluc 3.1; Hyty-3-O-gluc 4.3; total Ty 12.9; free Ty 3.3; Ty-4-O-gluc 9.6; Total HVAlc 4.5; free HVAlc 1.4; HVAlc-4-O-gluc: 3.1	HVAlc; Hyty, Ty and HVAlc gluc	Control of compliance with the dietary recommendations by analyzing the plasma and urine concentrations of Hyty, Ty and HVAlc	UPLC-ESI-MS and GC-MS	Khymenets et al., 2011

[a] Percentage of administered amount after 24 h, unless otherwise indicated

Abbreviations: apo-B: apolipoprotein-B; COO: common olive oil; DAD: diode array detector; DOPAC: 3,4-dihydroxyphenylacetic acid; DOPAL: 3,4-dihydroxyphenylacetaldehyde; DPPH: 2,2-diphenyl-1-picrylhydrazyl; EA: elenolic acid; EVOO: extra virgin olive oil; gluc: glucuronide; GC: gas chromatography; GI: gastrointestinal; HPLC: high performance liquid chromatography; Hyty: hydroxytyrosol; HVA: homovanillic acid; HVAlc: homovanillic alcohol; IV: intravenous; LC: liquid chromatography; LDL: low-density lipoproteins; Lig Agl: ligstroside aglycone; MS: mass spectrometry; OI: oleuropein; Papp: apparent permeability coefficient; ROO: refined olive oil; Ty: tyrosol; UV: ultraviolet; VOO: virgin olive oil

Table 2. Bioavailability of olive oil polyphenols in animals and humans.

A major limitation of the commented human studies is that they used phenolics extracts or olive oil samples artificially enriched with phenolics extracts, and therefore extrapolation of these results to typical olive oil consumption may not be realistic. Further studies have been performed administering VOO at doses close to that used in the Mediterranean countries (30-50 g/day) (Bonanome et al., 2000; de la Torre-Carbot et al., 2006, 2007; García-Villalba et al., 2010; Khymenets et al., 2011; Miró-Casas et al., 2001a, 2001b, 2003a, 2003b; Suárez et al, 2009). Results confirmed that Hyty and Ty are mainly excreted in their glucurono-conjugated form; in fact, the role of glucuronidation in metabolism of main olive oil phenols can be evaluated in about 65-75% of totally recovered in urine after dietary VOO consumption (Khymenets et al., 2011; Miró-Casas et al., 2003b), which suggests an extensive first-pass intestinal/hepatic metabolism of the compounds ingested. Suárez et al. (2009) considered for the first time the absorption and disposition of flavonoids and lignans after the ingestion of VOO. Besides the presence of those VOO polyphenols in their conjugated forms, an important variability in the concentrations was observed between the plasma samples obtained from different volunteers. This variability may be attributed to differences in the expression of metabolizing enzymes due to genetic variability within the population. The most comprehensive study regarding the identification of metabolites in human urine of practically all the olive oil polyphenols described was reported by García-Villalba et al. (2010). These authors were able to achieve the tentative identification of 60 metabolites; the most abundant were those containing a catechol group, such as Hyty and the secoiridoids Ol Agl and deacetoxy-Ol Agl. Phenolic compounds were subjected to various phase I and phase II reactions, mainly methylation and glucuronidation. The report suggests that most of the olive oil polyphenols are absorbed to a greater or lesser extent, although absorption and metabolism seems to differ greatly among the different compounds.

2.2 Conjugation and nature of metabolites

Low doses of polyphenols are delivered through human diet and, generally, do not escape first-pass metabolism. As a result, most olive oil polyphenols undergo structural modifications, i.e. conjugation process; in fact, conjugates are the predominant forms in plasma. Once absorbed, olive oil polyphenols are subjected to three main types of conjugation: methylation, glucuronidation and, to a lesser extent, sulfation, through the respective action of catechol-O-methyl transferases (COMT), uridine-5'-diphosphate glucuronosyltransferases (UDPGT) and sulfotransferases (SULT) (Manach et al., 2004).

Recently, García-Villalba et al. (2010) carried out a broad study of the metabolites of most olive oil phenolic compounds excreted in human urine, showing that most polyphenols were absorbed, metabolized and excreted to a lesser or greater extent. It was initially suggested in literature that Ol Agl and Lig Agl were hydrolyzed in the gastrointestinal tract (GI) tract and then, the resulting polar phenols, Hyty and Ty, were absorbed and metabolized (Vissers et al., 2002). Nevertheless, the results obtained in later experiments with Caco-2 cells (Pinto et al., 2011) and humans (García-Villalba et al., 2010), showed that, at least, part of the secoiridoids can be absorbed and metabolized; reduction (hydrogenation) is the most probable metabolic pathway of these compounds. Hydroxylation and hydration are also possible pathways for the secoiridoids. In fact, they can precede or follow the action of COMT on compounds such as Hyty, deacetoxy-Ol Agl, and Ol Agl. Some compounds can even suffer a double hydroxylation before or after the methylation (García-Villalba et al., 2010).

A notable metabolic pathway for Hyty is the methylation, giving rise to the formation of HVAlc (Caruso et al., 2001; Bazoti et al., 2010; Manna et al., 2000; Visioli et al, 2003). Oxidation and methylation-oxidation, rendering 3,4-dihydroxyphenilacetic acid (DOPAC) and HVA, respectively, have been also proposed (D'Angelo et al., 2001). It is noteworthy that many of the reported metabolites of Hyty are also the major molecular species deriving from dopamine metabolism (HVA, DOPAC, 3,4-dihydroxyphenyl acetaldehyde - DOPAL); in fact, Hyty can be also called DOPET, a well-known dopamine metabolite.

Besides, olive oil phenolic compounds and most of their corresponding phase I metabolites can be subsequently subjected to phase II reactions, preferentially glucuronoconjugation (García-Villalba et al., 2010). The presence of glucuronoconjugates of phenolic compounds belonging to most of chemical classes families described in olive oil has been widely detected in both urine and plasma, whereas the presence of sulfated metabolites has scarcely been reported in literature.

The metabolism of olive oil lignans has not been reported in detail so far and one of the few references appeared only recently (Soler et al., 2010). In this study, pinoresinol glucuronide and sulfate conjugates were identified after incubation of free pinoresinol using differentiated Caco-2/TC7 cell monolayers.

As far as flavonoids are concerned, products of methylation and glucuronidation have been observed (Soler et al., 2010; Suárez et al., 2009). Methyl-monoglucuronides of apigenin and luteolin have been identified as well (García-Villalba et al., 2010).

2.3 Binding of olive oil polyphenols to lipoproteins

Several reports converge on the *in vitro* ability of olive oil phenolic compounds to bind low density lipoproteins (LDL) and to protect them against oxidation (Covas et al., 2000; Visioli et al., 1995). Moreover, both animal and human *in vivo* studies (Coni et al., 2000; Marrugat et al., 2004) have provided evidence on the effects of olive oil ingestion on LDL composition and the incorporation of olive oil phenolics and their metabolites in LDL. In one of the first studies, Bonanome et al. (2000) determined the presence of Hyty and Ty in human lipoprotein fractions after olive oil ingestion. Both compounds were recovered in all of the fractions, except in the very low density lipoproteins one; concentrations peaked between 1 and 2 h. Covas et al. (2006) demonstrated that the postprandial oxidative stress can be modulated by the olive oil phenolic content and that the degree of LDL oxidation decreases in a dose-dependent manner with the phenol concentration of the olive oil ingested. They arrived to these conclusions administering a single dose of olive oil, but similar results were obtained in studies using sustained doses; olive oil consumption for 1 week led to an increase in the total phenolic content of LDL (Gimeno et al., 2002). In a later study, volunteers were requested to ingest virgin, common or refined olive oil daily for 3 weeks (Gimeno et al., 2007). The concentration of total phenolic compounds in LDL was directly correlated with the phenolic concentration of the oils and with the resistance of LDL to their *in vitro* oxidation.

De la Torre-Carbot et al. (2006, 2007) developed a rapid method for the determination in LDL of Ty, Hyty and several of their metabolites. The presence of these compounds in LDL strengthens claims that these compounds can act as *in vivo* antioxidants. The effect of the intake of virgin and refined olive oils after long-term ingestion of real-life doses on the content of the metabolites in LDL was examined as well (de la Torre-Carbot et al., 2010). The phenols in VOO modulated the LDL content of 3 phenolic metabolites, Hyty, Ty, and HVA sulfates; the concentration of these compounds increased significantly after the ingestion of VOO, in contrast to the refined one. In parallel, the ingestion of VOO significantly reduced LDL and plasma oxidative markers, which suggests that the metabolic activities of phenols can be related to the capacity of these compounds to remain bound to LDL.

2.4 Plasma concentration and tissue uptake

In 1998, Bai et al. reported for the first time the absorption of Hyty into the bloodstream. Hyty was administered orally to rats and appeared in plasma 2 min after, reaching the highest level at 5-10 min. Its concentration was low compared to the administered amount. The experiment, however, did not take into account the presence of metabolites.

After this first approach, different methods for the simultaneous detection of Ol and Hyty in rat plasma have been optimized. Ruiz-Gutiérrez et al. (2000) determined Hyty in overnight-fasted rat plasma after its oral administration. Ol and Hyty plasma concentrations were measured after oral administration of a single dose of Ol to rats using soya oil and distilled water as administration vehicle (Del Boccio et al., 2003). Analysis of plasma showed the presence of unmodified Ol, reaching a peak value within 2 h, with a small amount of Hyty. In another study, Ol and Hyty plasma levels were monitored in rats after intravenous dosing of Ol (Tan et al., 2003). The dosing profile showed that at 10 min both Ol and Hyty were rapidly distributed.

Studies in which phenolic ingestion is closer to typical dietary patterns may be more appropriate for estimating bioavailability than the administration of pure compounds. Recently (Bazoti et al., 2010), the simultaneous quantification of Ol and its major metabolites in rat plasma was carried out after a control diet of 80 days supplemented with Ol or with EVOO. Basal levels of HVAlc were detected in the blood stream after the enzymatic treatment of the samples with β-glucuronidase. Before the enzymatic treatment, HVAlc was detected below the limits of quantification in plasma samples of rats supplement with Ol. Hyty was not detected, which indicates that it was metabolized to HVAlc. Ol was detected below the LOQ before and after the enzymatic treatment. These results are in accordance with the study made by Del Boccio et al. (2003), who demonstrated that Ol was rapidly metabolized and eliminated.

Miró-Casas et al. (2003b) quantified Hyty and HVAlc in human plasma and urine after real-life doses of VOO. Both compounds appeared rapidly in plasma mainly as glucuronides, with peak concentrations at 30 min for Hyty and 50 min. for HVAlc after the ingestion, supporting the premise that the small intestine is the major site of absorption for these compounds (Vissers et al., 2002).

Most of the studies described so far have centered their attention on Hyty, Ty and Ol derivatives. In a recent work, the absorption and disposition of other olive oil polyphenols (flavonoids and lignans) have been considered (Suárez et al., 2009). Samples were obtained from healthy humans 1 and 2 h after the ingestion of VOO. The major compounds identified and quantified in plasma corresponded to metabolites of Ty, although Ty sulfate was only detected in one subject, and especially Hyty, as glucuronide and sulfate conjugates. HVA sulfate could be the direct product of the Hyty methylation, and vanillin sulfate and vanillic acid sulfate could be formed as products of alcohol dehydrogenase and aldehyde dehydrogenase activities. Suárez & co-workers also found hydroxybenzoic acid in all the plasma samples. The glucuronide metabolite of apigenin was tentatively quantified in all the samples analyzed, but showing a considerable inter-individual variation. Lignans (pinoresinol and acetoxypinoresinol) could not be detected in the plasma samples even in glucuronide or sulfate conjugated forms.

Once the polyphenols reach the bloodstream, they are able to penetrate tissues, particularly those in which they are metabolized. The nature of the tissular metabolites may be different from that of blood metabolites; data are still very scarce, even in animals, and their ability to accumulate within specific target tissues needs to be further investigated. An article written by D'Angelo et al. (2001) studied the fate of radiolabelled [14]C Hyty intravenously injected in rats in different biological fluids (plasma, urine and feces) and tissues (brain, heart, kidney, liver, lung, skeletal muscle and GI content). The pharmacokinetic analysis indicated a fast and extensive uptake of the molecule by the organs and tissues investigated, with a preferential renal uptake. Over 90% of the administered radioactivity was excreted in urine after 5 h and about 5% was detectable in feces and GI content. Less than 8% of the administered radioactivity was still present in the blood stream 5 min after injection. Regarding tissues, the time course analysis indicated that the highest level of radioactivity was detected 5 min after injection, followed by a rapid decrease. It is worth noting that Hyty is able to cross the blood-brain barrier, even though its brain uptake is lower compared with other organs. In all the investigated tissues, Hyty was enzimatically converted in four oxidized and/or methylated derivatives (HVAlc, HVA, DOPAC, DOPAL) and sulfoconjugated derivatives. Enzymatic methylation is presumably operative in the brain, HVAlc representing 41.9% of the detected, labeled species. This reflects the key role of COMT in the central nervous system. The occurrence in the analyzed organs of both labeled DOPAL and DOPAC implies a sequential oxidation of Hyty ethanol side chain catalyzed by alcohol, and aldehyde dehydrogenase, respectively. Labeled HVA, the product of both methylation and oxidation, was also identified. Sulfoconjugated metabolites were mainly found in plasma (43.3%) and urine (44.1%).

As data on plasma concentration of olive oil phenols are still scarce, an alternative is to look at olive oil phenols excreted in urine; these may provide information on the form in which phenols are present in plasma.

2.5 Elimination

The amount and form in which the olive oil phenols are excreted in urine may give an insight into their metabolism in the human body. The first experimental evidence of the absorption of Ty and Hyty from olive oil in humans was obtained by Visioli et al. (2000) from a single oral dose of 50 ml of phenolic-enriched olive oil. The proportions of Hyty and Ty recovered in glucuronidase-hydrolyzed urine, with respect to ingested dose, were in the ranges of 30–60% and 20–22%, respectively. This paper postulated that Hyty and Ty were dose-dependently absorbed in humans and excreted in urine as glucuronide conjugates.

Miró-Casas et al. (2001a) measured the urinary recovery of administered Ty during the 24 h after EVOO ingestion. Maximal Ty values were obtained in the 0-4 h urine samples and decrease to reach basal values after 8-12 h. Ty was excreted in urine mainly in its glucuro-conjugated form, with only 6-11% excreted in the free form. In a later study (Miró-Casas et al., 2001b), the simultaneous determination of Hyty and Ty in human urine after the intake of VOO was reported. Like the previous study, Hyty and Ty levels in urine rose after VOO consumption, reaching a peak at 0-4 h and returning to basal values at 12-24 h. After hydrolytic treatment, the amount of total compounds recovered in 24 h urine was also determined for Hyty and Ty. Recoveries ranged between 32-98% for Hyty and 12.1-52% for Ty. Both compounds were mainly excreted in conjugated form since only 5.9% Hyty and 13.8% Ty of the total amount excreted were in free form. The hydrolysis procedure applied in this study was limited because it did not provide specific information about the type of conjugation involved. This paper also postulated that Ol is not the main source of Hyty after ingestion of olive oil. The absorption of Hyty and Ty was later confirmed in an experiment using single and sustained doses of VOO (Miró-Casas et al., 2003a). Urinary recoveries of Ty were similar for both cases; however, mean recovery values for Hyty after ingestion of 25 ml/day VOO for one week, were 1.5-fold of those obtained after a 50 ml single dose.

Vissers et al. (2002) studied the absorption of Hyty, Ty and, for the first time, Ol and Lig Agl, in ileostomy subjects and in volunteers with a colon. The results showed that 55-66% of the ingested olive oil phenols were absorbed in ileostomy subjects, which implies that most phenols are absorbed in the small intestine. Excreted phenolics, mainly in the form of Hyty and Ty, were determined to be 5-16% of the total ingested. Similar levels of Hyty and Ty were found in the urine of subjects with and without a colon, confirming that olive oil phenols are absorbed mainly in the small intestine. The obtained values, lower than those reported by others, could be underestimations because metabolites of olive oil phenols were not considered. In this work it is also suggested that an important step in the metabolism of the Ol glycoside and Ol and Lig-aglycones is their transformation into Hyty or Ty. This was supported by finding that 15% of an Ol glycoside supplement administered to healthy human subjects was excreted in urine as Hyty and Ty.

Tuck et al. (2001) investigated the *in vivo* fate of tritium labeled Hyty and Ty after intravenous (in saline, tail vein) and oral dosing (in oil- and water-based solutions) to rats. For both Hyty and Ty, the elimination of radioactivity in urine within 24 h for the intravenously and orally administered oil-based dosing was significantly greater (95 and 75%, respectively) than the oral aqueous dosing method (74 and 53%, respectively). The majority of the excreted dose was eliminated from the body within 2 h, when intravenously dosed, and within 4 h for both oral dosings. Later, urine samples collected after 24 h were re-examined (Tuck et al., 2002). After oral oil dosing Hyty represented 4.10% of compound eliminated, monosulfate conjugate 48.42%, glucuronide conjugate 9.53%, HVA 10.26% and other possible metabolites 20.27%. Other study with rats supplemented with Ol (Ol rats) or with EVOO (EVOO rats) was developed for the simultaneous determination of Hyty, Ty and EA in rat urine (Bazoti et al., 2005). The urinary levels of free Ty and Hyty were higher in EVOO rats than in Ol rats. When the urine sample were treated with β-glucuronidase, the total amount of metabolites measured for the EVOO rat was higher for Ty but lower for Hyty than in Ol rats. EA was not detected, probably because of its further metabolism to simpler molecules. Nevertheless, as already mentioned, caution should be taken interpreting the results achieved from rats (Visioli et al., 2003).

The urinary excretion of HVAlc and HVA in humans was reported for the first time by Caruso et al. (2001) after the intake of different VOOs. HVAlc contributes to 22% of the total excretion of Hyty and its metabolites, and HVA 56%. The excretion of both metabolites correlated with the administered dose of Hyty. Even at low doses, HVAlc and HVA were excreted. In a later study, Miró-Casas et al. (2003b) observed how urinary amounts of Hyty and HVAlc increased in response to VOO ingestion, reaching the maximum peak at 0-2 h. Urinary recovery 12 h after olive oil ingestion was rather different depending on the hydrolytic treatment applied. Under acidic conditions recoveries were higher for both Hyty and HVAlc than with enzymatic hydrolysis. It was apparent that 65% of Hyty was in its glucuronoconjugated form and 35% in other conjugated forms.

To understand the impact of glucuronidation on the metabolic pathway of olive oil phenols, the simultaneous determination of Ty, Hyty, HVAlc and their corresponding O-glucuronides in human urine was carried out by Khymenets et al. (2011). It was the first time that the glucuronides of these compounds have been directly identified and quantified in urine samples of volunteers supplemented with EVOO, because previous methods measured either free or total phenols after hydrolysis. The maximum excretion of Hyty and Ty occurred during the first 6 h after administration, which is in agreement with earlier reported data. The free Ty and Hyty, as well as HVAlc, were detected at significant concentrations in all urine samples collected 6 h after EVOO acute intake. Concentrations of O-glucuronide metabolites (4-O-gluc-Ty, 4-O-gluc-Hyty, 3-O-gluc-Hyty, and 4-O-gluc-HVAlc) were substantially higher in 6 h postprandial samples when compared to their parent compounds. About 13% of the consumed olive oil polyphenols were recovered in 24 h urine, 75% of which were in the form of glucuronides and 25% as free compounds.

In another recent work, specific information about the type of conjugates in human urine was provided (García-Villalba et al., 2010). The authors were able to indentify more than 60 metabolites. This was the first report in which metabolites of simple phenols, flavonoids, lignans and secoiridoids have been found in human urine samples. Phenolic compounds were subjected to different phase I (hydrogenation, hydroxylation, methylation) and phase II (mainly glucuronidation) reactions. Ten metabolites were identified as possible biomarkers of olive oil intake and their levels in urine after the olive oil ingestion were monitored, finding the highest level of most of them 2 h after administration.

In summary, data on urinary excretion indicate that at least 5% of ingested olive oil phenols is recovered in urine mainly as glucuronidated Ty and Hyty. The remaining phenols are metabolized into other compounds, such as O-methylated Hyty. Monosulfate conjugates might be other metabolites, as shown in two rat studies; however, if olive oil phenols are also metabolized into these conjugates in humans remains to be elucidated.

3. Need and difficulties of carrying out bioavailability studies of polyphenols

In this section, we will discuss some common mistakes that can be made when bioavailability studies are carried out, the difficulties that the analyst can find and the limitations of some of the studies made so far. **Figure 1** gives a general idea of the most important topics commented in the section.

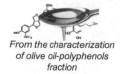 *Rigorous determination of bioavailability of polyphenols?*

From the characterization of olive oil-polyphenols fraction

To achieve evidence about health benefits and to understand their metabolism

Bioavailability ➠ Limits/difficulties of some of the studies carried out so far

Fraction of an ingested nutrient or compound that reaches the systemic circulation and the specific sites where it can exert its biological action

Three main steps are implicit

-Release from the carrier matrix
-Intestinal absorption
-Tissue uptake

In-vitro
- Great care is required when interpreting
- Are doses applied realistic?
- Compounds tested are available (or isolated) standards – i.e. polyphenol aglycones or their sugar conjugates - rather than active metabolites
Animal models
- Differences between human and animal genomes
In-vivo
- Intervention studies are difficult
- "Single-dose" design
- Bioavailability in target tissues?
Food matrix
- Polyphenols in food are more complex than often thought
Metabolomic studies
- Complex analytical tools
- Lack of standards (identification / quantization)

Fig. 1. Definition of bioavailability and the limits affecting bioavailability studies of polyphenols

Since intervention studies are very difficult to carry out, in many cases the researchers have to turn to *in vitro* or animal studies. *In vitro* studies are a pillar of pharmacological research and build the bases for future *in vivo* assays; however, the interpretation and extrapolation of the achieved data have to be made very carefully (Kroon et al., 2004). When the biological activity of polyphenols is assessed by using culture cells as tissue models, in most of the cases, cells are treated with aglycones or polyphenols-rich extract derived from plants or, in this case, from olive oil, and data are reported at concentrations which elicited a response. It is absolutely necessary to bear in mind that plasma and tissues are not exposed *in vivo* to polyphenols in these forms: the molecular forms reaching the peripheral circulation and tissues are different from those present in the olive oil (Day et al., 2001). Moreover, the polyphenols concentration tested should be of the same order as the maximum plasma concentration attained after a polyphenol-rich meal (0.1-10 µmol/l).

Matters of practicality determine the use of rats rather than humans as the model of choice for *in vivo* studies, although interspecies variability renders comparisons between the model species (animals, humans) complex and sometimes questionable (Visioli et al., 2003), since the rats and rodents in general are not the best model for the study of dietary problem of human metabolism.

When *in vivo* studies are carried out, we can say that most of researches have investigated the kinetics and extent of polyphenol absorption by measuring plasma concentrations and/or urinary excretion among adults after the ingestion of a single dose of polyphenol, provided as pure compound, plant extract, or whole food/beverage. Using this "single-dose" design, the increase in the blood concentration is transitional and reflects mainly the ability of the organism to take up the polyphenol from the food matrix. Consequently, most of the data from humans presented in the literature on the bioavailability refer only to the release of the polyphenols from the food matrix and their consequent absorption (D'Archivio et al., 2010; Vissers et al., 2004).

To address the bioavailability of olive oil phenols, we should exclude studies without a control diet and studies in which the amount of ingested phenols is not reported or could not be estimated (Miró-Casas et al., 2001a; Visioli et al., 2000a; Vissers et al., 2004). In other words, it is essential to characterize in depth the polyphenolic extract of the olive oil before starting bioavailability studies to assure their usefulness; since this fraction is quite complex and heterogeneous, it represents another requirement which difficults the whole process.

Advances in the understanding of olive oil polyphenols metabolism have been made possible by improvements in the analytical methodologies used, particularly high-resolution chromatographic systems with mass spectrometry as detector (Bai et al., 1998; Del Boccio et al., 2003; García-Villalba et al., 2010; Khymenets et al., 2011; Miró-Casas et al., 2003b). Performing metabolomic studies is challenging and requires measurements of a very high quality using powerful platforms. Even if the analyst uses proper tools, the fully structural assignment of the metabolites under study is sometimes very difficult due to the lack of the metabolite standards; fact which makes difficult the correct quantification too (D'Archivio et al., 2010; García-Villalba et al., 2010). The amount of information about the sample under study achieved in metabolomic studies is considerable, that is why for meaningful interpretation the appropriate statistical tools must be employed to manipulate the large raw data sets in order to provide understandable and workable information (Manach et al., 2009).

A very interesting review written by D'Archivio et al. (2010) gives a critical overview about the difficulties and the controversies surrounding the studies aimed at determining the bioavailability of polyphenols. Summarizing, there are some essential steps to be followed to establish conclusive evidences for the effectiveness of polyphenols in disease prevention and in human health improvement: 1) determination of the distribution of these compounds in our diet, estimating their content in each food; 2) identification of which of the existing polyphenols are likely to provide the greatest effects in the context of preventive nutrition, and 3) assesment of the bioavailability of polyphenols and their metabolites, to evaluate their biological activity in target issues.

Even though the bioavailability studies are properly designed, we have to be aware of how many different endogenous and exogenous variables are involved and the difficulties that have to be faced. The main factors recognized as affecting olive oil polyphenols bioavailability can be grouped in the following categories: factors related to the polyphenol characteristics, food/food processing related factors, external factors and factors related to the host, as it can be observed in **Figure 1.** An in-depth discussion of every factor influencing the bioavailability of olive oil polyphenols has been made by Manach et al. (2004) and Cicerale et al. (2009).

4. Conclusions

To explore and understand the mechanism of action of olive oil polyphenols and their role in disease prevention and human health improvement, extensive studies of absorption, metabolism, excretion, toxicity, and efficacy are needed. Although *in vitro* studies can be very useful and provide valuable information, they have to be completed with extensive *in vivo* research. The first requirement for a beneficiary dietary compound is that it enters into the blood circulation; therefore to demonstrate *in vivo* effects of olive oil phenolics it is necessary to assess first their *bioavailability*.

Analysis of plasma and urine provide valuable information on the identity and pharmacokinetics of circulating metabolites after ingestion. Since the metabolites sequestered in body tissues are not usually taken into account, results from urine samples could be an underestimation. There have been several studies which have determined the metabolites of the various olive oil polyphenols (mainly Hyty, Ty, and Ol) in human plasma and urine after oral intake, although the information is still scarce. The conjugation mechanisms that occur in the small intestine and later in the liver are highly efficient. The resulting metabolites are mainly glucuronate and sulfate conjugates with or without methylation across the catechol group (many are multiply conjugated).

Bioavailability studies are gaining increasing interest as food industries are continually involved in developing new products, defined as "functional" by virtue of the presence of specific antioxidants or phytochemicals. The difference between functional foods and medicines calls for moderation when the "medicinal" properties of individual food items, be it olive oil, are indicated. The correct message should be to select foods whose components have proven, albeit limited in magnitude, biological activities and build a balanced diet round them, to reduce several chronic diseases.

5. Acknowledgements

The authors are very grateful to Junta de Andalucía (Project P09-FQM-5469), to the International Campus of Excellence (CEI Granada 2009), to the Ministry of Education of Spain and to the Regional Government of Economy, Innovation and Science of Andalusia.

6. References

Bai, C.; Yan, X.; Takenaka, M.; Sekiya, K. & Nagata, T. (1998). Determination of Synthetic Hydroxytyrosol in Rat Plasma by GC-MS. *Journal of Agricultural and Food Chemistry*, Vol. 46, No. 10, (September 1998), pp. 3998-4001, ISSN 0021-8561

Bazoti, F.N.; Gikas, E.; Puel, C.; Coxam, V. & Tsarbopoulos, A. (2005). Development of a Sensitive and Specific Solid Phase Extraction-Gas Chromatography–Tandem Mass Spectrometry Method for the Determination of Elenolic Acid, Hydroxytyrosol, and Tyrosol in Rat Urine. *Journal of Agriculture and Food Chemistry*, Vol. 53, No. 16, (August 2005), pp. 6213–6221, ISSN 0021-8561

Bazoti, F.N.; Gicas, E. & Tsarbopoulos, A. (2010). Simultaneous Quantification of Oleuropein and its Metabolites in Rat Plasma by Liquid Chromatography Electrospray Ionization Tandem Mass Spectrometry. *Biomedical Chromatography*, Vol. 24, (May 2010), pp. 506-515, ISSN 0269-3879

Bendini, A.; Cerretani, L.; Carrasco-Pancorbo, A.; Gómez-Caravaca, A.M.; Segura-Carretero, A.; Fernández-Gutiérrez A. et al. (2007). Phenolic Molecules in Virgin Olive Oils: A Survey of Their Sensory Properties, Health Effects, Antioxidant Activity and Analytical Methods. An Overview of the Last Decade. *Molecules*, Vol. 12, (August 2007), pp. 1679-1719, ISSN 1420-3049

Bonanome, A.; Pagnan, A.; Caruso, D; Toia, A.; Xamin, A.; Fedeli, E. et al. (2000). Evidence of Postprandial Absorption of Olive Oil Phenols in Humans. *Nutrition, Metabolism & Cardiovascular Diseases*, Vol. 10, No. 3, (June 2000), pp. 111-120, ISSN 0939-4753

Boskou, D. (2000) Olive oil, In: *Mediterranean Diets (World Review of Nutrition and Dietetics, Vol. 87)*, A.P. Simopoulos, F. Visioli, (Eds.), pp. 56-77, Karger Press, ISBN 978-3-8055-7066-4, Basel, Switzerland

Boskou, D.; Tsimidou, M. & Blekas, D. (2006). Polar phenolic compounds, In: *Olive Oil, Chemistry and Technology*, D. Boskou, (Ed.), pp. 73-92, AOCS Press, ISBN 978-1-893997-88-2, Champaign, IL, USA

Boskou, D. (2009). Phenolic compounds in olives and olive oil, In: *Olive Oil. Minor Constituents and Health*, D. Boskou, (Ed.), pp. 11-44, CRC press, ISBN 978-1-4200-5993-9, NY, USA

Brenes, M.; García, A.; García, P.; Rios, J.J. & Garrido, A. (1999). Phenolic Compounds in Spanish Olive Oils. *Journal of Agricultural and Food Chemistry*, Vol. 47, No. 9, (September 1999), pp. 3535-3540, ISSN 0021-8561

Brenes, M.; García, A.; Rios, J.J.; García, P. & Garrido, A. (2002). Use of 1-acetoxypinoresinol to authenticate Picual olive oils. *International Journal of Food Science & Technology*, Vol. 37, (November 2002), pp. 615-625, ISSN 0950-5423

Carrasco-Pancorbo, A.; Gómez-Caravaca, A.M.; Cerretani, L.; Bendini, A.; Segura-Carretero, A. & Fernández-Gutiérrez, A. (2006). Rapid Quantification of the Phenolic Fraction of Spanish Virgin Olive Oils by Capillary Electrophoresis with UV Detection. *Journal of Agricultural and Food Chemistry*, Vol. 54, No. 21, (October 2006), pp. 7984-7991, ISSN 0021-8561

Caruso, D.; Visioli, F.; Patelli, R.; Galli, C. & Galli, G. (2001). Urinary Excretion of Olive Oil Phenols and their Metabolites in Humans, *Metabolism: Clinical and Experimental*, Vol. 50, No. 12, (December 2001), pp. 1426–1428, ISSN 0026-0495

Cicerale, S.; Conlan, X.A.; Sinclair, A.J. & Keast, R.S.J. (2009). Chemistry and Health of Olive Oil Phenolics. *Critical Reviews in Food Science and Nutrition*, Vol. 49, No. 3, (March 2009), pp. 218-236, ISSN 1040-8398

Coni, E.; Di Benedetto, R.; Di Pasquale, M.; Masella, R.; Modesti, D.; Mattei, R. et al. (2000). Protective Effect of Oleuropein, an Olive Oil Biophenol, on Low Density Lipoprotein Oxidizability in Rabbits. *Lipids*, Vol. 35, No. 1, (January 2000), pp. 45-54, ISSN 0024-4201

Corona, G.; Tzounis, X.; Dessì, M.A.; Deiana, M.; Debnam, E.S.; Visioli, F. et al. (2006). The Fate of Olive Oil Polyphenols in the Gastrointestinal Tract: Implications of Gastric and Colonic Microflora-Dependent Biotransformation. *Free Radical Research*, Vol. 40, No. 6, (June 2006), pp. 647-658, ISSN 1071-5762

Corona, G; Spencer, J.P.E. & Dessì, M.A. (2009). Extra Virgin Olive Oil Phenolics: Absorption, Metabolism, and Biological Activities in the GI Tract. *Toxicology and Industrial Health*, Vol. 25, (May-June 2009), pp. 285-293, ISSN 0748-2337

Covas, M.I.; Fitó, M.; Lamuela-Raventós, R.M.; Sebastiá, N.; de la Torre-Boronat, C. & Marrugat, J. (2000). Virgin Olive Oil Phenolic Compounds: Binding to Human Low Density Lipoprotein (LDL) and Effect on LDL Oxidation. *International Journal of Clinical Pharmacology Research*. Vol. 20, No. 3-4, (n.d.), pp. 49-54, ISSN 0251-1649

Covas, M.I.; de la Torre, K.; Farre-Albaladejo, M.; Kaikkonen, J.; Fito, M.; Lopez-Sabater, C. et al. (2006). Postprandial LDL Phenolic Content and LDL Oxidation are modulated by Olive Oil Phenolic Compounds in Humans. *Free Radical Biology & Medicine*, Vol. 40, No. 4 (February 2006), pp. 608-616, ISSN 0891-5849.

Covas, M.I.; Khymenets, O.; Fitó, M. & de la Torre, R. (2009). Bioavailability and Antioxidant Effect of Olive Oil Phenolic Compounds in Humans, In: *Olive Oil. Minor Constituents and Health*, D. Boskou, (Ed.), pp. 109-128, CRC press, ISBN 978-1-4200-5993-9, NY, USA.

D'Angelo, S.; Manna, C.; Migliardi, V.; Mazzoni, O.; Morrica, P.; Capasso, G. et al. (2001). Pharmacokinetics and Metabolism of Hydroxytyrosol, a Natural Antioxidant from Olive Oil. *Drug Metabolism and Disposition*, Vol. 29, No. 11, (November 2001), pp. 1492-1498, ISSN 0090-9556

D'archivio, M.; Filesi, C.; Vari, R.; Scazzocchio, B. & Masella, R. (2010). Bioavailability of the Polyphenols: Status and Controversies. *International Journal of Molecular Sciences*, Vol. 11, No. 4, (March 2010), pp 1321-1342, ISSN 1422-0067

Day, A.J.; Mellon, F.A.; Barron, D.; Sarrazin, G.; Morgan, M.R. & Williamson, G. (2001). Human Metabolism of Flavonoids: Identification of Plasma Metabolites of Quercetin. *Free Radical Research*, Vol. 35, No. 6, (December 2001), pp. 941-952, ISSN 1071-5762

Dilis, V. & Trichopolou, A. (2009). Mediterranean diet and olive oil consumption-Estimations of daily intake of antioxidants from Virgin Olive Oil and olives, In: *Olive Oil. Minor Constituents and Health*, D. Boskou, (Ed.), pp. 201-210, CRC press, ISBN 978-1-4200-5993-9, NY, USA.

De la Torre-Carbot, K.; Jauregui, O.; Castellote, A.I.; Lamuela-Raventos, R.M.; Covas, M.I.; Casals, I. et al. (2006). Rapid High–Performance Liquid Chromatography-Electrospray Ionization Mass Spectrometry Method for Qualitative and Quantitative Analysis of Virgin Olive Oil Phenolic Metabolites in Human Low-Density Lipoproteins. *Journal of Chromatography A*, Vol. 1116, No. 1-2, (May 2006), pp. 69-75, ISSN 0021-9673

De la Torre-Carbot, K.; Chavez-Servin, J.L.; Jáuregui, O.; Castellote, A.I.; Lamuela-Raventos, R.M.; Fito, M., et al. (2007). Presence of Virgin Olive Oil Phenolic Metabolites in Human Low Density Lipoprotein Fraction: Determination by High–Performance Liquid Chromatography-Electrospray Ionization Tandem Mass Spectrometry. *Analytica Chimica Acta*, Vol. 583, No. 2, (February 2007), pp. 402-410 ISSN 0003-2670

De la Torre, R. (2008). Bioavailability of Olive Oil Phenolic Compounds in Humans. *Inflammopharmacology*, Vol. 16 (September 2008), pp. 245-247, ISSN 0925-4692

De la Torre-Carbot, K.; Chávez-Servín, J.L.; Jaúregui, O.; Castellote, A.I.; Lamuela-Raventós, R.M.; Nurmi, T. et al. (2010). Elevated Circulating LDL Phenol Levels in Men who Consumed Virgin rather than Refined Olive Oil are Associated with Less Oxidation of Plasma LDL. *The Journal of Nutrition*, Vol. 140, No. 3, (March 2010), pp. 501-508, ISSN 0022-3166

Del Boccio, P.; Di Deo, A.; De Curtis, A.; Celli, N.; Iacoviello, L. & Rotilio, D. (2003). Liquid Chromatography–Tandem Mass Spectrometry Analysis of Oleuropein and its Metabolite Hydroxytyrosol in Rat Plasma and Urine after Oral Administration. *Journal of Chromatography B*, Vol. 785, No. 1, (February 2003), pp. 47-56, ISSN 1570-0232

Edgecombe, S.C.; Stretch, G.L. & Hayball, P.J., (2000). Oleuropein, an Antioxidant Polyphenol from Olive Oil, Is Poorly Absorbed from Isolated Perfused Rat

Intestine. *The Journal of Nutrition*, Vol. 130, No. 12, (December 2000), pp. 2996-3002, ISSN 0022-3166

Food and Agricultural Organization [FAO], (2000). Food Balance Sheets, 1999. Rome.

Fitó, M.; de la Torre, R.; Farré-Albadalejo, M.; Khymenets, O.; Marrugat, J. & Covas, M.I. (2007). Bioavailability and Antioxidant Effects of Olive Oil Phenolic Compounds in Humans: A Review. *Annali dell'Istituto Superiore di Sanità*, Vol. 43, No. 4, (n.d.), pp. 375-381 ISSN 0021-2571

García-Villalba, R.; Carrasco-Pancorbo, A.; Nevedomskaya, E.; Mayboroda O.A,; Deelder, A.M.; Segura-Carretero, A. et al. (2010). Exploratory Analysis of Human Urine by LC-ESI-TOF MS after High Intake of Olive Oil: Understanding the Metabolism of Polyphenols. *Analytical and Bioanalytical Chemistry*, Vol. 398, No. 1, (September 2010), pp. 463-75, ISSN 1618-2642

Gimeno, E.; Fito, M.; Lamuela-Raventos, R.M.; Castellote, A.I.; Covas, M.; Farre, M. et al. (2002). Effect of Ingestion of Virgin Olive Oil on Human Low Density Lipoprotein Composition. *European Journal of Clinical Nutrition*, Vol. 56, No. 2, (February 2002), pp. 114-120, ISSN 0954-3007

Gimeno, E.; de la Torre-Carbot, K.; Lamuela-Raventós, R.M.; Castellote, A.I.; Fitó, M.; de la Torre, R. et al. (2007). Changes in the Phenolic Content of Low Density Lipoprotein after Olive Oil Consumption in Men. A Randomized Crossover Controlled Trial. *British Journal of Nutrition*, Vol. 98, No. 6, (December 2007), pp. 1243-1250, ISSN 0007-1145

Hillestrom, P.R.; Covas, M.I. & Poulsen, H.E. (2006). Effect of Dietary Virgin Oil on Urinary Excretion of Etheno-DNA Adducts. *Free Radical Biology & Medicine*, Vol. 41, No. 7, (October 2006), pp. 1133-1138, ISSN 0891-5849

Khymenets, O.; Farré, M.; Pujadas, M.; Ortiz, E; Joglar, J.; Covas, M.I et al. (2011). Direct Analysis of Glucuronidated Metabolites of Main Olive Oil Phenols in Human Urine after Dietary Consumption of Virgin Olive Oil. *Food Chemistry*, Vol. 126, No. 1, (May 2011), pp. 306-314, ISSN 0308-8146

Kroon, P.A.; Clifford, M.N.; Crozier, A.; Day, A.J.; Donovan, J.L.; Manach, C. et al. (2004). How Should we Asses the Effects of Exposure to Dietary Polyphenols *In Vitro*? *The American Journal of Clinical Nutrition*, Vol. 80 No. 1, (July 2004), pp. 15-21, ISSN 0002-9165

Manach, C.; Scalbert, A.; Morand, C.; Rémésy, C. & Jiménez, L. (2004). Polyphenols: Food Sources and Bioavailability. *The American Journal of Clinical Nutrition*, Vol. 79, No. 5, (May 2004), pp. 727-747, ISSN 0002-9165

Manach, C.; Hubert, J.; Llorach, R. & Scalbert, A. (2009). The Complex Links between Dietary Phytochemicals and Human Health Deciphered by Metabolomics. *Molecular Nutrition & Food Research*, Vol. 53, No. 10, (October 2009), pp. 1303-1315, ISSN 1613-4125

Manna, C.; Galletti, P.; Maisto, G.; Cucciolla, V.; D'Angelo, S. & Zappia, V. (2000). Transport Mechanism and Metabolism of Olive Oil Hydroxytyrosol in Caco-2 Cells. *FEBS Letters*, Vol. 470, No. 3, (March 2000), pp. 341-344, ISSN 0014-5793

Marrugat, J.; Covas, M.I.; Fito, M.; Schroder, H.; Miró-Casas, E.; Gimeno, E. et al. (2004). Effects of Differing Phenolic Content in Dietary Olive Oils on Lipids and LDL Oxidation--A Randomized Controlled Trial. *European Journal of Nutrition*, Vol. 43, No. 3, (2004), pp. 140-147, ISSN 1436-6207

Mateos, R.; Goya, L. & Bravo, L. (2005). Metabolism of the Olive Oil Phenols Hydroxytyrosol, Tyrosol, and Hydroxytyrosyl Acetate by Human Hepatoma HepG2 Cells. *Journal of Agricultural and Food Chemistry*, Vol.53, No. 26, (December 2005), pp. 9897-9905, ISSN 0021-8561

Mateos, R.; Pereira-Caro, G.; Saha, S.; Cert, R.; Redondo-Horcajo, M.; Bravo, L. et al. (2011). Acetylation of Hydroxytyrosol Enhances its Transport across Differentiated Caco-2 Cell Monolayers. *Food Chemistry*, Vol. 125, No. 3, (April 2011), pp. 865-872, ISSN 0308-8146

Miró-Casas, E.; Farre-Albaladejo, M.; Covas-Planells, M.I.; Fito-Colomer, M.; Lamuela-Raventos, R.M. & de la Torre-Fornell, R. (2001a). Tyrosol Bioavailability in Humans after Ingestion of Virgin Olive Oil. *Clinical Chemistry*, Vol. 47, No. 2, (February 2001), pp. 341-343, ISSN 0009-9147

Miró-Casas, E.; Farre-Albaladejo, M.; Covas, M.I.; Ortuno-Rodriguez, J.O.; Colomer, E.; Lamuela-Raventos, R.M. et al. (2001b). Capillary Gas Chromatography-Mass Spectrometry Quantitative Determination of Hydroxytyrosol and Tyrosol in Human Urine after Olive Oil Intake. *Analytical Biochemistry*, Vol. 294 No. 1, (July 2001), pp. 63-72, ISSN 0003-2697

Miró-Casas, E.; Covas, M.I.; Fitó, M.; Farre-Albaladejo, M.; Marrugat, J. & de la Torre, R. (2003a). Tyrosol and Hydroxytyrosol are Absorbed from Moderate and Sustained Doses of Virgin Olive Oil in Humans. *European Journal of Clinical Nutrition*, Vol. 57, No. 1, (January 2003), pp. 186-190, ISSN 0954-3007

Miró-Casas, E.; Covas, M.I.; Farre, M.; Fito, M.; Ortuno, J.; Weinbrenner, T. et al. (2003b). Hydroxytyrosol Disposition in Humans. *Clinical Chemistry*, Vol. 49, No. 6, (June 2003), pp. 945-952, ISSN 0009-9147

Murkovic, M. ; Lechner, S. ; Pietzka, A. ; Bratakos, M. & Katzogiannos, M. (2004). Analysis of Minor Constituents in olive Oil, *Journal of Biochemical and Biophysical Methods*, Vol. 61, No. 1-2, (October 2004), pp. 155-160, ISSN 0165-022X

Pinto, J.; Paiva-Martins, F.; Corona, G.; Debnam, E.S.; Oruna-Concha, J.M.; Vauzour, D. et al. (2011). Absorption and Metabolism of Olive Oil Secoiridoids in the Small Intestine. *British Journal of Nutrition*, Vol. 105, No. 11, (June 2011), pp. 1607-1618, ISSN 0007-1145

Romero, M.P.; Tovar, M.J.; Girona, J. & Motilva, M.J. (2002). Changes in the HPLC Phenolic Profile of Virgin Olive Oil from Young Trees (*Olea europaea* L. cv Arbequina) Grown Under Different Deficit Irrigation Strategies. *Journal of Agricultural and Food Chemistry*, Vol. 50, No. 19, (September 2002), pp. 5349-5354, ISSN 0021-8561

Ruiz-Gutiérrez, V.; Juan, M.E.; Cert, A. & Planas, M. (2000). Determination of Hydroxytyrosol in Plasma by HPLC. *Analytical Chemistry*, Vol. 72, No. 18, (September 2000), pp. 4458-4461, ISSN 0003-2700

Scalbert, A. & Williamson, G. (2000). Dietary Intake and Bioavailability of Polyphenols. *The Journal of Nutrition*, Vol. 130, No. 8 Suppl., (August 2000), pp.2073S-2085S, ISSN 0022-3166

Servili, M. & Montedoro, G., (2002). Contribution of Phenolic Compounds to Virgin Olive Oil Quality. *European Journal of Lipid Science and Technology*, Vol. 104, No. 9-10, (October 2002), pp. 602-613, ISSN 1438-7697

Servili, M.; Selvaggini, R.; Esposto, S.; Taticchi, A.; Montedoro, G. & Morozzi. G. (2004). Health and Sensory Properties of Virgin Olive Oil Hydrophilic Phenols: Agronomic and Technological Aspects of Production that Affect Their Occurrence in the Oil. *Journal of Chromatography A*, Vol. 1054, No. 1-2, (October 2004), pp. 113-127, ISSN 0021-9673

Soler, A.; Romero, M.P.; Macià, A.; Saha, S.; Furniss, C.S.S.; Kroon, P.A. et al. (2010). Digestion Stability and Evaluation of the Metabolism and Transport of Olive Oil Phenols in the Human Small-Intestinal Epithelial Caco-2/TC7 Cell Line. *Food Chemistry*, Vol. 119, No. 2, (March 2010), pp. 703-714, ISSN 0308-8146

Stahl, W.; Van Den Berg, H.; Arthur, J.; Bast, A.; Dainty, J.; Faulks, R.M. et al. (2002). Bioavailability and Metabolism. *Molecular Aspects of Medicine*, Vol. 23, (February 2002), pp.39-100, ISSN 0098-2997

Stark, A.H. & Madar, Z. (2002). Olive Oil as a Functional Food: Epidemiology and Nutritional Approaches. *Nutrition Reviews*, Vol. 60, No. 6, (June 2002), pp. 170-176, ISSN 0029-6643

Suárez, M.; Romero, M.P.; Macià, A.; Valls, R.M.; Fernández, S.; Solà, R. et al. (2009). Improved Method for Identifying and Quantifying Olive Oil Phenolic Compounds and Their Metabolites in Human Plasma by Microelution Solid-Phase Extraction Plate and Liquid Chromatography-Tandem Mass Spectrometry. *Journal of Chromatography B*, Vol. 877, No. 32, (December 2009), pp. 4097-4106, ISSN 1570-0232

Tan, H.-W.; Tuck, K.L.; Stupans, I. & Hayball, P.J. (2003). Simultaneous Determination of Oleuropein and Hydroxytyrosol in Rat Plasma Using Liquid Chromatography with Fluorescence Detection. *Journal of Chromatography B*, Vol. 785, No. 1, (February 2003), pp. 187-191, ISSN 1570-0232

Tovar, M.J.; Motilva, M.J. & Romero, M.P. (2001). Changes in the Phenolic Composition of Virgin Olive Oil from Young Trees (*Olea europaea* L. cv Arbequina) Grown Under Linear Irrigation Strategies. *Journal of Agricultural and Food Chemistry*, Vol. 49, No. 11, (November 2001), pp. 5502-5508, ISSN 0021-8561

Tuck, K.L.; Freeman, M.P.; Hayball, P.J.; Stretch, G.L. & Stupans, I. (2001). The *In Vivo* Fate of Hydroxytyrosol and Tyrosol, Antioxidant Phenolic Constituents of Olive Oil, after Intravenous and Oral Dosing of Labeled Compounds to Rats. *The Journal of Nutrition*, Vol. 131, No. 7, (July 2001), pp. 1993–1996, ISSN 0022-3166

Tuck, L.K. & Hayball, P.J. (2002). Major Phenolic Compounds in Olive Oil: Metabolism and Health Effects. *Journal of Nutritional Biochemistry*, Vol. 13, No. 11, (July 2002), pp. 636-644, ISSN 0955-2863

Tuck, K.L.; Hayball, P.J. & Stupans, I. (2002). Structural Characterization of the Metabolites of Hydroxytyrosol, the Principal Phenolic Component in Olive Oil, in Rats. *Journal of Agriculture and Food Chemistry*, Vol. 50, No. 8, (April 2002), pp. 2404–2409, ISSN 0021-8561

Uceda, M.; Hermoso, M.; García-Ortiz, A.; Jiménez, A. & Beltran, G. (1999). Intraspecific Variation of Oil Contents and the Characteristics of Oils in Olive Cultivars. *Acta Horticulturae (ISHS)*, Vol. 474, No. 2, (April 1999), pp. 659–652, ISSN 0567-7572

Visioli, F.; Bellomo, G.; Montedoro, G. & Galli, C. (1995). Low Density Lipoprotein Oxidation is Inhibited In Vitro by Olive Oil Constituents. *Atherosclerosis*, Vol. 117, No. 1, (September 1995), pp. 25-32, ISSN 0021-9150

Visioli, F.; Galli, C.; Bornet, F.; Mattei, A.; Patelli, R.; Galli, G. et al. (2000). Olive Oil Phenolics are Dose-Dependently Absorbed in Humans, *FEBS Letters*, Vol. 468, No. 2-3, (February 2000), pp. 159-160, ISSN 0014-5793

Visioli, F.; Caruso, D.; Plasmati, E.; Patelli, C.; Mulinacci, N.; Romani, A. et al. (2001). Hydroxytyrosol, as a Component of Olive Mill Waste Water, is Dose-Dependently Absorbed and Increases the Antioxidant Capacity of Rat Plasma. *Free Radical Research*, Vol. 34, No. 3, (March 2001), pp. 301-305, ISSN 1071-5762

Visoli, F.; Galli, C.; Galli, G. & Caruso, D. (2002). Biological Activities and Metabolic Fate of Olive Oil Phenols. *European Journal of Lipid Science and Technology*, Vol. 104, No. 9-10, (October 2002), pp. 677-684, ISSN 1438-7697

Visioli, F.; Galli, C.; Grande, S.; Colonnelli, K.; Patelli, C.; Galli, G. et al. (2003). Hydroxytyrosol Excretion Differs Between Rats and Humans and Depends on the Vehicle of Administration. *The Journal of Nutrition*, Vol. 133, No. 8, (August 2003), pp. 2612-2615, ISSN 0022-3166

Visioli, F. & Bernardini, E. (2011). Extra Virgin Olive Oil's Polyphenols: Biological Activities. *Current Pharmaceutical Design*, Vol. 17, No. 8, (n.d.), pp. 786-804, ISSN 1381-6128

Vissers, M.N.; Zock, P.L.; Roodenburg, A.J.C.; Leenen, R. & Katan, M.B. (2002). Olive Oil Phenols are absorbed in Humans? *The Journal of Nutrition*, Vol. 132, No. 3, (March 2002), pp. 409-417 ISSN 0022-3166

Vissers, M.N.; Zock, P.L. & Katan, M.B. (2004). Bioavailability and Antioxidant Effect of Olive Oil Phenols in Humans: A Review. *European Journal of Clinical Nutrition*, Vol. 58, No. 6, (June 2004), pp. 955-965, ISSN 0954-3007

Weinbrenner, T.; Fitó, M.; Farré-Albaladejo, M.; Saez, G.T.; Rijken, P.; Tormos, C. et al. (2004). Bioavailability of Olive Oil Phenolic Compounds from Olive Oil and Oxidative/Antioxidative Status at Postprandial State in Humans. *Drugs under Experimental and Clinical Research*, Vol. 30, No. 5-6, (n.d.), pp. 207-212, ISSN 0378-6501

8

Biological Properties of Hydroxytyrosol and Its Derivatives

José G. Fernández-Bolaños, Óscar López,
M. Ángeles López-García and Azucena Marset
University of Seville
Spain

1. Introduction

Polyphenols are a wide family of compounds found in fruits and vegetables, wine, tea, cocoa, and extra-virgin olive oil, which exhibit strong antioxidant activity by scavenging different families of Reactive Oxygen Species (ROS). One of the most effective members of the polyphenol family in terms of free radical scavenging is hydroxytyrosol, 2-(3,4-dihydroxyphenyl)ethanol (Fernández-Bolaños et al., 2008), a simple phenol found predominantly in olive tree (*Olea europaea*).

Hydroxytyrosol (HT) can be found in leaves and fruits of olive, extra virgin olive oil and it is specially abundant in olive oil mill wastewaters from where it can be recovered (Fernández-Bolaños et al., 2008; Sabatini, 2010). Hydroxytyrosol is a metabolite of oleuropein (Fig. 1), another major phenolic component of olive products; they both give to extra-virgin olive oil its bitter and pungent taste (Omar, 2010a). Hydroxytyrosol shows a broad spectrum of biological properties due to its strong antioxidant and radical-scavenging properties. More active than antioxidant vitamins and synthetic antioxidants, hydroxytyrosol exerts its antioxidant activity by transforming itself into a catechol quinone (Rietjens, 2007).

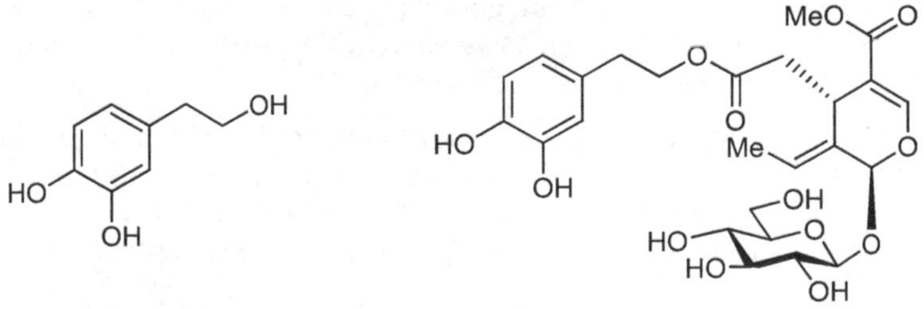

Hydroxytyrosol   Oleuropein

Fig. 1. Structures of hydroxytyrosol and oleuropein

2. Biological activity of hydroxytyrosol

Historically, olive tree leaves were used for traditional therapy by ancient civilizations. Extracts from olive tree leaves were found to have a positive effect on hypertension by the middle of last century (Scheller, 1955; Perrinjaquet-Moccetti et al. 2008; Susalit et al., 2011), and, since then, the benefits of minor olive components have been extensively investigated (Tripoli et al., 2005).

2.1 Antioxidant activity

The antioxidant activity is the most studied property of olive phenolic compounds. The interest of hydroxytyrosol is based on its remarkable pharmacological and antioxidant activities. Reactive oxygen species, which are continuously being formed as a result of metabolic processes in the organism, may cause oxidation and damage of cellular macromolecules, and therefore, may contribute to the development of degenerative diseases, such as atherosclerosis, cancer, diabetes, rheumatoid arthritis and other inflammatory diseases (Balsano & Alisi, 2009).

The high antioxidant efficiency of HT, attributed to the presence of the o-dihydroxyphenyl moiety, is due to its high capacity for free radical scavenging during the oxidation process and to its reducing power on Fe^{3+} (Torres de Pinedo et al., 2007). The antioxidant properties of the o-diphenols are associated with their ability to form intramolecular hydrogen bonds between the hydroxyl group and the phenoxy radical (Visioli et al., 1998); therefore, the catechol avoids the chain propagation by donating a hydrogen radical to alkylperoxyl radicals (ROO˙) formed in the initiation step of lipid oxidation (Scheme 1).

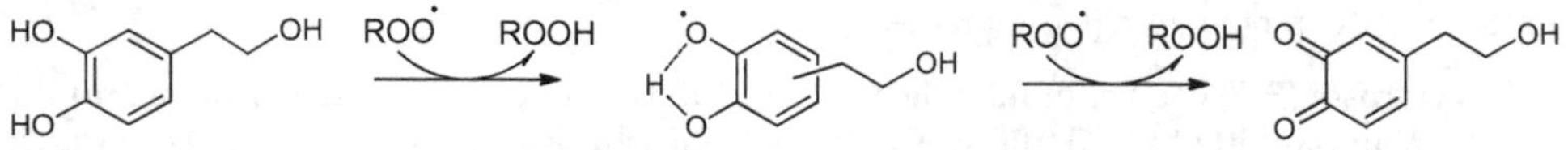

Scheme 1. Mechanism of free radical scavenging by hydroxytyrosol

Oxidation of low-density lipoproteins (LDLs) is a lipid peroxidation chain reaction, which is initiated by free radicals. It has been shown that hydroxytyrosol can inhibit LDL oxidation efficiently due to its capacity to scavenge peroxyl radicals (Arouma et al., 1998; Turner et al., 2005). Hydroxytyrosol reduces oxidation of the low-density lipoproteins carrying cholesterol (LDL-C), which is a critical step in the development of atherosclerosis and other cardiovascular diseases (Gonzalez-Santiago et al., 2010; Vázquez-Velasco et al., 2011); hydroxytyrosol has also a potential protective effect against oxidative stress induced by *tert*-butyl hydroperoxide (Goya et al., 2007).

It has been reported that hydroxytyrosol enhances the lipid profile and antioxidant status preventing the development of atherosclerosis. This compound may also reduce the expression of vascular cell adhesion molecules (Carluccio et al., 2007) and inhibit platelet aggregation in rats (González-Correa et al., 2008a) and hypercholesterolaemia in humans (Ruano et al., 2007).

2.2 Anticancer activity

Numerous studies about the relationship between olive oil consumption and cancer prevention have been carried out (Pérez-Jiménez et al., 2005). Antioxidant compounds supplied in the diet can reduce the risk of cancer due to the fact that they can minimize DNA damage, lipid peroxidation and the amount of ROS generated (Omar, 2010a; Hillestrom, 2006; Manna, 2005).

It has been reported that HT may exert a pro-apoptotic effect by modulating the expression of genes involved in tumor cell proliferation of promyelocytes (HL60 cells) (Fabiani et al., 2006, 2008, 2009, 2011). Moreover, it has been shown that HT inhibits proliferation of human MCF-7 breast cancer cells (Siriani et al., 2010; Bulotta et al. 2011; Bouallagui et al., 2011a), human HT29 colon carcinoma cells (Guichard et al., 2006), human M14 melanoma cells (D'Angelo et al., 2005) and human PC3 prostate cells (Quiles et al., 2002).

Pre-treatment of HepG2 cells with hydroxytyrosol prevented cell damage, what could be due to the fact that hydroxytyrosol may prepare the antioxidant defense system of the cell to face oxidative stress conditions (Goya et al., 2007, 2010).

2.3 Osteoporosis

Hydroxytyrosol may have critical effects on the formation and maintenance of bone, and could be used as an effective remedy in the treatment of osteoporosis symptoms, as it can stimulate the deposition of calcium and inhibit the formation of multinucleated osteoclasts in a dose-dependent manner. HT also suppressed the bone loss of spongy bone in femurs of ovariectomized mice (Hagiwara et al., 2011).

2.4 Antimicrobial activity

Antimicrobial activity of oleuropein, tyrosol and hydroxytyrosol has been studied *in vitro* against bacteria, viruses and protozoa (Bisignano et al., 1999).

The *in vitro* antimycoplasmal activity of HT has been investigated, concluding that this compound might be considered as an antimicrobial agent for treating human infections caused by bacterial strains or casual agents of intestinal or respiratory tract (Furneri et al., 2004).

It has been shown that polyphenols from olive oil are powerful anti-*Helicobacter Pylori* compounds *in vitro* (Romero et al., 2007), a bacteria linked to a majority of peptic ulcers and to some types of gastric cancer.

2.5 Antiinflammatory activity

Inflammation and its consequences play a crucial role in the development of atherosclerosis and cardiovascular diseases. Polyphenols have been shown to decrease the production of inflammatory markers, such as leukotriene B4, in several systems (Biesalski, 2007).

The effect of hydroxytyrosol on platelet function has been tested. Hydroxytyrosol was proven to inhibit the chemically induced aggregation, the accumulation of the pro-aggregant agent thromboxane in human serum, the production of the pro-inflammatory molecules leukotrienes and the activity of arachidonate lipoxygenase (Visioli et al., 2002).

Recently, it has been described that HT-20, an olive oil extract containing about 20% of hydroxytyrosol, inhibits inflammatory swelling and hyperalgesia, and suppresses proinflammatory cytokine in a rat inflammation model (Gong et al., 2009).

2.6 Antiviral activity

Hydroxytyrosol and oleuropein have been identified as a unique class of HIV-1 inhibitors that prevent HIV from entering into the host cell and binding the catalytic site of the HIV-1 integrase. Thus, these agents provide an advantage over other antiviral therapies in which both, viral entry and integration, are inhibited (Lee-Huang et al., 2007a, 2007b, 2009).

HT and its derivatives are also useful, when applied topically, as microbicide for preventing HIV-infection, as well as other sexually transmitted diseases caused by fungi, bacteria or viruses (Gómez-Acebo et al., 2011). Furthermore, it has been reported that hydroxytyrosol inactivated influenza A viruses, suggesting that the mechanism of the antiviral effect of HT might require the presence of a viral envelope (Yamada et al., 2009).

2.7 Hydroxytyrosol as an antinitrosating agent

The antinitrosating properties of hydroxytyrosol and other plant polyphenols of dietary relevance have been investigated (De Lucia et al., 2008). It has been shown that HT reacts with sodium nitrite at pH 3 to give 2-nitrohydroxytyrosol, supporting a protective role of HT as an efficient scavenger of nitrosating species (Fig. 2).

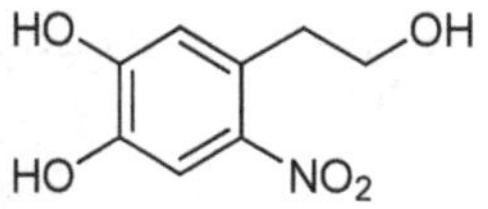

Fig. 2. 2-Nitrohydroxytyrosol formed by nitrosation of HT

3. Hydroxytyrosol derivatives

3.1 Lipophilic hydroxytyrosol esters

Many different hydroxytyrosol lipophilic analogues occur naturally in olive fruit and in virgin olive oil. The amount of these compounds is related to olive variety and ripeness, climate, location, type of crushing machine and oil extraction procedures. As an example, the concentration of hydroxytyrosyl acetate is similar to that of HT in some olive oil varieties such as Arbequina, twice as high in the Picual variety, and between one third and one fourth in the Manzanilla and Hojiblanca oils (Romero et al., 2007).

Due to the limited solubility of HT in lipid media, the search for new lipophilic hydroxytyrosol esters with enhanced properties is of great interest, both in food industry and in medicine. Studies on olive polyphenols have shown the importance of the lipophilicity of the antioxidants on the cell uptake and membrane crossing, and on the substrate to be protected (membrane constituents or LDL), (Grasso et al., 2007). These facts explain the efforts made in the development of new synthetic analogues with increased lipophilicity.

3.1.1 Synthetic approaches

Phenolic acids, such as caffeic acid, have been esterified with good chemoselectivity in the presence of strong protic acids (Fischer esterification), but the severe reaction conditions together with the large excess of alcohol required make this strategy of limited applicability (Burke et al., 1995). Under basic catalysis, phenols can be easily deprotonated, so the esterification of phenolic alcohols and phenolic acids via acyl nucleophilic substitution requires previous protection of the phenolic hydroxyl groups, due to the competition between aliphatic and phenolic hydroxyl groups (Appendino et al., 2002; Gambacorta et al., 2007).

3.1.1.1 Protection of phenolic hydroxyl groups

As an example, benzyl groups have been used to carry out the HT esterification under basic conditions, followed by catalytic hydrogenation to remove the protective groups (Gordon et al., 2001), as depicted in Scheme 2.

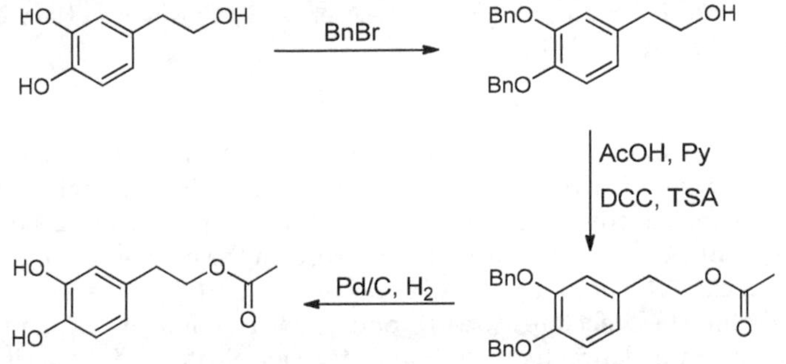

Scheme 2. Synthesis of hydroxytyrosyl acetate via benzylation of phenolic hydroxyls

A two-step procedure involving the reaction of methyl orthoformate-protected hydroxytyrosol with acetyl chloride, and hydrolytic deprotection in phosphate buffer under very mild conditions (pH=7.2) to get hydroxytyrosyl acetate (87% overall yield) (Scheme 3) was also described as a successful procedure for the preparation of HT-derived esters (Gambacorta et al., 2007). The key synthetic orthoester intermediate was also used for the synthesis of HT upon reduction with LiAlH$_4$ and acidic deprotection.

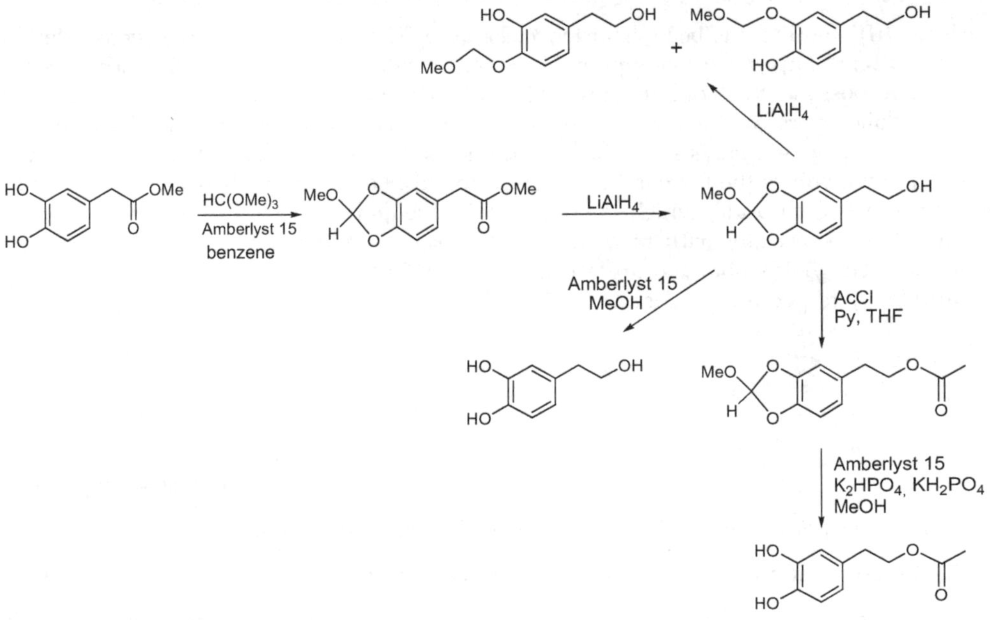

Scheme 3. Synthesis of hydroxytyrosyl acetate via methyl orthoformate-protected hydroxytyrosol.

In order to overcome the problems associated to the protection and deprotection steps of the phenolic hydroxyl groups, different methods for the preparation of hydroxytyrosyl esters by reaction of hydroxytyrosol with various acylating agents have been described, such as esterification with free acids (Appendino et al., 2002), transesterification with methyl or ethyl esters (Alcudia et al., 2004; Trujillo et al., 2006), acyl chlorides (Torregiani et al., 2005) and the use of enzymatic methodologies (Grasso et al., 2007; Mateos et al., 2008; Torres de Pinedo et al., 2005; Buisman, 1998).

3.1.1.2 Acid catalyzed transesterification

HT transesterification using methyl or ethyl esters and p-toluenesulfonic acid as catalyst has been described as a method without the need of protection of the aromatic hydroxyl groups due to its total chemoselectivity (Alcudia et al., 2004; Trujillo et al., 2006). This method involves heating a solution solution of hydroxytyrosol in the corresponding ethyl or methyl ester, containing a catalytic amount of p-toluenesulfonic acid (Scheme 4). This protocol has been optimized for HT acetate (86%), and also for longer aliphatic chains like hydroxytyrosyl butyrate, laureate, palmitate, stearate, oleate and linoleate, obtained in acceptable to good yields (62-76%) (Mateos et al., 2008).

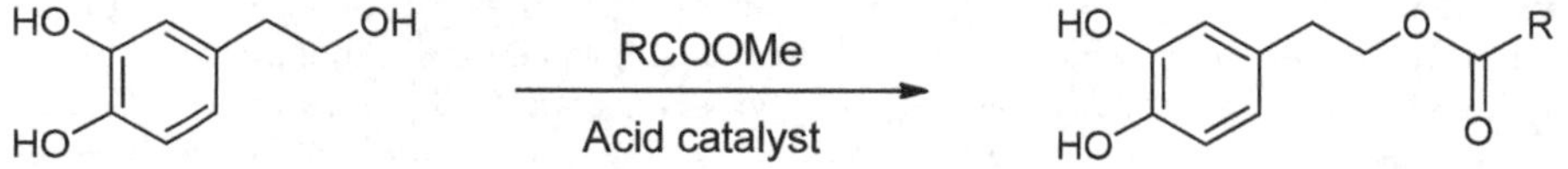

Scheme 4. General procedure of acid-catalyzed transesterification

3.1.1.3 Acylation of polyphenolic alcohols with the couple CeCl₃–RCOCl

Cerium (III) chloride has been reported to be an efficient promoter for the chemoselective esterification of unprotected polyphenolic alcohols with acyl halides as acyl donors, thereby making it possible to avoid the protection of phenolic hydroxyl groups and providing polyphenolic esters of interest (Torregiani, 2005). This reaction is one example of the so-called Lewis acid catalysis by lanthanide salts (Ishihara et al., 1995). The reaction presumably involves the formation of an electrophilic Lewis adduct between acyl chlorides and cerium (III) chloride, which is quenched by the more nucleophilic aliphatic hydroxyl group of the substrate, with formation of the ester, and regeneration of the lanthanide promoter. The yields obtained are acceptable for HT using nonanoyl and oleoyl chlorides (53 and 52%), respectively (Scheme 5).

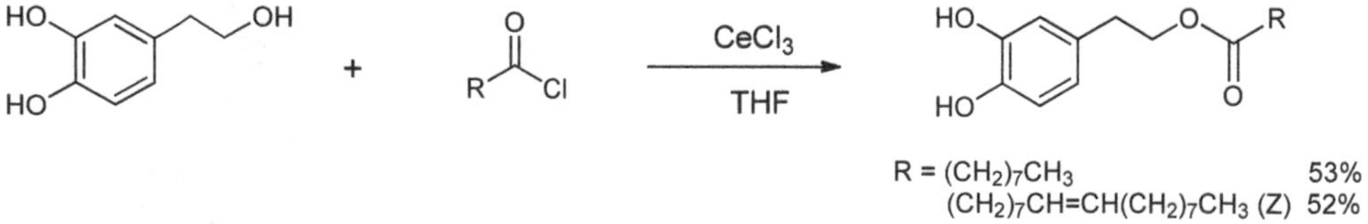

R = (CH₂)₇CH₃   53%
(CH₂)₇CH=CH(CH₂)₇CH₃ (Z) 52%

Scheme 5. Acylation of hydroxytyrosol with acyl chlorides and Ce(III)

3.1.1.4 Esterification with free acids: Mitsunobu esterification

The Mitsunobu reaction has been also applied to the chemoselective esterification of phenolic acids with phenolic alcohols (Appendino et al., 2002) as demonstrated by the condensation of hydroxytyrosol with gallic acid, and of vanillyl alcohol with caffeic acid in a one step procedure with 48% and 50% yields, respectively. The esterification is carried out using DIAD (diisopropyl azodicarboxylate) and TPP (triphenylphosphine) in THF (Scheme 6). The removal of byproducts arising during the Mitsunobu reaction, a major problem of this type of reactions, could be solved by gel-permeation chromatography on Sephadex LH-20.

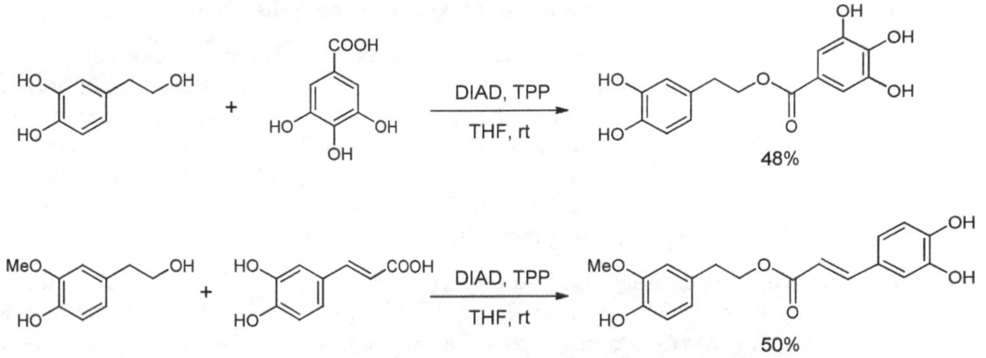

Scheme 6. Mitsunobu esterification of hydroxytyrosol and vanillyl alcohol

3.1.1.5 Syntheses of hydroxytyrosol esters from tyrosol and homovanillyl alcohol

The syntheses previously described in the previous sections had all in common hydroxytyrosol as a precursor of its esters, but some efforts have also been done to get hydroxytyrosyl esters starting from different and cheaper reagents. In this context, the syntheses of hydroxytyrosol esters from tyrosol and homovanillyl alcohol have been proposed (Bernini et al., 2008b). This procedure involves the selective esterification of tyrosol and homovanillyl alcohol with acyl chlorides in dicholoromethane as solvent, to give tyrosyl and homovanillyl acetates in 90% and quantitative yields, respectively, by using only a little excess of acetyl chloride in dichloromethane without any catalysts. The authors suggested acid catalysed acylation due to traces of hydrochloric acid derived from the hydrolysis of the acetyl chloride under the experimental conditions. A similar selectivity was observed by using several saturated or unsaturated acyl chlorides with longer chains such as hexanoyl, palmitoyl, oleoyl and linoleoyl chlorides.

The subsequent oxidation with 2-iodoxybenzoic acid (IBX) or Dess-Martin periodinane reagent (DMP) and *in situ* reduction with sodium dithionite ($Na_2S_2O_4$) of tyrosyl and homo-vanillyl esters led to the corresponding hydroxytyrosol derivatives. In general, the oxidation of tyrosol derivatives proceeded with higher yields (92-77%) compared to those of homovanillyl derivatives (88-58%). The use of DMP gave similar results to those obtained with IBX. The procedure of oxidation/reduction with IBX/$Na_2S_2O_4$ to obtain the different esters is under protection of two patents (Bernini et al. 2007, 2008c).

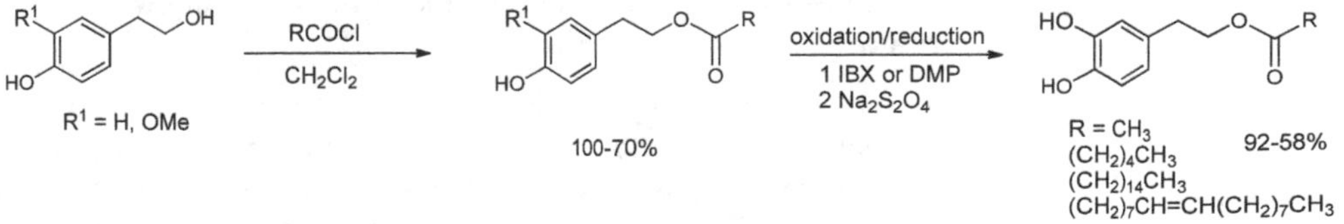

Scheme 7. Synthesis of hydroxytyrosol esters from tyrosol and homovanillyl alcohol

3.1.1.6 Lipase-catalyzed transesterification

The use of enzymes, like lipases, as catalysts in non-aqueous solvents to prepare lipophilic derivatives directly from HT has been widely described in the last few years (Grasso et al., 2007; Torres de Pinedo et al., 2005; Mateos et al., 2008; Buisman et al., 1998). This procedure avoids the use of toxic reagents and allows mild reaction conditions.

The esterification of phenols with carboxylic fatty acids and lipases as biocatalysts was firstly investigated by Buisman et al., (1998), using hydroxytyrosol, octanoic acid in hexane, and immobilized lipases from *Candida antartica* (CAL-B). Furthermore, a strong dependence of the yield on the solvent used was observed; so, in diethyl ether a conversion of 85% was obtained within 15 hours (35 °C), while conversions of roughly 20% were found in the case of solvents like chloroform, dichloromethane or THF. Yields of 70–80% were observed using *n*-pentane and *n*-hexane, in spite of the poor solubility of HT in such solvents.

Different enzymes have been tested on hydroxytyrosol (Grasso et al., 2007) including lipases from *A. niger, C. cylindracea, M. javanicus, P. cepacia, M. miehei, C. viscosum, P. fluorescens, R. arrhizus, R. niveus, C. antarctica*, porcine pancreas and wheat germ, using vinyl acetate as reagent and *tert*-butyl methyl ether as solvent. The best results were obtained with *C. antarctica* in terms of short reaction time, chemioselective conversion and good yield. *C. antarctica* lipase (CAL) was selected for acylation of hydroxytyrosol and homovanillic alcohol with vinyl esters of different acyl chains on a preparative scale, as shown in Table 1. The use of *C. antarctica* with increasing alkyl chain length required longer reaction times. The homovanillyl alcohol and its esters were found to exhibit scarce effectiveness both as radical scavengers and antioxidant agents.

Transesterification of HT with ethyl saturated, mono- and poly-unsaturated fatty acid esters, catalized by Novozym® 435 (immobilized *C. antarctica* lipase B), in vacuum under solventless conditions, has been successfully developed (Torres de Pinedo et al. 2005). This procedure gave hydroxytyrosyl esters in 59-98% yield for the saturated fatty acid esters, and 32-97% yield for the mono- and poly-unsaturated fatty acid esters.

3.1.2 Biological activity

3.1.2.1 Antioxidant activity

The antioxidant activity of hydroxytyrosyl esters has been measured with different methods, including DPPH (1,1-diphenyl-2-picrylhydrazyl radical), ABTS (2,2'-azino-bis(3-etilbenzotiazolin-6-sulfonic acid), FRAP (ferric reducing antioxidant power) and Rancimat (Mateos et al., 2008; Gordon et al., 2001; Bouallagui et al., 2011b). The Rancimat test is a method commonly used to evaluate the antioxidant power in lipophilic food matrices, such as oils and fats, while the ABTS and FRAP assays are used for the evaluation of antioxidant activity in hydrophilic medium; the ABTS assay evaluating the radical-scavenging capacity, and the FRAP method determining the reducing activity.

The Rancimat test revealed a lower activity for ester derivatives compared to HT, in agreement with the so-called polar paradox, according to which hydrophilic antioxidants are more effective in less polar media, such as bulk oils, whereas lipophilic antioxidants are more effective in relatively more polar media, such as in oil-in-water emulsions or liposomes (Frankel et al., 1994; Shahidi & Zhong, 2011).

Phenol	Acylating agent	Product	Time (min)	Yield (%)
			35	95.0
			35	96.5
			75	93.3
			180	92.3
			60	96.8
			90	90.9
			90	97.5
			240	98.0

Phenol: acylating agent 1:20, *C. antarctica* lipase, *t*-BuOMe, 40 °C

Table 1. Enzymatic esterification of HT and homovanillyl alcohol (Grasso et al., 2007)

The order of the scavenging activities toward the ABTS radical was hydroxytyrosyl esters ≥ α-tocopherol > hydroxytyrosol > tyrosyl >tyrosyl esters ≅ BHT. In a similar trend, comparison of FRAP values obtained for the free hydroxytyrosol and tyrosol with the corresponding esters revealed that while hydroxytyrosyl esters showed a significantly higher reducing activity than their precursor, all the tyrosyl esters showed a lower antioxidant activity than that of tyrosol. The same conclusion was obtained from DPPH assay of the radical scavenging activity (Grasso et al., 2007).

In connection with the size of the acyl chain, the reported literature seems to conclude that the antioxidant capacity of hydroxytyrosyl esters is better for medium-sized (C4–C9) alkyl chains in comparison with HT, whereas further elongation of the acyl chain does not improve the antioxidant activity. This confirms that antioxidant capacity does not depend only on lipophilicity. A possible explanation could be related to the fact that the conformational freedom of the ester chain increases with the acyl chain length, and this could result in folded structures in which catechol hydroxyls are shielded (Tofani et al., 2010; Pereira-Caro et al., 2009; Medina et al., 2009).

This antioxidant activity has also been proved in biological assays, in order to check the ability of hydroxytyrosyl esters to protect proteins and lipids against oxidation caused by peroxyl radicals, using a brain homogenate as an *ex vivo* model (Trujillo et al., 2006) and cumene hydroperoxide to induce oxidation. The results obtained showed a protective effect in these systems, which was more effective in preventing the generation of carbonyl groups in proteins than the generation of malondialdehyde in lipid; hydroxytyrosyl linoleate showed the greatest activity. This fact proves that the introduction of a lipophilic chain in the hydroxytyrosol molecule increases both protein and lipid protection.

Dichlorodihydrofluorescein (DCF) fluorometric assay on whole cells, carried out to check the antioxidant activity of a large serie of hydroxytyrosyl esters (Tofani et al., 2010) on rat muscle cells, showed that hydroxytyrosol esters had a better antioxidant activity compared to HT due to the better penetration into the cells of the lipophilic derivatives.

Hydroxytyrosol fatty acid esters have shown a nonlinear tendency in antioxidant capacity in fish oil-in-water emulsions (Lucas et al., 2010), where a maximum of antioxidant efficiency appeared for hydroxytyrosol octanoate in a study of hydrosytyrosyl esters with alkyl chains varying from C2 to C18. These results seem to be in disagreement with the antioxidant polar paradox.

3.1.2.2 Cardiovasvular diseases

Platelet aggregation is considered one of the main events in arterial thrombosis; therefore aggregation prevention is a major goal of cardiovascular research. It has been proved that hydroxytyrosol acetate inhibits platelet aggregation induced by ADP, collagen or arachidonic acid and stimulates nitric oxide production, more efficiently than hydroxytyrosol, and as effectively as acetylsalicylic acid; the latter is the most widely used drug in the world to prevent ischaemic cardiovascular diseases because of its antiplatelet aggregating action. This conclusion has been achieved *in vivo* in a study of oral administration of this ester to rats (González-Correa et al., 2008b), and *in vitro* in both human whole blood and platelet-rich plasma (González-Correa et al., 2009).

3.1.2.3 DNA damage oxidative protection

The atypical Comet test on whole blood cells has been applied to several hydroxytyrosyl esters to check their capacity to counteract the oxidative stress caused by H_2O_2 and the basal DNA damage.

The results obtained show that antidamaging properties on DNA of HT acetate and propanoate are comparable to those of HT, whereas the protective effect progressively decreases in the order butanoate < decanoate $\cong$ estearate (Fig. 3). This behavior was not observed for the lipophilic analogues of homovanillyl alcohol which appear to be scarcely protective, indicating that o-diphenols are more effective antioxidants than simple phenols (Grasso et al., 2007).

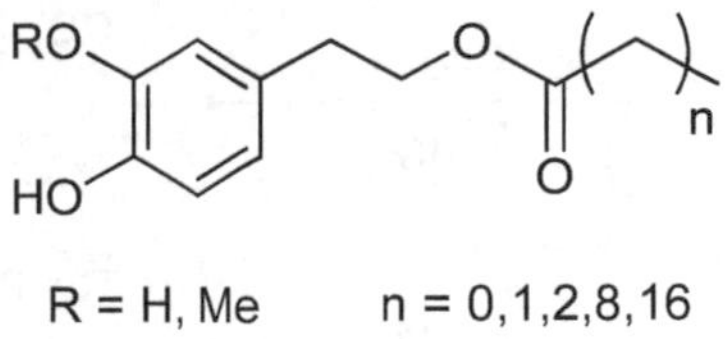

R = H, Me   n = 0,1,2,8,16

Fig. 3. Hydroxytyrosol lipophilic analogues

3.1.2.4 Prevention of oxidative stress

The ability of hydroxytyrosol and its esters to prevent iron-induced oxidative stress has been studied on human cervical cells (HeLa cells) by the TBARS protocol (Bouallagui et al., 2011b). Pre-incubation of HeLa cells in the presence of 100 µM phenolic compounds led to a significant improvement of the oxidative status. In fact, thiobarbituric acid-reactive substance (TBARS) production was decreased by 30%, 36% and 38% with hydroxytyrosol, hydroxytyrosyl acetate and hydroxytyrosyl oleate, respectively.

3.1.2.5 Transport, absorption and metabolism

The study of the metabolism of hydroxytyrosol, tyrosol, and hydroxytyrosyl acetate has been carried out using human hepatoma cells (HepG2) as a model system of the human liver (Mateos et al., 2005). The results showed extensive uptake and metabolism of hydroxytyrosol and scarce metabolism of tyrosol, while hydroxytyrosyl acetate showed an interesting behavior, with formation of deacetylated hydroxytyrosol after only 2 h. Because hydroxytyrosyl acetate was stable in the culture medium, the hydroxytyrosol detected in the extracellular medium should be attributed to the action of the hepatic cells.

3.1.2.6 Neuroprotective effect of hydroxytyrosyl and hydroxytyrosol acetate

Neuroprotection exerted by HT derivatives has been investigated in rat brain slices subjected to hypoxia-reoxygenation, both *in vitro* and after oral administration (González-Correa et al., 2008). This study was carried out to confirm to the previously demonstrated neuroprotective effects of virgin olive oil in rats (González-Correa et al., 2007). Although the studies gave positive results in the neuroprotective activity of both HT and hydroxytyrosyl acetate, mechanisms that underlie this effect are still unknown.

3.2 Lipophilic hydroxytyrosyl alkyl ethers

3.2.1 Synthetic approaches

Hydroxytyrosyl alkyl ethers have been obtained (Madrona et al., 2009) in a three-step procedure starting from hydroxytyrosol isolated from olive oil waste waters (Scheme 8). This procedure requires first the selective protection of the aromatic hydroxyl groups *via* benzylation with benzyl bromine in the presence of K_2CO_3, and then the addition of an alkyl iodide under basic conditions, and the subsequent deprotection by catalytic hydrogenation (Pd/C) to obtain the corresponding ethers.

The yield for the alkylation step varies depending on the length of the alkyl chain; as depicted in Scheme 8, the yields decrease as the length of the alkyl chain increases, due to the reduced solubility of the corresponding long chain alkyl iodides in the solvent (DMSO).

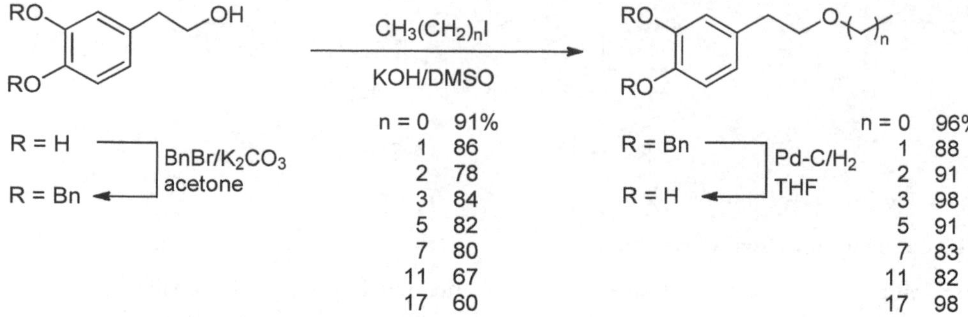

Scheme 8. Synthesis of hydroxytyrosyl alkyl ethers by alkylation with alkyl iodides

The oxidative stability of lipid matrix in the presence of these compounds, measured by the Rancimat method, has shown that these derivatives retain the high protective capacity of free hydroxytyrosol and similar induction times, having higher induction times than butylhydroxytoluene (BHT) and α-tocopherol (Madrona et al., 2009). These results are in agreement with those obtained in the case of hydroxytyrosyl esters, covered in the previous section (Mateos et al., 2008). The antioxidant activity has been checked by the DPPH, FRAP and ABTS assays in a hydrophilic medium (Pereira-Caro et al., 2009). The antioxidant activity of the lipophilic hydroxytyrosyl ethers was slightly lower in bulk oils and higher in hydrophilic media in comparison with their reference HT, supporting the polar paradox. The length of the alkyl chain did have a positive influence in hydrophilic medium for ethers with a short alkyl chain (methyl, ethyl, propyl), while ethers with longer alkyl chains (from butyl to octadecyl) maintained or decreased their antioxidant activity, probably due to the steric effect of the hydrocarbon chains.

3.2.2 Biological activity

In order to evaluate the safety and potential biological activity of these ethers, studies of their transport, absorption and metabolism in cellular and animal models have been developed (Pereira-Caro et al., 2010a, 2010b) using a human hepatoma cell line (HepG2) as a model system of the human liver and human enterocyte-like Caco-2/TC7 cells, which are commonly used to characterize the intestinal absorption of a range of drugs, nutrients, and other xenobiotics.

The results showed a direct relationship between the lipophilic nature of each compound and the level of metabolization; as an example, hydroxytyrosyl butyl ether biotransformation was complete after 18 h, whereas small amounts of the others remained after the same time. Furthermore, an intestinal absorption increase was observed from methyl to *n*-butyl ethers.

Protective effects against oxidative stress have also been studied (Pereira-Caro et al., 2011) using HepG2 cells, the ones previously employed to assess the metabolism of the synthesized HT ethers. The results obtained show the potential to prevent cell damage induced by *tert*-butyl hydroperoxide (*t*-BuOOH) and the ability to maintain unaltered cellular redox status, partially after 2 hours of pretreatment and almost completely after 20 hours. These results are in accordance with those obtained with hydroxytyrosol (Martín et al., 2010), but they also show the relevance of the role of the lipophilic character of the phenolic compounds on their antioxidant potential against cell damage: HT methyl and ethyl ethers are less effective than HT propyl and butyl ethers.

3.3 Hydroxytyrosol-derived isochromans

Isochroman fragment is a ubiquitous scaffold that can be found in natural products, drugs and agrochemicals (Larghi & Kaufman, 2006). Access to dihydroxyisochromans derived from HT can be achieved by using the oxa-Pictet-Spengler reaction, by reaction of arylethanols with aldehydes, ketones or masked-carbonyl derivatives (Guiso et al. 2001). The reaction is highly regioselective, as intramolecular cyclization takes place mainly in the less hindered position, as it can be deduced from its reaction mechanism, shown in Scheme 9. Two of the synthetized isochromans (Fig. 4) have been detected in olive oil (Bianco et al., 2001).

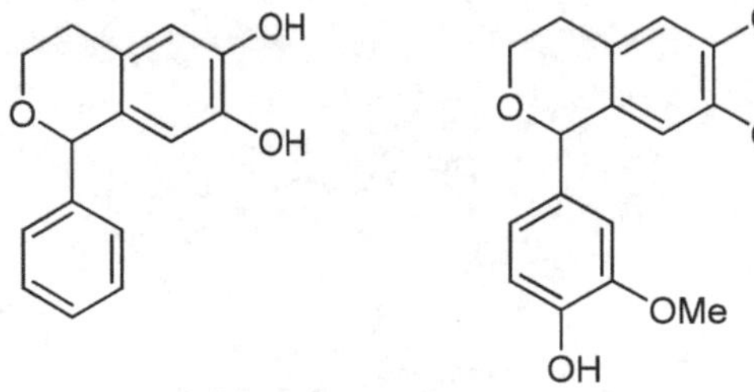

Fig. 4. Isochromans naturally present in olive oil

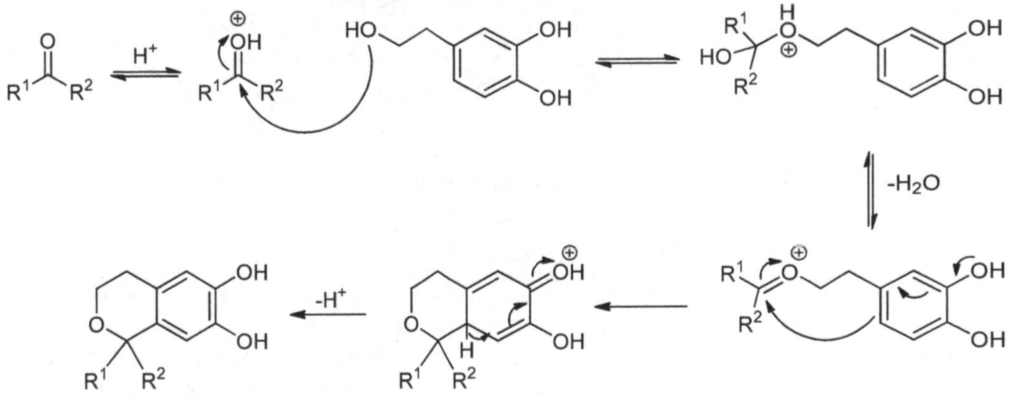

Scheme 9. Synthesis of hydroxytyrosol isochroman derivatives

Hydroxytyrosol isochroman derivatives shown in Fig. 4 were effective free radical scavengers able to inhibit platelet aggregation and thromboxane release (Togna, 2003).

3.4 Hydroxytyrosol glucoronide derivatives

One of the major metabolic pathways found *in vivo* for dietary phenolic compounds such as hydroxytyrosol is *O*-conjugation via glucuronidation and sulfation. Therefore, it is of interest to study these metabolites and their biological activities.

Biocatalyzed syntheses of hydroxytyrosol and other phenolic glucuronides have been developed using porcine liver microsomes (Khymenets et al., 2006, 2010). This type of glucuronides has also been synthesised stereoselectively (Lucas et al., 2009) in the phenolic or aliphatic hydroxyl groups using efficient chemical method from *O*-partially protected hydroxytyrosol and glucuronosyl trichloroacetimidate donors (Scheme 10).

The antioxidant activities of hydroxytyrosol conjugates have been evaluated, concluding that none of these glucuronides displayed significant antioxidant activities at the concentration tested (Khymenets et al., 2010).

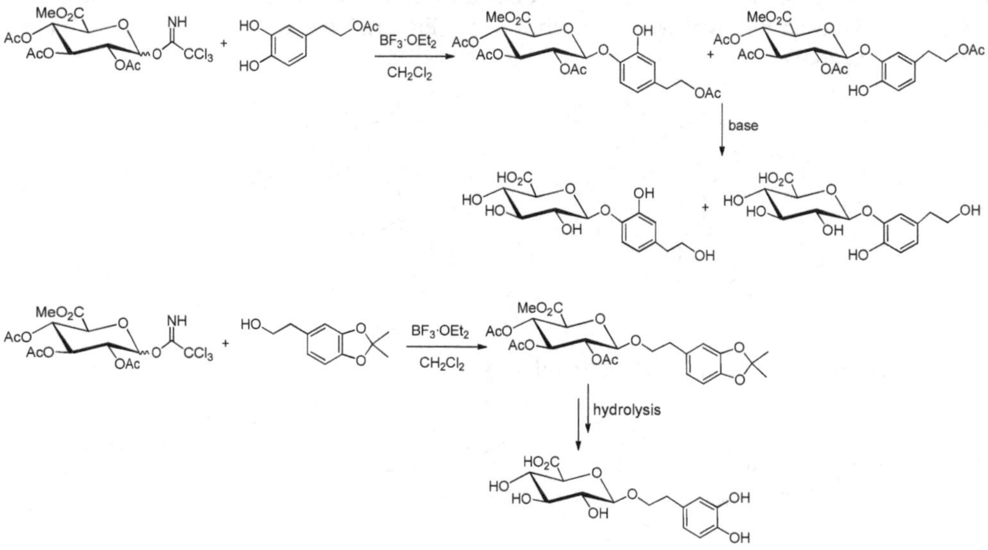

Scheme 10. Synthesis of hydroxytyrosol glucuronides

3.5 Hydroxytyrosol glucosides

The three isomers of hydroxytyrosol β-D-glucopyranosides (Fig.5) have been reported to be present in olives (Bianco et al., 1998). The 4-glucoside (Romero et al., 2004) and the 1-glucosides (Medina et al., 2007) have been found in table olive brines, and have been analysed as antimicrobial compounds against *Lactobacillus pentosus* with negative results. It has been recently shown that hydroxytyrosol 4-glucoside was the main phenolic compound in the aqueous phase of fresh alpeorujo, followed by hydroxytyrosol, and hydroxytyrosol 1-glucoside.

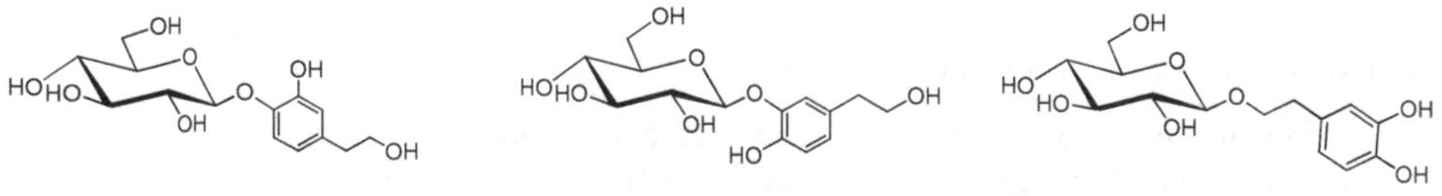

Fig. 5. The three isomers of hydroxytyrosol · −Δ−γλυχοπψρανοσιδεσ

3.6 Arylhydroxytyrosol derivatives

The synthesis of 2-arylhydroxytyrosols from 2-halohydroxytyrosol derivatives has been described (Bernini et al., 2008a). The reaction of the corresponding 2-chloro precursors via Suzuki-Miyaura cross-coupling reaction with arylboronic acids containing electron-donating, electron-withdrawing, as well as *ortho* substituents, yielded this family of compounds in high to excellent yields (Scheme 11).

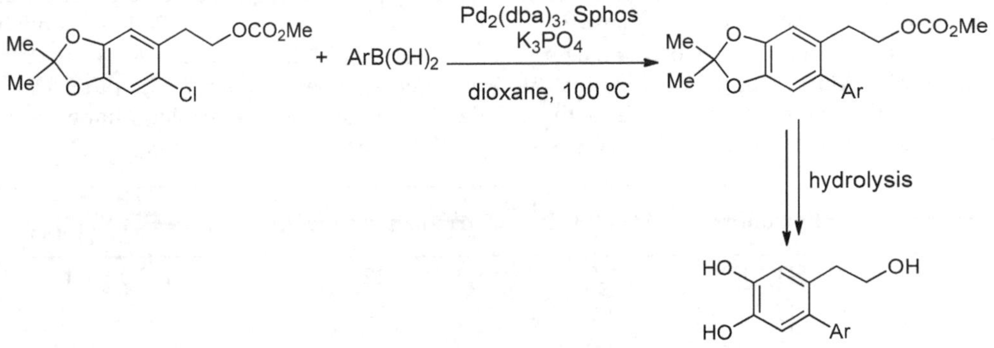

Scheme 11. Synthesis of 2-arylhydroxytyrosol derivatives

3.7 Complexation of hydroxytyrosol with β-cyclodextrins

The complexation of hydroxytyrosol with commercially-available β-cyclodextrin (β-CD) (López-García et al., 2010; Rescifina et al. 2010) and hydroxypropyl-β-cyclodextrin (HP-β-CD) (López-García et al., 2010) in aqueous solutions has been studied. The stoichiometries, the association constants and the geometry of the complexes have been determined by NMR techniques. The stoichiometries of both complexes are 1:1 and the association constants are 93 ± 7 M^{-1} for HT/β-CD complex and 43 ± 1 M^{-1} for HT/HP-β-CD complex (López-García et al., 2010). In both cases, the insertion of the catechol moiety took place by directing the hydroxyalkyl chain to the primary rim. The postulated geometry of the 1:1 HT/β-CD inclusion complex is depicted in Fig. 6.

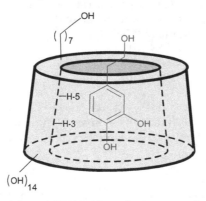

Fig. 6. Postulated geometry of the 1:1 HT/β-CD inclusion complex

Moreover, the antioxidant activity of encapsulated HT, together with the photoprotection effect of β-CD on HT, has been evaluated by scavenging of the stable DPPH radical. It has been proven that β-Cyclodextrin acts as a secondary antioxidant and provides a moderate improvement of the radical scavenging activity of HT measured by the DPPH assay.

β-Cyclodextrin exerts a strong photoprotection of HT upon UV irradiation, which could be deduced from the EC_{50} values (Table 2). For equimolecular mixtures of HT and β-CD at 1.2 mM, the observed degradation after 24 h and 48 h is similar to the degradation found for HT at the same concentration and time (entries 4 and 5) showing no protection at 24 h and only a slight protection after 48 h. However, using 1:4 mixtures of HT (1.2 mM) and β-CD (4.8 mM), a remarkable reduction of the degradation rate was observed when compared with pure HT. In this way, the complexation of HT with cyclodextrins might enhance stability, improve its performance as antioxidant and extend its storage life (López-García et al., 2010).

Entry	Antioxidant	[HT] (mM)	Irradiation time (h)	EC_{50} (g HT / kg DPPH)
1	HT	1.2	12	119.0 ± 1.4
2	HT	1.2	24	353.6 ± 23.4
3	HT	1.2	48	1436.1 ± 73.2
4	HT-βCD (1:1)	1.2	24	357.2 ± 30.7
5	HT-βCD (1:1)	1.2	48	1011.6 ± 171.9
6	HT-βCD (1:4)	1.2	12	112.4 ± 6.4
7	HT-βCD (1:4)	1.2	24	198.0 ± 4.6
8	HT-βCD (1:4)	1.2	48	387.2 ± 13.3

Table 2. Effect of the encapsulation of HT on its photostability

4. Conclusions

Hydroxytyrosol is a phenolic compound that can be isolated from olive oil mill wastewaters. The remarkable biological properties of this compound, mainly due to its strong antioxidant activity, has stimulated the synthesis of a series of derivatives, some of them are also naturally-occurring in the olive tree. Among these derivatives hydroxytyrosyl esters and ethers are of great interest, as some of them show strong antioxidant activity and improved bioavailability.

5. Acknowledgement

We thank the Junta de Andalucía (P08-AGR-03751 and FQM 134) and Dirección General de Investigación of Spain (CTQ2008-02813) for financial support. M.A.L.G. and A.M. thanks Ministerio de Educación and Junta de Andalucía, respectively, for their fellowships.

6. References

Alcudia F, Cert A, Espartero JL, Mateos R, & Trujillo M. (2004) Method of preparing hydroxytyrosol esters, esters thus obtained and use of same. PCT WO 2004/005237.

Appendino G, Minassi A, Daddario N, Bianchi F, & Tron GC. (2002) Esterification of phenolic acids and alcohols. Org Lett 4: 3839–3841.

Aruoma O, Deiane M, Jenner A, Halliwell B, Kaur M, Banni S, et al. (1998) Effect of hydroxytyrosol found in extra virgin olive oil on oxidative DNA damage and on low-density lipoprotein oxidation. J Agric Food Chem 46: 5181–5187.

Balsano C & Alisi A. (2009) Antioxidant effects of natural bioactive compounds. Curr Pharm Des 15: 3063-3073.

Bernini R, Mincione E, Barontini M, & Crisante F. (2007) Procedimento per la preparazione di derivati dell'idrossitirosolo e di idrossitirosolo via demetilazione ossidativa MI2007 A001110.

Bernini R, Cacchi S, Fabrizi G, & Filisti E. (2008a) 2-Arylhydroxytyrosol derivatives via Suzuki-Miyaura cross-coupling. Org Lett, 10: 3457–3460.

Bernini R, Mincione E, Barontini M, & Crisante F. (2008b) Convenient synthesis of hydroxytyrosol and its lipophilic derivatives from tyrosol or homovanillyl alcohol. J Agric Food Chem 56: 8897–8904.

Bernini R, Mincione E, Barontini M, & Crisante, F. (2008c) Method for preparing hydroxytyrosol and hydroxytyrosol derivatives. PCT/IB2008/000598.

Bianco A, Mazzei RA, Melchioni C, Romeo G, Scarpati ML, Soriero A, et al. (1998) Microcomponents of olive oil III. Glucosides of 2(3,4-dihydroxy-phenyl)ethanol. Food Chem 63, 461–464.

Bianco, A., Coccioli, F., Guiso, M. & Marra, C. (2001) The occurrence in olive oil of a new class of phenolic compounds: hydroxy-isochromans. Food Chem. 77: 405–411.

Biesalski HK. (2007) Polyphenols and inflammation: basic interactions. Curr Opin Clin Nutr Metab Care 10: 724–728.

Bisignano G, Tomaino A, Lo Cascio R, Crisafi G, Uccella N, & Saija A. (1999) On the in-vitro antimicrobial activity of oleuropein and hydroxytyrosol. J Pharm Pharmacol 51: 971–974.

Bouallagui Z, Han J, Isoda H, & Sayadi S. (2011a) Hydroxytyrosol rich extract from olive leaves modulates cell cycle progression in MCF-7 human breast cancer cells. Food Chem Toxicol 49: 179–184.

Bouallagui Z, Bouaziz M, Lassoued S, Engasser JM, Ghoul M, & Sayadi S. (2011b) Hydroxytyrosol acyl esters: Biosynthesis and Activities. Appl Biochem Biotechnol 163: 592–599.

Buisman GJH, van Helteren CTW, Kramer GFH, Veldsink JW, Derksen JTP, & Cuperus FP. (1998) Enzymatic esterifications of functionalized phenols for the synthesis of lipophilic antioxidants. Biotechnol Lett 20: 131–136.

Bulotta S, Corradino R, Celano M, D'Agostino M, Maiuolo J, Oliverio M, et al. (2011) Antiproliferative and antioxidant effects on breast cancer cells of oleuropein and its semisynthetic peracetylated derivatives. Food Chem 127: 1609–1614.

Burke TR Jr, Fesen MR, Mazumder A, Wang J, Carothers AM, Grunberger D, et al. (1995) Hydroxylated aromatic inhibitors of HIV-1 integrase. J Med Chem 38: 4171–4178.

Carluccio MA, Ancora MA, Massaro M, Carluccio M, Scoditti E, Distante A, et al. (2007) Homocysteine induces VCAM-1 gene expression through NF-kB and NAD(P)H oxidasa activation: protective role of Mediterranean diet polyphenolic antioxidants. Am J Physiol Heart Circ Physiol 293: 2344–2354.

D'Angelo S, Ingrosso D, Migliardi V, Sorrentino A, Donnarumma G, Baroni A, et al. (2005) Hydroxytyrosol, a natural antioxidant from olive oil, prevents protein damage induced by long-wave ultraviolet radiation in melanoma cells. Free Radic Biol Med 38: 908–919.

De Lucia M, Panzella L, Pezzella A, Napolitano A, & D'Ischia M. (2008) Plant catechols and their S-glutathionyl conjugates as antinitrosating agents: expedient synthesis and remarkable potency of 5-S-Glutathionylpiceatannol Chem Res Toxicol 21: 2407–2413.

Fabiani R, De Bartolomeo A, Rosignoli P, Servili M, Selvaggini R, Montedoro GF, et al. (2006) Virgin olive oil phenols inhibit proliferation of human promyelocytic leukemia cells (HL60) by inducing apoptosis and differenciation. J Nutr 136: 614–619.

Fabiani R, Rosignoli P, De Bartolomeo A, Fuccelli R, Servili M, Montedoro GF, et al. (2008) Oxidative DNA damage is prevented by extracts of olive oil, hydroxytyrosol, and other olive phenolic compounds in human blood mononuclear cells and HL60 cells. J Nutr 138: 1411–1416.

Fabiani R, Fuccelli R, Pieravanti F, De Bartolomeo A, & Morozzi G. (2009) Production of hydrogen peroxide is responsible for the induction of apoptosis by hydroxytyrosol on HL60 cells. Mol Nutr Food Res 53: 887–896.

Fabiani R, Rosignoli P, De Bartolomeo A, Fuccelli R, Servili M, & Morozzi G. (2011) The production of hydrogen peroxide is not a common mechanism by which olive oil phenol compounds induce apoptosis on HL60 cells. Food Chem 125: 1249–1255.

Fernández-Bolaños JG, López O, Fernández-Bolanos J, & Rodríguez- Gutiérrez G. (2008) Hydroxytyrosol and derivatives: Isolation, synthesis, and biological properties. Curr Org Chem 12: 442–463.

Frankel EN, Huang SW, Kanner J, & German JB. (1994) Interfacial phenomena in the evaluation of antioxidants: Bulk oils versus emulsions. J Agric Food Chem 42:1054–1059.

Furneri PM, Piperno A, Sajia A, & Bisignano G. (2004) Antimycoplasmal activity of hydroxytyrosol. Antimicrob Agents Chemother 48: 4892–4894.

Gambacorta A, Tofani D, & Migliorini A. (2007) High-yielding synthesis of methyl orthoformate-protected hydroxytyrosol and its use in preparation of hydroxytyrosyl acetate. Molecules 12: 1762–1770.

Gómez-Acebo E, Alcami Pertejo J, & Aunon Calles D. (2011) Topical use of hydroxytyrosol and derivatives for the prevention of HIV infection. Pat. Appl. Publ. WO 2011067302 A1.

Gong D, Geng C, Jiang L, Cao J, Yoshimura H, & Zhong L. (2009) Effects of hydroxytyrosol-20 on carrageenan-induced acute inflammation and hyperalgesia in rats. Phytoth Res 23: 646–650.

González-Correa JA, Muñoz-Marín J, Arrebola MM, Guerrero A, Carbona F, López-Villodres J, et al. (2007) Dietary virgin olive oil reduces oxidative stress and cellular damage in rat brain slices subjected to hypoxia-reoxygenation Lipids 42: 921–929.

González-Correa JA, Navas MD, Lopez-Villodres JA, Trujillo M, Espartero JL, & de la Cruz, JP. (2008a) Neuroprotective effect of hydroxytyrosol and hydroxytyrosol acetate in rat brain slices subjected to hypoxia-reoxygenation. Neurosci Lett 446: 143–146.

González-Correa JA, Navas MD, Muñoz-Marín J, Trujillo M, Fernádez-Bolaños J, & de la Cruz JP. (2008b) Effects of hydroxytyryosol and hydroxytyrosol acetate administration to rats on platelet function compared to acetylsalicylic acid. J Agric Food Chem 56: 7872–7876.

González Correa JA, López-Villodres JA, Asensi R, Espartero JL, Rodríguez-Gutiérrez G, & de la Cruz JP. (2009) Virgin olive oil polyphenol hydroxytyrosol acetate inhibits in vitro platelet aggregation in human whole blood: comparison with hydroxytyrosol and acetylsalicylic acid. Br J Nutr 101: 1157–1164.

Gonzalez-Santiago M, Fonolla J, & Lopez-Huertas E. (2010) Human absorption of a supplement containing purified hydroxytyrosol, a natural antioxidant from olive oil, and evidence for its transient association with low-density lipoproteins. Pharmacol Res 61: 364–370.

Gordon MH, Paiva-Martins F, & Almeida M. (2001) Antioxidant activity of hydroxytyrosol acetate compared with other olive oil polyphenols. J Agric Food Chem 49: 2480–2485.

Goya L, Mateos R, & Bravo L. (2007) Effect of the olive oil phenol hydroxytyrosol on human hepatoma HepG2 cells protection against oxidative stress induced by tert-butylhydroxyperoxide. Eur J Nutr 46: 70–78.

Goya L, Mateos R, Martin MA, Ramos S, & Bravo L. (2010) Uptake, metabolism and biological effect of the olive oil phenol hydroxytyrosol by human HepG2 cells, In: Olives and olive oil in health and disease prevention. Preedy VR and Watson RR, (Eds.), Oxford: Academic Press, Elsevier (USA), pp. 1157-1165, ISBN 978-0-12-374420-3.

Grasso S, Siracusa L, Spatafora C, Renis M, & Tringali C. (2007) Hydroxytyrosol lipophilic analogues: Enzymatic synthesis, radical scavenging activity and DNA oxidative damage protection. Bioorg Chem 35: 137–152.

Guichard C, Pedruzzi E, Fay M, Marie JC, Braut-Boucher F, Daniel F, et al. (2006) Dihydroxyphenylethanol induces apoptosis by activating serine/threonine protein phosphatase PP2A and promotes the endoplasmic reticulum stress response in human colon carcinoma cells. Carcinogenesis 27: 1812–1827.

Guiso, M., Marra, C. & Cavarischia, C. (2001) Isochromans from 2-(3,4-dihydroxy)phenylethanol. Tetrahedron Lett 42: 6531–6534.

Hagiwara K, Goto T, Araki M, Miyazaki H, & Hagiwara H. (2011) Olive polyphenol hydroxytyrosol prevents bone loss. Eur J Pharmacol 662: 78–84.

Hillestrom PR, Covas MI, & Poulsen HE. (2006) Effect of dietary virgin olive oil on urinary excretion of etheno-DNA adducts. Free Radic Biol Med 41: 1133–1138.

Ishihara K, Kubota M, Kurihara H, & Yamamoto H. (1995) Scandium trifluoromethanesulfonate as an extremely active acylation catalyst. J Am Chem Soc 117: 4413–4414.

Khymenets O, Joglar J, Clapés P, Parella T, Covas M-I, & de la Torre R. (2006) Biocatalyzed synthesis and structural characterization of monoglucuronides of hydroxytyrosol, tyrosol homovanillic alcohol and 3-(4'-hydroxyphenyl)propanol. Adv Synth Catal 34: 2155–2162.

Khymenets O, Clapés P, Parella T, Covas M-I, de la Torre R, & Joglar J. (2009) Biocatalyzed synthesis of monoglucuronides of hydroxytyrosol, tyrosol homovanillic alcohol and 3-(4'-hydroxyphenyl)propanol using liver cells microsomal frations. In: Practical methods for biocatalysis and biotransformations, Wittall J, Sutton P (eds.), John Wiley& Sons, Ltd, pp. 245-250.

Khymenets O, Fito M, Taurino S, Muñoz-Aguayo D, Pujadas M, Torres JL, et al. (2010) Antioxidant activities of hydroxytyrosol main metabolites do not contribute to beneficial health effects after olive oil ingestion. Drug Metab Dispos 38: 1417-1421.

Larghi EL & Kaufman TS. (2006) The oxa-Pictet-Spengler cyclization: Synthesis of isochromans and related pyran-type heterocycles. Synthesis 187-220.

Lee-Huang S, Lin Huang P, Zhang D, Wook Lee J, Bao J, Sun Y, et al. (2007a) Discovery of small-molecule HIV-1 fusion and integrase inhibitors oleuropein and hydroxytyrosol: Part I. Integrase inhibition. Biochem Biophys Res Commun 354: 872-878.

Lee-Huang S, Lin Huang P, Zhang D, Wook Lee J, Bao J, Sun Y, et al. (2007b) Discovery of small-molecule HIV-1 fusion and integrase inhibitors oleuropein and hydroxytyrosol: Part II. Integrase inhibition. Biochem Biophys Res Commun. 354: 879-884.

Lee-Huang S, Huang PL, Huang PL, Zhang D, Zhang JZH, Chang YT, et al. (2009) Compositions and methods for treating obesity, obesity related disorders and for inhibiting the infectivity of human immunodeficiency virus. US Pat Appl Publ US 20090061031 A1 20090305.

López-García MÁ, López Ó, Maya I, & Fernández-Bolaños JG. (2010) Complexation of hydroxytyrosol with β-cyclodextrins: an efficient photoprotection. Tetrahedron 66: 8006-8011.

Lucas R, Alcantara D, & Morales JC. (2009) A concise synthesis of glucuronide metabolites of urolithin-B, resveratrol and hydroxytyrosol. Carbohydr Res 344: 1340-1346.

Lucas R, Comelles F, Alcántara D, Maldonado OS, Curcuroze M, Parra JL, et al. (2010) Tyrosol and hydroxytyrosol fatty acid esters: a potential explanation for the nonlinear hypothesis of the antioxidant activity in oil-in-water emulsions. J Agric Food Chem 58: 8021-8026.

Madrona A, Pereira-Caro G, Mateos R, Rodríguez G, Trujillo M, Fernández-Bolaños J, et al. (2009) Synthesis of hydroxytyrosyl alkyl ethers from olive oil waste waters. Molecules, 14: 1762-1772.

Manna C, Migliardi V, Sannino F, De Artino A, & Capasso R. (2005) Protective effects of synthetic hydroxytyrosol acetyl derivatives against oxidative stress in human cells. J Agric Food Chem 53: 9602-9607.

Martín MA, Ramos S, Granado-Serrano AB, Rodríguez-Ramiro I, Trujillo M, Bravo L, & et al. (2010) Hydroxytyrosol induces antioxidant/detoxificant enzymes and Nrf2 translocation via extracellular regulated kinases and phosphatidylinositol-3-kinase/protein kinase B pathways in HepG2 cells. Mol Nutr Food Res 54: 1-11.

Mateos R, Goya L, & Bravo L. (2005) Metabolism of the olive oil phenols hydroxytyrosol, tyrosol, and hydroxytyrosyl acetate by human hepatoma hepG2 cells. J Agric Food Chem 53: 9897-9905.

Mateos R, Trujillo M, Pereira-Caro G, Madrona A, Cert A, & Espartero JL. (2008) New lipophilic tyrosyl esters. Comparative antioxidant evaluation with hydroxytyrosyl esters. J Agric Food Chem 56:10960-10966.

Medina E, Romero C, de los Santos B, de Castro A, García A, Romero F, & Brenes M. (2011) Antimicrobial activity of olive solutions from stored alpeorujo against plant pathogenic microorganisms. J Agric Food Chem 59: 6927-6932.

Medina E, Brenes M, Romero C, Garcia A, & de Castro A. (2007) Main antimicrobial compounds in table olives. J Agric Food Chem 55: 9817-9823.

Medina I, Lois, S, Alcántara D, Lucas R, & Morales JC. (2009) Effect of lipophilization of hydroxytyrosol on its antioxidant activity in fish oils and fish oil-in-water emulsions. J Agric Food Chem 57: 9773–9779.

Omar SH. (2010a) Cardioprotective and neuroprotective roles of oleuropein in olive. Saudi Pharm J 18: 111–121.

Omar SH. (2010b) Oleuropein in olive and its pharmacological effects. Sci Pharm 78: 133-154.

Pereira-Caro G, Madrona A, Bravo L, Espartero JL, Alcudia F, Cert A, & Mateos A. (2009) Antioxidant activity evaluation of alkyl hydroxytyrosyl ethers, a new class of hydroxytyrosol derivatives. Food Chem 115: 86–91.

Pereira-Caro G, Bravo L, Madrona A, Espartero JL, & Mateos R. (2010a) Uptake and metabolism of new synthetic lipophilic derivatives, hydroxytyrosyl ethers, by human hepatoma HepG2 cells. J Agric Food Chem 58: 798–806.

Pereira-Caro G, Mateos R, Saha, S, Madrona A, Espartero JL, Bravo L, et al. (2010b) Transepithelial transport and metabolism of new lipophilic ether derivatives of hydroxytyrosol by enterocyte-like Caco-2/TC7 cells. J Agric. Food Chem 58: 11501–11509.

Pereira-Caro G, Sarriá B, Madrona A, Espartero JL, Goya L, Bravo L, & et al. (2011) Alkyl hydroxytyrosyl ethers show protective effects against oxidative stress in HepG2 cells. J Agric Food Chem 59: 5964–5976.

Pérez-Jiménez F, Álvarez de Cienfuegos G, Badimon L, Barja G, Battino M, Blanco A, et al. (2005) International conference on the healthy effect of virgin olive oil. Eur J Clin Invest 35: 421–424.

Perrinjaquet-Moccetti T, Busjahn A, Schmidlin C, Schmidt A, Bradl B, & Aydogan C. (2008) Food supplementation with an olive (Olea europaea L.) leaf extract reduces blood pressure in borderline hypertensive monozygotic twins. Phytother Res 22: 1239–1242.

Quiles JL, Farquharson AJ, Simpson DK, Grant I, & Wahle KWJ. (2002) Olive oil phenolics: effects on DNA oxidation and redox enzyme mRNA in prostate cells. Br J Nutr. 88: 225–234.

Rescifina A, Chiacchio U, Iannazzo D, Piperno A, & Romeo G. (2010) β-Cyclodextrin and caffeine complexes with natural polyphenols from olive and olive oils: NMR, thermodynamic, and molecular modeling studies. J Agric Food Chem 58: 11876–11882.

Rietjens SJ, Bast A, & Haenen GRMM. (2007) New insights into controversies on the antioxidant potential of the olive oil antioxidant hydroxytyrosol. J Agric Food Chem 55: 7609–7614.

Romero C, Brenes M, García P, García A, Garrido A. (2004) Polyphenol changes during fermentation of naturally black olives. J Agric Food Chem 52:1973–1979.

Romero C, Medina E, Vargas J, Brenes M, & de Castro A. (2007) In vitro activity of olive oil polyphenols against Helicobacter pylori. J Agric Food Chem 55: 680–686.

Ruano J, López-Miranda J, de la Torre R, Delgado-Lista J, Fernández J, Caballero J, et al. (2007) Intake of phenol-rich virgin olive oil improves the postprandial prothrombotic profile in hypercholesterolemic patients. J Clin Nutr 86: 341–346.

Sabatini N. (2010) Recent patents in olive oil industry: new technologies for the recovery of phenols compounds from olive oil, olive oil industrial by-products and waste waters. Recent Pat Food Nutr Agric 2: 154–159.

Scheller EF. (1955) Treatment of hypertension with standardized olive leaf extract. Med Klin 50: 327–329.

Shahidi F & Zhong Y. (2011) Revisiting the polar paradox theory: A critical overview. J Agric Food Chem 59: 3499–3504.

Siriani R, Chimento A, De Luca A, Casaburi I, Rizza P, Onofrio A, et al. (2010) Oleuropein and hydroxytyrosol inhibit MCF-7 breast cancer cell proliferation interfering with ERK1/2 activation. Mol Nutr Food Res 54: 833–840.

Susalit E, Agus N, Effendi I, Tjandrawinata RR, Nofiarny D, Perrinjaquet-Moccetti T, et al. (2011) Olive (*Olea europaea*) leaf extract effective in patients with stage-1 hypertension: Comparison with captopril. Phytomedicine 18: 251–258.

Tofani, D, Balducci V, Gasperi T, Incerpi S, & Gambacorta, A. (2010) Fatty acid hydroxytyrosyl esters: structure/antioxidant activity relationship by ABTS and in cell-culture DCF assays. J Agric Food Chem 58: 5292–5299.

Togna GI, Togna AR, Franconi M, Marra C, & Guiso M. (2003) Olive oil isochromans inhibit human platelet reactivity. J Nutr 2532–2536.

Torregiani E, Seu G, Minassi A, & Appendino G. (2005) Cerium(III) chloride-promoted chemoselective esterification of phenolic alcohols. Tetrahedron Lett 46: 2193–2196.

Torres de Pinedo A, Peñalver P, Rondón P & Morales JC. (2005) Efficient lipase-catalyzed synthesis of new lipid antioxidants based on a catechol structure. Tetrahedron 61: 7654–7660.

Torres de Pinedo A, Peñalver P, & Morales JC. (2007) Synthesis and evaluation of new phenolic-based antioxidants: structure-activity relationship. Food Chem 103: 55-61.

Tripoli E, Giammanco M, Tabacchi G, Di Majo D, Giammanco S, & La Guardia M. (2005) The phenolic compounds of olive oil: structure, biological activity and beneficial effects on human health. Nutr Res Rev 18: 98–112.

Trujillo M, Mateos R, Collantes de Terán L, Espartero JL, Cert R, Jover M, et al. (2006) Lipophilic hydroxytyrosyl esters. Antioxidant activity in lipid matrices and biological systems. J Agric Food Chem 54: 3779–3785.

Turner R, Etienne N, Alonso MG, de Pascual-Teresa S, Minihane AM, Weinberg PD, et al. (2005) Antioxidant and anti-atherogenic activities of olive oil phenolics. Int J Vitam Nutr Res. 75: 61–70.

Vázquez-Velasco M, Esperanza Díaz L, Lucas R, Gómez-Martínez S, Bastida S, Marcos A, et al. (2011) Effects of hydroxytyrosol-enriched sunflower oil consumption on CVD risk factors. Br J Nutr 105: 1448–1452.

Visioli F, Bellomo GF, & Galli C. (1998) Free radical-scavenging properties of olive oil polyphenols. Biochem Biophys Res Commun 247: 60–64.

Visioli F, Poli A, & Galli C. (2002) Antioxidant and other biological activities of phenols from olives and olive oil. Med Res Rev 22: 65–75.

Yamada K, Ogawa H, Hara A, Yoshida Y, Yonezawa Y, Karibe K, et al. (2009) Mechanism of the antiviral effect of hydroxytyrosol on influenza virus appears to involve morphological change of the virus. Antiviral Res 83: 35–44.

9

Differential Effect of Fatty Acids in Nervous Control of Energy Balance

Christophe Magnan, Hervé Le Stunff and Stéphanie Migrenne
Université Paris Diderot, Sorbonne Paris Cité, Biologie Fonctionnelle et Adaptative,
Equipe d'accueil conventionnée Centre National de la Recherche Scientifique, Paris,
France

1. Introduction

Energy homeostasis is kept through a complex interplay of nutritional, neuronal and hormonal inputs that are integrated at the level of the central nervous system (CNS). A disruption of this regulation gives rise to life-threatening conditions that include obesity and type-2 diabetes, pathologies that are strongly linked epidemiologically and experimentally. The hypothalamus is a key integrator of nutrient-induced signals of hunger and satiety, crucial for processing information regarding energy stores and food availability. Much effort has been focused on the identification of hypothalamic pathways that control food intake but, until now, little attention has been given to a potential role for the hypothalamus in direct control of glucose homeostasis and nergy balance. Recent studies have cast a new light on the role of the CNS in regulating peripheral glucose via a hypothalamic fatty acid (FA)-sensing device that detects nutrient availability and relays, through the autonomic nervous system, a negative feedback signal on food intake, insulin sensitivity and insulin secretion. Indeed, accumulating evidences suggest that FA are used in specific areas of CNS not as nutrients, but as cellular messengers which inform "FA sensitive neurons" about the energy status of the whole body (Blouet & Schwartz, 2010; Migrenne et al., 2006; Migrenne et al., 2011). Thus it has been described that up to 70% of hypothalamic arcuate nucleus (ARC) and ventromedian nucleus (VMN) neurons are either excited or inhibited by long chain fatty acids such as oleic acid (Jo et al., 2009; Le Foll et al., 2009; Migrenne et al., 2011). Within the VMN, 90% of the glucosensing neurons also have their activity altered by FA. In a large percentage of these neurons, glucose and FA have opposing effects on neuronal activity, much as they do on intracellular metabolism in many other cells (Randle et al., 1994). Neuronal FA sensing mechanisms include activation of the K_{ATP} channel by long chain fatty acid acyl CoA (Gribble et al., 1998) or inactivation by generation of ATP or reactive oxygen species during mitochondrial β-oxidation (Jo et al., 2009; Le Foll et al., 2009; Migrenne et al., 2011; Wang et al., 2006). Many fatty acid sensing neurons are activated by interaction of long chain fatty acids with the fatty acid transporter/receptor, FAT/CD36, presumably by activation of store-operated calcium channels by a mechanism that is independent of fatty acid metabolism (Jo et al., 2009). Importantly, most neurons utilize FA primarily for membrane production rather than as a metabolic substrate (Rapoport et al., 2001; Smith & Nagura, 2001) and only nanomolar concentrations of fatty acid are required to alter the activity of fatty acid sensing neurons in the absence of astrocytes (Jo et al., 2009). While cerebral lipids are both produced in the brain and transported into it from the periphery (Rapoport et al., 2001; Smith & Nagura, 2001), the mechanism of this transport and the actual levels of various FA in the extracellular space in the brain remains largely unknown. As mentioned above, hypothalamic FA sensing may be involved in the control of feeding behaviour, hepatic glucose production and insulin secretion. It seems also that intracellular FA metabolism is important to relay their effects (β oxidation has been showed to be involved in oleate effect in hypothalamus) (Cruciani-Guglielmacci et al., 2004; Obici et al., 2003). In addition differential effect of FA in regard to feeding behaviour or glucose production may be related to their chain length and degree of saturation. For exemple, it has been showed in rodents that oleate both inhibits food intake and hepatic glucose production whereas octanoate has no effect on these parameters (Obici et al., 2002). In another study we showed that intracerebroventricular infusion of palmitate induced an hepatic insulin resistance and an impaired insulin signaling in hypothalamus (Benoit et al., 2009). In contrast oleate has no deleterious effect in this parameter (Benoit et al., 2009). Poly-unsaturated fatty acids (PUFA) such as n-3 or n-6 may have also different effects in neuronal activity and cognitive function such as memorization. The present work was aimed at studying differential effect of FA or triglycerides emulsion infused in rats in glucose homeostasis. In addition, in order to identify molecular mechanisms involved in specific effects of FA, mRNA expression of key genes involved in FA metabolism as well as ceramides and diacylglycerol (DAG) content have been measured in hypothalamus. Regarding physiopathology aspects it must be pointed out that dysfunction of central FA sensing could be a contributing factor to the early development of type 2 diabetes mellitus and/or obesity which leads to further dysfunction in predisposed subjects. A better understanding of these mechanisms, as well as further characterization of FA sensitive neurons and their role in physiological and pathological processes, might lead to identification of novel pharmacological targets for the prevention and treatment of diabetes and obesity.

2. FA sensing in hypothalamus

There is now growing amount of evidence suggesting, at least in rodents models, that some neurons located in hypothalamus (and brainstem) are sensitive to FA, ie their electrical activity is either increased or decreased in presence of variations of FA concentration. This has been evidenced both in vivo and in vitro. A key point is the transport of FA across the blood brain barrier (BBB). It cannot be excluded that FA may be produce directly in neurons from hydrolysis of intracellular troglycerides (TG).

2.1 Transport of FA uptake into the brain and neurons

Cerebral lipids are an essential component of both membranes and intracellular signalling pathways. They represent 50% of brain dry weight; the highest organ lipid content after adipose tissue (Edmond, 2001; Watkins et al., 2001). However, the mechanism by which FA are transported into the brain remains poorly understood. A growing body of evidence suggests that cerebral lipids are derived both from local synthesis and uptake from the blood (Rapoport et al., 2001). Several studies show that some poly-unsaturated FA (PUFA) have the ability to cross the BBB (Rapoport et al., 2001; Smith & Nagura, 2001). The question of whether brain FA uptake occurs by passive diffusion or involves a protein which facilitates the transport is still matter of debate. However, once across the BBB, it is likely that neurons can take up FA since some neurons do appear to have FA transporters. For example, dissociated neurons from the VMN of rats express mRNA's for FA transport proteins (FATP)-1 and 4 and the FA transporter/receptor FAT/CD36 (Le Foll et al., 2009). Also, while it is unlikely that neurons derive much of their energy supply from FA, these same neurons do express mRNA's for the intracellular metabolism of FA such as long chain acyl-CoA synthetase, carnitine palmitoyltransferase-1a and 1c and uncoupling protein-2 (Le Foll et al., 2009). They also express enzymes for de novo FA synthesis such as FA synthetase (Le Foll et al., 2009). But, it seems likely that much of the reported oxidation of FA such as palmitate in the brain probably occurs in astrocytes (Escartin et al., 2007) whereas other FA such as arachidonate are largely incorporated into phospholipids (Rapoport et al., 2001).

2.2 Some hypothalamic neurons are lipid responsive

The presence of neurons sensitive to variations in extracellular glucose levels is clearly demonstrated in the brain (Gilbert et al., 2003) and in particular in the hypothalamus (review in (Luquet & Magnan, 2009; Migrenne et al., 2011; Penicaud et al., 2002). Thirty-five years ago Oomura and colleagues first showed that FA activated lateral hypothalamic neurons which suggested a role for FA as neuronal signaling molecules (Oomura et al., 1975). As shown in Figure 1, FA also modify neuronal firing rate in hypothalamic arcuate nucleus (ARC) (Wang et al., 2006). Both FA "excited" (around 20% of arcuate neurons) and "inhibited" neurons (about 12%) are detected in arcuate nucleus of rat using this patch clamp technique (Wang et al., 2006). These FA sensitive neurons are also detected *in vivo* using multi-unit recording approaches (Wang et al., 2006). Therefore we demonstrated that single injection of oleic acid (OA) through carotid artery induced either increased or decreased neuronal activity depending on location of microelectrode in hypothalamus. It seems that some areas are mainly composed with FA "excited" neurons whereas others are mainly composed with FA "inhibited" neurons. Such data also suggest that physiological variations of plasma FA concentrations (reflecting the metabolic state and energy availability) can be detected and integrated by FA sensing neurons in critical brain areas involved in the regulation of feeding behaviour, glucose and lipid metabolism (Clement et al., 2002; Obici et al., 2002). Indeed increased plasma FA concentration during fasted or starvation may be detected by FA excited neurons which in turn may have an impact on nervous control of energy balance. On the contrary decreased plasma FA concentration during a meal could be also detected by these sensitive neurons which may act a satiety signal like insulin and glucose do during a meal when acting on hypothalamus sensitive neurons (Gilbert et al., 2003).

The physiological relevance of brain FA sensing is supported by various studies showing that local increases in brain and hypothalamic FA levels are associated with changes in insulin secretion and hepatic glucose output with variable effects on food intake (Clement et al., 2002; Obici et al., 2002; Ross et al., 2010; Schwinkendorf et al., 2010). For example, a 6 hour intracerebroventricular (icv) infusion of the monounsaturated FA, oleic acid (OA), reduced food intake as well as hepatic glucose production (HGP) (Obici et al., 2002). Reducing hypothalamic FA oxidation by inhibition of carnitine palmitoyl transferase -1 (CPT1), the enzyme that promotes β-oxidation by facilitating transport of medium- and long-chain FA into mitochondria, mimicked these effects on food intake and HGP induced by icv infusion of OA (Obici et al., 2003). In another study a direct bilateral infusion of OA into the mediobasal hypothalamus decreased hepatic glucose production (Ross et al., 2010). In addition, it seems that the hypothalamus differentially senses FA. For example, icv infusions of OA or docosahexanoic acid, but not palmitic acid, reduce food intake and body weight (Schwinkendorf et al., 2010). However, icv and direct infusions of FA into the brain are not physiological. Thus, they might produce non-specific effects by evoking an inflammatory response by irritating ependymocytes and tanycytes lining the ventricles or by exciting microglia and astrocytes in the brain parenchyma.

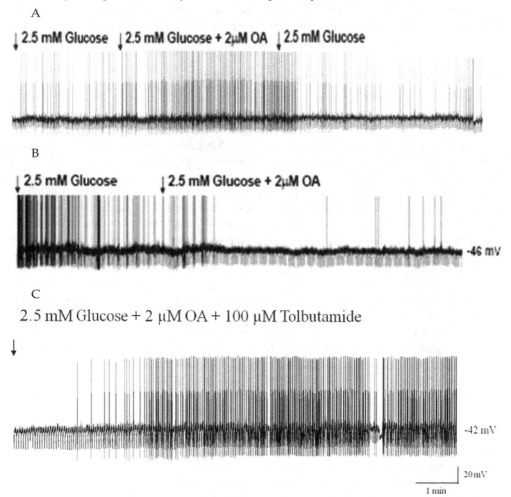

Fig. 1. Fragments of whole cell current clamp recordings of oleic acid (OA) excited (A) and inhibited (B) neuron in arcuate nucleus of rat (Adapted from Wang et al, 2006 et Migrenne et al, 2006).The inhibitory effect of OA on neuronal activity is inhibited by tolbutamide (C), suggesting involvement of K_{ATP} channels in OA effect.

More physiological routes include elevating systemic levels of FA or infusing them directly into the carotid arteries, the major route by which FA reach the forebrain. For example, a two-fold increase in plasma triglycerides produced by a two day systemic infusion of triglycerides was associated with decreased sympathetic activity. This reduced sympathetic tone, which is also produced by central FA infusions (Magnan et al., 1999), might contribute to the associated FA-induced exaggeration of glucose-induced insulin secretion (GIIS), a condition which is similar to what occurs in the prediabetic state (Magnan et al., 1999). Also, this exaggerated GIIS and a reduction in HGP were mimicked by infusing triglycerides into the carotid artery (Cruciani-Guglielmacci et al., 2004). These exaggerated responses were reduced by central inhibition CPT1 (Magnan et al., 1999). Similarly, central CPT1 inhibition was associated with an increase in the acyl CoA intracellular pool which was postulated to be the "final" satiety signal rather than FA themselves (review in (Lam et al., 2005; Luquet & Magnan, 2009).

However, there are at least two potential problems involved in the interpretation of such in vivo data. First, the idea that increases in brain FA levels act as a satiety signal to inhibit feeding (Obici et al., 2003) is counterintuitive given the fact that plasma FA levels do not rise substantially after food ingestion, but do rise significantly during fasting (Ruge et al., 2009). Second, the vast majority of FA oxidation in the brain occurs in astrocytes rather than neurons (Escartin et al., 2007). While a select group of neurons in the hypothalamus clearly responds directly to changes in ambient FA levels by altering their activity (Le Foll et al., 2009; Oomura et al., 1975), only a relatively small percentage of these responses depend upon neuronal FA metabolism (Le Foll et al., 2009). Furthermore, although β-oxidation and formation of malonyl-CoA and FA metabolites such as acyl-CoA may be mediators of the in vivo effects produced by FA infusions (Dowell et al., 2005; Migrenne et al., 2011) it is likely that most of these occur at the level of the astrocyte. If so, then there must be a mechanism by which alterations in astrocyte FA metabolism can provide a signal to those neurons which regulate HGP and food intake. We suggest that this communication between astrocyte FA metabolism and neuronal FA sensing involves the production and export of ketone bodies from astrocytes (Escartin et al., 2007) and subsequent uptake by neurons. Finally another important issue is the nature of the FA and its effect on sensitive neurons. As previously mentioned OA and octanoate have differential effect regarding food intake or hepatic glucose production, suggesting that medium or long chain fatty acids may have different effects (Obici et al., 2002). Thus, the aim of the present study was to test whether different FA may have different effect on glucose homeostasis when infused in rats brains through carotid artery. Triglyceride emulsion either enriched in ω3, ω6 FA or saturated FA (lard oil) have been also tested.

3. Methods

All animal care and experimental procedures were approved by the animal ethics committee of the university Paris-Diderot. Four weeks-old male wistar rats were purchased from Charles Rivers (Lyon, France) and housed at 21°C with normal light/dark cycle and free access to water and food.

3.1 First serie of experiments

Rats received an intracerebroventricular (icv) infusion of FA during 3 days. Briefly, rats anesthetized with isoflurane were stereotactically implanted with a chronic stainless steel cannula in the right lateral cerebral ventricle. The cannula was connected via a polyethylene catheter to a subcutaneously osmotic minipump filled up with FA (oleate, octanoate or linolenate) or saline. Infusions started 6h after surgery. The rate of infusion was 0.5 µl/h. Blood was daily removed (~ 80 µl) from caudal vessels for measurement of plasma substrate (FA and glucose) and insulin. Food intake was daily measured. At day 3 of infusion glucose-induced insulin secretion (GIIS) was measured in response to a single intraperitoneal injection of glucose (0.5g/kg bw). was made in overnight fasted rats. The glycemia was determined by a glucometer (AccuChek, Rabalot, France) from 2 µl collected from the tip of the tail vein at time 0, 5, 10, 15, 20, 30 and 60 min. In addition 20 µl of blood was sampled at the same time for insulin measurement (RIA, Diasorin, France). In another serie of experiments, etomoxir (CPT1 inhibitor) was concomitantly infused with FA. At the end of experiment brain were removed and five hypothalamus nuclei (arcuate, lateral, ventromedian, paraventricular and dorsomedian) were micropunched in order to measure gene expression (acetylCoA carboxylase, ACC, carnitine palmitoyl transferase, CPT1, FA synthase, FAS, G protein related peptide GPR41). Briefly, total RNA was isolated from the hypothalamus using RNeasy Lipid kit (Qiagen). To remove residual DNA contamination, the RNA samples were treated with DNAse RNAse-free (Qiagen). 4 µg of total RNA from each sample was reverse transcribed with 40 U of M-MLV Reverse Transcriptase (Invitrogen, life technologies) using random hexamer primers.

3.2 Second serie of experiments

In a second serie of experiments our goal was to test the effect of different triglyceride emulsion on glucose tolerance and both diacylglycerol and ceramides content in hypothalamus. To that end rats received an intracarotid infusion during 24 h of lard oil, mainly composed of saturated FA (SFA), ω3-enriched (Omegaven, Santec, France) or ω6-enriched (Ivelip, Rabalot, France) polyunsaturated FA (PUFA) triglyceride emulsion. The long-term unrestrained infusion technique was used, as previously described (Gilbert et al., 2003). Briefly, 5 days before the beginning of the infusion, rats were anaesthetized with isoflurane for catheterization of right carotid artery, towards the brain. Catheter was then exteriorized at the vertex of the head, and animals were allowed to recover for 5 days. For infusion, catheter is connected to a swiveling infusion device, allowing the animal free access to water and food and infused with a triglyceride emulsion. Food intake was measured after the 24h infusion period. In another set of experiments, oral glucose tolerance test (3g/kg, OGTT) was also performed.

3.3 Extraction and analysis of ceramids and DAG content in the hypothalamus

Diacylglycerol and ceramide levels in tissues extracts were measured by the diacylglycerol kinase enzymatic method as previously described (Escalante-Alcalde et al., 2003; Le Stunff et al., 2002). Briefly, aliquots of the chloroform phases from cellular lipid extracts were re-suspended in 7.5% (w/v) octyl-β-D-glucopyranoside/5 mM cardiolipin in 1 mM DETPAC/10 mM imidazole (pH 6.6). The enzymatic reaction was started by the addition of 20 mM DTT, 0.88 U/ml E. coli diacylglycerol kinase, 5 µCi/10 mM [γ-^{32}P]ATP and the reaction buffer (100 mM imidazole (pH 6.6), 100 mM NaCl, 25 mM MgCl2, and 2 mM EGTA). After incubation for 1 h at room temperature, lipids were extracted with chloroform/methanol/HCl (100:100:1, v/v) and 1 M KCl. [γ-32P]-phosphatidic acid was resolved by TLC with chloroform/acetone/methanol/acetic acid/water (10:4:3:2:1, v/v) and quantified with a Molecular Dynamics Storm PhosphorImager. Known amounts of diacylgycerol and ceramide standards were included with each assay. Ceramide and diacylglycerol levels were expressed as pmol by nmol of phospholipid (PL) levels. Total phospholipids present in cellular lipid extracts used for ceramide analysis were quantified as described previously (Escalante-Alcalde et al., 2003; Le Stunff et al., 2002) with minor modifications. Briefly, a mixture of 10N H_2SO_4/70% perchloric acid (3:1, v/v) was added to lipid extracts which were incubated for 30 min at 210°C. After cooling, water and 4.2% ammonium molybdate in 4 N HCl/0.045% malachite green (1:3 v/v) was added. Samples were incubated at 37°C for 30 min, and absorbance was measured at 660 nm.

4. Results

In the first serie of experiment, whatever the FA ie oleate or linolenate), there was no change in food intake during experiment (data not shown). Basal plasma glucose, FA, and insulin concentrations were also similar in all groups. As displayed in figure 2, in response to glucose load in linoleate group time course of glycemia was similar to control but was associated with an increased glucose induced insulin secretion (GIIS), suggesting an insulin resistance state which was compensated by this increased GIIS. In oleate infused group there was no change in plasma glucose or GIIS compared to controls.

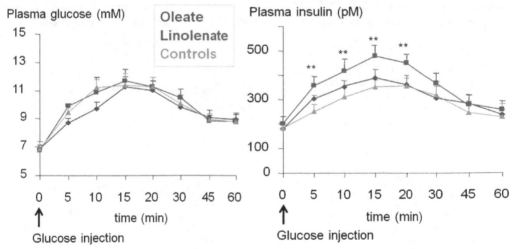

Fig. 2. Time course of plasma glucose and insulin concentration in response to glucose injection in oleate, linolenate, and NaCl (controls) icv 24h infused rats. **, p < 0.01 vs oleate and controls.

In order to test whether β oxidation was required to relay FA effect, GIIS was measured in presence or not of etomoxir a specifix inhibitor of CPT1 activity, a rate-limiting enzyme of β oxidation (figure 3). Results are expressed as insulinogenic index (ie ratio of areas under the curve of insulin to glucose during GIIS). Effect of linolenate on GIIS was reversed by etomoxir.

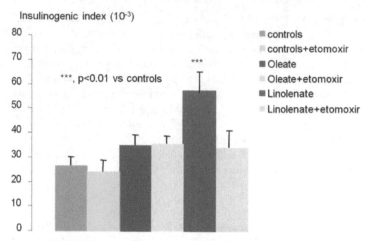

Fig. 3. Insulinogenic index in control rats, oleate and linolenate infused rats with or without etomoxir, a specific inhibitor of CPT1.*** p<0.01 vs controls

Hypothalamic gene expression have been measured in five areas (figure 4) involved in nervous control of energy balance: arcuate nucleus (ARC), ventromedian (VMH), lateral (LH), dorsomedian hypothalamus (DMH) and paraventricular nucleus (PVN). Studied genes were CPT1 (carnitine palmitoyl transferase 1), FAS (fatty acid synthase), GPR41 (G protein related receptor 41, GPR40 and 43 have been also tested but were not detected in our models), ACCβ (Acetyl carboxylase β) and AMPKα2. Results are displayed in figure 5.

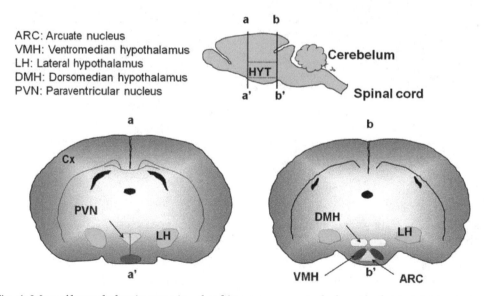

Fig. 4. Map of hypothalamic areas involved in nervous control of energy homeostasis. Arcuate nucleus (ARC), ventromedian (VMH), lateral (LH), dorsomedian hypothalamus (DMH) and paraventricular nucleus (PVN).

Hypothalamic gene expression were also modified in some areas depending on FA (figure 5). For example, FAS expression was inhibited in all nuclei except in VMH of linolenate infused rats. GPR41 was up-regulated in ARC of linolenate infused rats.

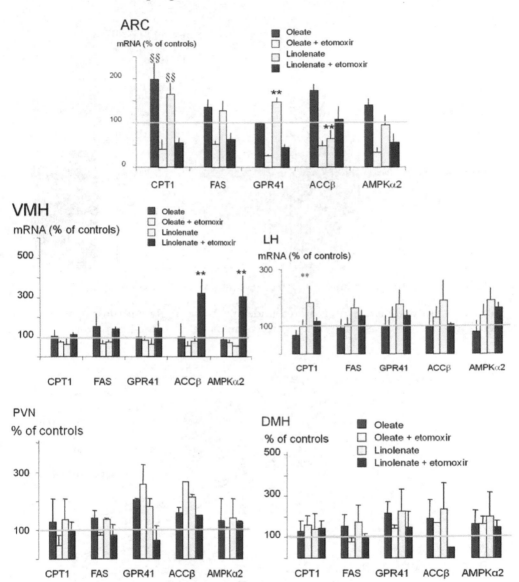

Fig. 5. mRNA expressionof target gene in different hypothalamic areas in oleate or linolenate +/- etomoxir. The green line represents gene expression in control rats (infused with NaCl). **, p <0.01 vs controls. §§, p <0.01 vs controls. ARC: arcuate nucleus; VMH: ventromedian hypothalamus; LH: lateral hypothalamus; DMH: dorsomedian hypothalamus; PVN: paraventricular nucleus.

In the second serie of experiments, we first measured food intake (figure 6). As depicted, there was a decreased in food intake with omegaven and ivelip infusion but not with lard oil.

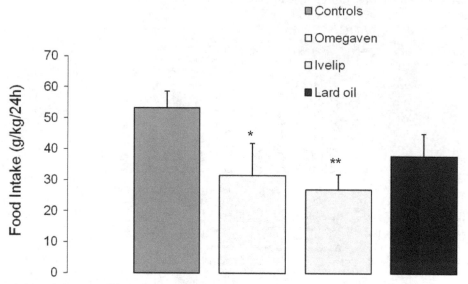

Fig. 6. Measurement of food intake. *, p<0.05 vs controls, **p<0.01 vs controls.

Figure 7 depicted time course and area under the curse of glycemia in response to oral glucose tolerance test. Lard oil induced glucose intolerance compared to controls.

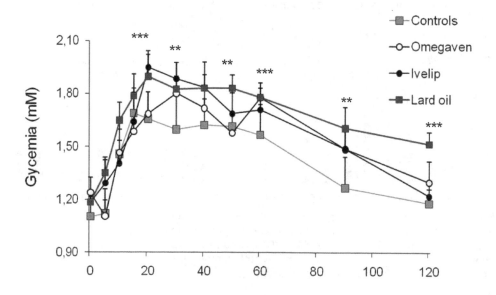

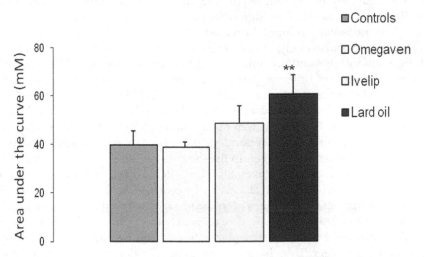

Fig. 7. Time course of glycemia (left) and area under the curve of glycemia during oral glucose tolerance test (right). **p<0.01, *** p<0.001 vs controls.

Interestingly, the effect of lard oil was associated with an accumulation of ceramide in the hypothalamus (figure 8). Therefore, our data suggest that ceramide accumulation in the hypothalamus following icv infusion of saturated fatty acid could contribute to the installation of an insulin resistant state by altering nervous output and consequently nervous control of insulin secretion and action.

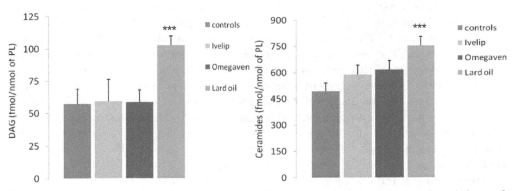

Fig. 8. Hypothalamic content of diacylglycerol (DAG) and ceramides in rats infused through carotid artery with NaCl (controls), Ivelip, Omegaven or Lard oil during 24h. ***p<0.001 vs controls.

4. Discussion

It is now clearly evidenced that hypothalamic FA sensing is an important regulator of nervous control of energy balance. In the present study we highlighted the differential effects of FA regarding their chain length and degree of saturation. We firstly evidenced here that oleate and linolenate have differential effects in regard to glucose homeostasis and GIIS. Especially linolenate induced increased GIIS compared to both oleate and control group whereas time course of glycemia remained similar. Thus there is a difference between effect of monounsaturated and polyunsaturated fatty acids when infused toward the brain. This suggest activation of different pathways. It must be pointed out that we previously demonstrated that short term infusion of oleate (6h and 24h) induced an increased in insulin secretion induced by glucose compared to control rats (Migrenne et al., 2006; Wang et al., 2006). In the present study infusion was made during 3 days which can explain different effect in short vs long term infusion peridod. Indeed we cannot exclude an adaptation to oleate effect when infusion stay longer. In the same way of ideas, inhibitory effect of central infusion of oleate on food intake was also lost after 3 days of infusion as previously evidenced by obici et al (Obici et al., 2002). In contrast, in the present study effect of linolenate was still present after 3 days of infusion. Linolenate effect may induce an insulin-resistance state and increased GIIS could be an adaptation to this insulin resistance state. By acting on FA sensitive neurons, linolenate may affect nervous output from CNS, especially autonomic nervous system. This change in autonomic nervous system balance will in turn modify nervous control of insulin secretion and action. We previously demonstrated that lipid infusion induced changes in both sympathetic and parasympathetic nervous activity in both rodents (Magnan et al., 1999) and humans (Magnan et al., 2001). In both studies decreased sympathetic nervous activity induced an increased insulin secretion in response to glucose and insulin resistance. In addition, in the present study we showed that linolenate effect involved its metabolism since it had no more effect in presence of etomoxir an inhibitor of β oxidation. The involvement of β oxidation to relay FA effect on sensitive neurons have been also described in different models by us (Cruciani-Guglielmacci et al., 2004) and others (Obici et al., 2003). Finally that specific effects of linolenate compared to oleate could be, at least in part, related to differential gene transcription involved in FA metabolism such as CPT1, FAS or ACC in key areas of hypothalamus. More precisely in ARC and LH CPT1 expression was increased in linolenate infused rats and etomoxir induced a decreased in this gene and its return to basal value. However, it is difficult to further analyze these results since in other nuclei, there is no change of CPT1 expression. In addition, in some area others genes are differently expressed such as AMPKα2 or ACCβ, both key enzymes of glucose and FA metabolism. Altogether these data suggest that oleate or linolenate may act on different subpopulations of neurons (or astrocytes) thus highlighting the fact that FAs may have different effect in regard of the area in which they act. It is also interesting that expression of gene such as GPR41 can be also modified by linolenate infusion. Indeed it has been recently evidenced that short-chain fatty acids and ketones directly regulate sympathetic nervous system via GPR41 at the level of sympathetic ganglion (Kimura et al., 2011). Thus changes in hypothalamic GRP41 gene expression may have an impact during starvation, a situation in which ketone bodies production is increased. thereby control body energy expenditure in maintaining metabolic homeostasis.

In the second part of our work we demonstrated a differential role of PUFA vs saturated FA (SFA) regarding induction of insulin resistance and ceramides production in hypothalamus by using triglyceride emulsion infusion, in order to mimic a more "physiological approach".

Indeed 24h of lard oil infusion in carotid which had no effect on plasma TG or FA concentrations (data not shown) induced a glucose intolerance suggesting a deregulation of insulin sensitivity and or secretion. This deleterious effect of lard oil in nervous control of glucose homeostasis was associated with an increased in DAG and ceramides content in hypothalamus. An important role for ceramides has emerged from research on the pathogenesis of metabolic diseases associated with obesity, such as diabetes (Holland & Summers, 2008). Indeed, ceramides appear to be particularly deleterious components of the lipid milieu that accrues in obesity, and levels of ceramides are often elevated in skeletal muscle, liver, and/or serum of obese humans and rodents (Adams et al., 2004; Clement et al., 2002). DAG and ceramides are known to activate kinase such as PKC, which phosphorylate insulin receptor substrate and Akt leading to an inhibition of the insulin signaling (Mullen et al., 2009; Newton et al., 2009). A recent study also evidenced that sphingolipids such as ceramide might be key components of the signaling networks that link lipid-induced inflammatory pathways to the antagonism of insulin action that contributes to diabetes (Holland et al., 2011). We also recently demonstrated that the atypical protein kinase C, PKCϴ, is expressed in discrete neuronal populations of the ARC and the dorsal medial hypothalamic nucleus (Benoit et al., 2009). CNS exposure to saturated palmitic acid via direct infusion or by oral gavage increased the localization of PKCϴ to hypothalamic cell membranes in association impaired hypothalamic insulin and leptin signaling (Benoit et al., 2009). This finding was specific for palmitic acid, as the monounsaturated FA, OA, neither increased membrane localization of PKCϴ nor reduced insulin signaling. Finally, ARC-specific knockdown of PKCϴ attenuated diet-induced obesity and improved hypothalamic insulin signaling (Benoit et al., 2009). These results suggest that many of the deleterious effects of high fat diets, specifically those enriched with palmitic acid, are CNS mediated via PKCϴ activation, resulting in reduced insulin activity. Therefore, our data suggest that ceramide accumulation in the hypothalamus following icv infusion of saturated fatty acid could contribute to the installation of an insulin resistant state by altering nervous output and consequently nervous control of insulin secretion and action.

Further studies are needed to clearly identify molecular mechanism relaying ceramides production. However there is now several experiments highlighting some of these mechanisms in FA sensitive neurons as described below.

4.1 Molecular mechanisms involved in neuronal FA sensing

In FA sensitive neurons, exposure to long chain FA can alter the activity of a wide variety of ion channels including Cl-, $GABA_A$ (Tewari et al., 2000), potassium, K^+-Ca^{2+} (Honen et al., 2003) or calcium channels (Oishi et al., 1990). Additionally, FA inhibit the Na^+-K^+ ATPase pump (Oishi et al., 1990). For example, OA activates ARC POMC neurons by inhibiting ATP-sensitive K^+ (K_{ATP}) channel activity (Jo et al., 2009) and the effect of OA on HGP is abolished by icv administration of a K_{ATP} channel inhibitor (Jo et al., 2009). However, K_{ATP} channels are ubiquitously expressed on neurons throughout the brain, not only in FA sensing neurons, making the mechanism and site of such in vivo manipulations difficult to discern (Dunn-Meynell et al., 1998). Using in vivo and in vitro electrophysiological approaches, OA sensitive-neurons have been characterized using whole cell patch clamp records in ARC slices from 14 to 21 day old rats (Wang et al., 2006). Of these 13 % were excited by OA and 30% were inhibited by OA (Oomura et al., 1975). The excitatory effects of OA appeared to be due to closure of chloride channels leading to membrane depolarization and increased action potential frequency (Migrenne et al., 2006). On the other hand, inhibitory effect of OA may involve the K_{ATP} channels since this inhibition was reversed by the K_{ATP} channel blocker tolbutamide (Migrenne et al., 2006). Using fura-2 Ca^{2+} imaging in dissociated neurons from the ventromedial hypothalamic nucleus (VMN) neurons, we found that OA excited up to 43% and inhibited up to 29% of all VMN neurons independently of glucose concentrations (Le Foll et al., 2009). However, in these neurons, inhibition of the K_{ATP} channel mediated FA sensing in only a small percentage of FA sensing neurons. Importantly, although a relatively large percentage of hypothalamic neurons are FA sensors, a select population that also sense glucose are highly dependent upon ambient glucose concentrations for the resultant effect of FA on the activity of these neurons (Le Foll et al., 2009). Such data suggest that the responses of hypothalamic FA sensitive neurons are dependent upon the metabolic state of the animal and thus might be expected to respond differently during fasting (when FA levels rise and glucose levels fall) vs. the overfed state when glucose levels rise while free FA levels remain relatively unchanged (Le Foll et al., 2009). However, it must be pointed out that FA are naturally complexed to serum albumin in the blood and the concentration of circulating free FA is less than 1% of total FA levels. All the studies investigating FA sensing in the hypothalamus either use non-complexed FA or cyclodextrin-complexed FA *in vitro* or *in vivo*. The concentration of free FA in cyclodextrin-complexed FA preparation is unknown. Whether or not the FA concentration used mimics FA levels in physiological states needs to be determined.

4.2 Metabolic-dependent FA sensing effects

The effects of FA on activity of some neurons are dependent upon intracellular metabolism of FA. Enzymes involved in FA metabolism such as FA synthase (FAS), CPT1 and acetyl-CoA carboxylase (ACC) are expressed in some hypothalamic neurons as well as in glial cells (reviewed in (Blouet & Schwartz; Le Foll et al., 2009). Malonyl-CoA may be an important sensor of energy levels in the hypothalamus. It is derived from either glucose or FA metabolism via the glycolysis or β-oxidation, respectively. The steady-state level of malonyl-CoA is determined by its rate of synthesis catalysed by ACC relative to its rate of turnover catalysed by FAS. The synthesis of malonyl-CoA is the first committed step of FA synthesis and ACC is the major site of regulation in that process. Thus, when the supply of glucose is increased, malonyl CoA levels increase in keeping with a decreased need for FA oxidation. This increase in both malonyl CoA and acyl CoA levels is associated with reduced food intake. Central administration of C75, an inhibitor of FAS, also increases malonyl-CoA concentration in the hypothalamus, suppresses food intake and leads to profound weight loss (Proulx & Seeley, 2005). It has been proposed that centrally, C75 and cerulenin (another inhibitor of FAS) alter the expression profiles of feeding-related neuropeptides, often inhibiting the expression of orexigenic peptides such as neuropeptide Y (Proulx et al., 2008). Whether through centrally mediated or peripheral mechanisms, C75 also increases energy expenditure, which contributes to weight loss (Clegg et al., 2002; Tu et al., 2005). In vitro and in vivo studies demonstrate that at least part of C75's effects are mediated by the modulation of AMP-activated kinase, a known energy-sensing kinase (Ronnett et al., 2005). Indeed, icv administration of 5-aminoimidazole-4-carboxamide ribonucleoside (AICAR), a 5'-AMP kinase activator, rapidly lowers hypothalamic malonyl-CoA concentration and increases food intake (Tu et al., 2005). These effects correlate closely with the phosphorylation-induced inactivation of ACC, an established target of AMP kinase. Collectively, these data suggest a role for FA metabolism in the perception and regulation of energy balance. However, it must be also pointed out that C75 and AICAR may also have non-specific or even opposite effects. For example, a major effect of C75 is to activate CPT-1 rather than lead to its inhibition *in vitro* (Aja et al., 2008). Finally the route of administration and the type of FA used are also critical. For example, bolus intracerebroventricular injections of OA, but not palmitic acid, reduce food intake and body weight, possibly mediated through POMC/MC4R signaling (Schwinkendorf et al., 2010). Again, such bolus icv injections could cause non-specific effects related to inflammation of ependymocytes and tanycytes. Also because so much of FA metabolism takes place in astrocytes, such manipulations done in vivo and in slice preparations are likely to alter FA metabolism that takes place in astrocytes which could then indirectly alter neuronal FA sensing (Escartin et al., 2007).

4.3 Non metabolic-dependent neuronal FA sensing

While intracellular FA metabolism may be responsible for altering neuronal activity in some FA sensitive neurons such as ARC POMC neurons (Jo et al., 2009) it accounts for a relatively small percent of the effects of OA on dissociated VMN neurons (Le Foll et al., 2009). In those neurons, inhibition of CPT1, reactive oxygen species formation, long-chain acyl CoA synthetase and K_{ATP} channel activity or activation of uncoupling protein 2 (UCP2) accounts for no more than 20% of the excitatory or approximately 40% of the inhibitory effects of OA (Le Foll et al., 2009). On the other hand, pharmacological inhibition of FAT/CD36, a FA transporter/receptor that can alter cell function independently of intracellular FA metabolism reduced the excitatory and inhibitory effects of OA by up to 45% (Le Foll et al., 2009). Thus, in almost half of VMN FA sensing neurons, CD36 may act primarily as receptor, rather than a transporter, for long chain FA as it does on taste cells on the tongue where it activates store-operated calcium channels to alter membrane potential and release of serotonin (Gaillard et al., 2008). These effects all occur in the presence of nanomolar concentrations of OA, whereas micromolar concentrations are generally required to effect similar changes in neuronal activity in brain slice preparations (Jo et al., 2009; Migrenne et al., 2011; Wang et al., 2006). Thus, in the absence of astrocytes, OA can directly affect VMN neuronal activity through both metabolic and non-metabolic pathways. Alternatively, FA might act as signaling molecules by covalent attachment to proteins (N-terminal acylation) to alter the function of membrane and intracellular signaling molecules. For example, palmitoylation facilitates the targeting and plasma membrane binding of proteins which otherwise would remain in the cytosolic compartment (Resh, 1999). Some membrane proteins (TGFα, synaptosomal associated protein of 25KDa (required for exocytosis) and plasma membrane receptors (seven transmembrane receptors such as α_{2a}- and β_2- adrenoceptors) are typically palmitoylated on one or several cysteine residues located adjacent to or just within the transmembrane domain (Resh, 1999) Such mechanisms might also modulate neuronal FA sensing.

4.4 Which neurotransmitters or neuropeptides?

The ultimate consequence of the activation or inactivation of a neuron is the release of neurotransmitters and neuropeptides. Since FA decrease food intake, they might be expected to alter activity neurons specifically involved in the regulation of feeding. In fact, OA activates catabolic POMC neurons directly, apparently via ß-oxidation and inactivation of the K_{ATP} channel in hypothalamic slice preparations (Jo et al., 2009). In vivo, Obici et al. (Obici et al., 2003) reported that icv administration of OA markedly inhibits glucose production and food intake, accompanied by a decrease in the hypothalamic expression of the anabolic peptide, neuropeptide Y. This decrease in the expression of such a critical anabolic peptide might contribute to the reduced food intake associated with direct central administration of OA. On the other hand, an n-3 FA enriched diet increases food intake in anorexic tumor-bearing rats, in association with reduced tumor appearance, tumor growth and onset of anorexia (Ramos et al., 2005). In these treated rats, neuropeptide Y immunoreactivity increased 38% in ARC and 50% in paraventricular nucleus, whereas α-melanocyte stimulating hormone (a catabolic peptide cleavage product of POMC) decreased 64% in the ARC and 29% in the paraventricular nucleus (Ramos et al., 2005). Finally, in the hippocampus, docosahexaenoic acid (22:6(n-3) increased the spontaneous release of acetylcholine (Aid et al., 2005).

4.5 Pathological implications of excess FA

Besides physiological regulation of energy balance by hypothalamic neuronal FA sensing, impaired regulation of such sensing might contribute to the development of metabolic diseases such as obesity and type 2 diabetes in predisposed subjects exposed to a chronic lipid overload (Luquet & Magnan, 2009; Migrenne et al., 2011). Excessive brain lipid levels may indeed alter control of glucose and lipid homeostasis through changes of autonomic nervous system activity. Increasing brain FA levels reduces sympathetic activity and increases GIIS in rats (Clement et al., 2002; Obici et al., 2003) a condition which would exacerbate the development of type 2 diabetes mellitus. Also, a lipid overload due to high-fat diet intake alters both hypothalamic monoamine turnover (Levin et al., 1983) and peripheral sympathetic activity in rats (Young & Walgren, 1994). In humans, overweight is often associated with an altered sympathetic tone (Peterson et al., 1988) suggesting a relationship between lipids and autonomic control centers in brain.

5. Conclusion

In conclusion, there is now increasing evidence that specialized neurons within hypothalamus and other areas such as the brainstem or hippocampus can detect changes in plasma FA levels by having FA directly or indirectly alter the of FA sensitive neurons involved in the regulation of energy and glucose homeostasis. Central FA effects on insulin secretion and action are related to their chain length or degree of saturation. Such effects are also mediated through differential changes in gene expression.

The neuronal networks of these FA sensitive neurons that sense and respond to FA are likely very complex given the fact that FA can either inhibit or excite specific neurons. In addition, many of these neurons also utilize glucose as a signaling molecule and there is often an inverse responsiveness of such "metabolic sensing" neurons to FA vs. glucose.

Thus, these neurons are ideally suited to respond differentially under a variety of metabolic conditions such as fasting, feeding, hypo- or hyperglycemia. However, while it is clear that specific neurons can respond to changes in ambient FA levels, many questions remain. We still do not know for certain how FA are transported into the brain, astrocytes or neurons and whether those FA that are transported are derived from circulating free FA or triglycerides. Since most studies suggest that rising FA levels reduce food intake, then we must explain why plasma FA levels are most elevated during fasting when the drive to seek and ingest food should be at its strongest. Another major issue relates to the interaction between astrocytes and neurons with regard to the metabolism and signaling of FA. Also, we still know little about the basic mechanisms utilized by neurons to sense FA, where such FA sensitive neurons reside throughout the brain and what neurotransmitters and peptides they release when responding to FA.

Finally, it has been postulated that diabetes may be a disorder of the brain (Elmquist & Marcus, 2003). If so, dysfunction of these FA sensitive neurons could be, at least in part, one of the early mechanisms underlying impairment of neural control of energy and glucose homeostasis and the development of obesity and type 2 diabetes in predisposed subjects. A better understanding of this central nutrient sensing, including both FA and glucose, could provide clues for the identification of new therapeutic targets for the prevention and treatment of both diabetes and obesity.

6. Acknowledgements

This work was partially supported by an award from European Foundation for Study of Diabetes (EFSD)/GSK 2007 (Stéphanie Migrenne).

7. References

Adams JM, 2nd; Pratipanawatr T; Berria R; Wang E; DeFronzo RA; Sullards MC & Mandarino LJ. (2004). Ceramide content is increased in skeletal muscle from obese insulin-resistant humans. *Diabetes*, Vol. 53, No. 1, pp 25-31, 0012-1797 (Print) 0012-1797 (Linking)

Aid S; Vancassel S; Linard A; Lavialle M & Guesnet P. (2005). Dietary docosahexaenoic acid [22: 6(n-3)] as a phospholipid or a triglyceride enhances the potassium chloride-evoked release of acetylcholine in rat hippocampus. *J Nutr*, Vol. 135, No. 5, pp 1008-1013,

Aja S; Landree LE; Kleman AM; Medghalchi SM; Vadlamudi A; McFadden JM; Aplasca A; Hyun J; Plummer E; Daniels K; Kemm M; Townsend CA; Thupari JN; Kuhajda FP; Moran TH & Ronnett GV. (2008). Pharmacological stimulation of brain carnitine palmitoyl-transferase-1 decreases food intake and body weight. *Am J Physiol Regul Integr Comp Physiol*, Vol. 294, No. 2, pp R352-361, 0363-6119 (Print) 0363-6119 (Linking)

Benoit SC; Kemp CJ; Elias CF; Abplanalp W; Herman JP; Migrenne S; Lefevre AL; Cruciani-Guglielmacci C; Magnan C; Yu F; Niswender K; Irani BG; Holland WL & Clegg DJ. (2009). Palmitic acid mediates hypothalamic insulin resistance by altering PKC-theta subcellular localization in rodents. *J Clin Invest*, Vol. 119, No. 9, pp 2577-2589,

Blouet C & Schwartz GJ. (2010). Hypothalamic nutrient sensing in the control of energy homeostasis. *Behav Brain Res*, Vol. 209, No. 1, pp 1-12,

Clegg DJ; Air EL; Woods SC & Seeley RJ. (2002). Eating elicited by orexin-a, but not melanin-concentrating hormone, is opioid mediated. *Endocrinology*, Vol. 143, No. 8, pp 2995-3000,

Clement L; Cruciani-Guglielmacci C; Magnan C; Vincent M; Douared L; Orosco M; Assimacopoulos-Jeannet F; Penicaud L & Ktorza A. (2002). Intracerebroventricular infusion of a triglyceride emulsion leads to both altered insulin secretion and hepatic glucose production in rats. *Pflugers Arch*, Vol. 445, No. 3, pp 375-380,

Cruciani-Guglielmacci C; Hervalet A; Douared L; Sanders NM; Levin BE; Ktorza A & Magnan C. (2004). Beta oxidation in the brain is required for the effects of non-esterified fatty acids on glucose-induced insulin secretion in rats. *Diabetologia*, Vol. 47, No. 11, pp 2032-2038,

Dowell P; Hu Z & Lane MD. (2005). Monitoring energy balance: metabolites of fatty acid synthesis as hypothalamic sensors. *Annu Rev Biochem*, Vol. 74, No. pp 515-534,

Dunn-Meynell AA; Rawson NE & Levin BE. (1998). Distribution and phenotype of neurons containing the ATP-sensitive K+ channel in rat brain. *Brain Res*, Vol. 814, No. 1-2, pp 41-54, 0006-8993 (Print) 0006-8993 (Linking)

Edmond J. (2001). Essential polyunsaturated fatty acids and the barrier to the brain: the components of a model for transport. *J Mol Neurosci*, Vol. 16, No. 2-3, pp 181-193; discussion 215-121,

Elmquist JK & Marcus JN. (2003). Rethinking the central causes of diabetes. *Nat Med*, Vol. 9, No. 6, pp 645-647,

Escalante-Alcalde D; Hernandez L; Le Stunff H; Maeda R; Lee HS; Jr Gang C; Sciorra VA; Daar I; Spiegel S; Morris AJ & Stewart CL. (2003). The lipid phosphatase LPP3 regulates extra-embryonic vasculogenesis and axis patterning. *Development*, Vol. 130, No. 19, pp 4623-4637, 0950-1991 (Print) 0950-1991 (Linking)

Escartin C; Boyer F; Bemelmans AP; Hantraye P & Brouillet E. (2007). IGF-1 exacerbates the neurotoxicity of the mitochondrial inhibitor 3NP in rats. *Neurosci Lett*, Vol. 425, No. 3, pp 167-172, 0304-3940 (Print) 0304-3940 (Linking)

Escartin C; Pierre K; Colin A; Brouillet E; Delzescaux T; Guillermier M; Dhenain M; Deglon N; Hantraye P; Pellerin L & Bonvento G. (2007). Activation of astrocytes by CNTF induces metabolic plasticity and increases resistance to metabolic insults. *J Neurosci*, Vol. 27, No. 27, pp 7094-7104, 1529-2401 (Electronic) 0270-6474 (Linking)

Gaillard D; Laugerette F; Darcel N; El-Yassimi A; Passilly-Degrace P; Hichami A; Khan NA; Montmayeur JP & Besnard P. (2008). The gustatory pathway is involved in CD36-mediated orosensory perception of long-chain fatty acids in the mouse. *FASEB J*, Vol. 22, No. 5, pp 1458-1468, 1530-6860 (Electronic) 0892-6638 (Linking)

Gilbert M; Magnan C; Turban S; Andre J & Guerre-Millo M. (2003). Leptin receptor-deficient obese Zucker rats reduce their food intake in response to a systemic supply of calories from glucose. *Diabetes*, Vol. 52, No. 2, pp 277-282,

Gribble FM; Proks P; Corkey BE & Ashcroft FM. (1998). Mechanism of cloned ATP-sensitive potassium channel activation by oleoyl-CoA. *J Biol Chem*, Vol. 273, No. 41, pp 26383-26387, 0021-9258 (Print) 0021-9258 (Linking)

Holland WL; Bikman BT; Wang LP; Yuguang G; Sargent KM; Bulchand S; Knotts TA; Shui G; Clegg DJ; Wenk MR; Pagliassotti MJ; Scherer PE & Summers SA. (2011). Lipid-induced insulin resistance mediated by the proinflammatory receptor TLR4 requires saturated fatty acid-induced ceramide biosynthesis in mice. *J Clin Invest*, Vol. 121, No. 5, pp 1858-1870, 1558-8238 (Electronic) 0021-9738 (Linking)

Holland WL & Summers SA. (2008). Sphingolipids, insulin resistance, and metabolic disease: new insights from in vivo manipulation of sphingolipid metabolism. *Endocr Rev*, Vol. 29, No. 4, pp 381-402, 0163-769X (Print) 0163-769X (Linking)

Honen BN; Saint DA & Laver DR. (2003). Suppression of calcium sparks in rat ventricular myocytes and direct inhibition of sheep cardiac RyR channels by EPA, DHA and oleic acid. *J Membr Biol*, Vol. 196, No. 2, pp 95-103,

Jo YH; Su Y; Gutierrez-Juarez R & Chua S, Jr. (2009). Oleic acid directly regulates POMC neuron excitability in the hypothalamus. *J Neurophysiol*, Vol. 101, No. 5, pp 2305-2316, 0022-3077 (Print) 0022-3077 (Linking)

Kimura I; Inoue D; Maeda T; Hara T; Ichimura A; Miyauchi S; Kobayashi M; Hirasawa A & Tsujimoto G. (2011). Short-chain fatty acids and ketones directly regulate sympathetic nervous system via G protein-coupled receptor 41 (GPR41). *Proc Natl Acad Sci U S A*, Vol. 108, No. 19, pp 8030-8035, 1091-6490 (Electronic) 0027-8424 (Linking)

Lam TK; Schwartz GJ & Rossetti L. (2005). Hypothalamic sensing of fatty acids. *Nat Neurosci*, Vol. 8, No. 5, pp 579-584,

Le Foll C; Irani BG; Magnan C; Dunn-Meynell AA & Levin BE. (2009). Characteristics and mechanisms of hypothalamic neuronal fatty acid sensing. *Am J Physiol Regul Integr Comp Physiol*, Vol. 297, No. 3, pp R655-664,

Le Stunff H; Galve-Roperh I; Peterson C; Milstien S & Spiegel S. (2002). Sphingosine-1-phosphate phosphohydrolase in regulation of sphingolipid metabolism and apoptosis. *J Cell Biol*, Vol. 158, No. 6, pp 1039-1049, 0021-9525 (Print) 0021-9525 (Linking)

Levin BE; Triscari J & Sullivan AC. (1983). Altered sympathetic activity during development of diet-induced obesity in rat. *Am J Physiol*, Vol. 244, No. 3, pp R347-355,

Luquet S & Magnan C. (2009). The central nervous system at the core of the regulation of energy homeostasis. *Front Biosci (Schol Ed)*, Vol. 1, No. pp 448-465,

Magnan C; Collins S; Berthault MF; Kassis N; Vincent M; Gilbert M; Penicaud L; Ktorza A & Assimacopoulos-Jeannet F. (1999). Lipid infusion lowers sympathetic nervous activity and leads to increased beta-cell responsiveness to glucose. *J Clin Invest*, Vol. 103, No. 3, pp 413-419,

Magnan C; Cruciani C; Clement L; Adnot P; Vincent M; Kergoat M; Girard A; Elghozi JL; Velho G; Beressi N; Bresson JL & Ktorza A. (2001). Glucose-induced insulin hypersecretion in lipid-infused healthy subjects is associated with a decrease in plasma norepinephrine concentration and urinary excretion. *J Clin Endocrinol Metab*, Vol. 86, No. 10, pp 4901-4907,

Migrenne S; Cruciani-Guglielmacci C; Kang L; Wang R; Rouch C; Lefevre AL; Ktorza A; Routh V; Levin B & Magnan C. (2006). Fatty acid signaling in the hypothalamus and the neural control of insulin secretion. *Diabetes*, Vol. 55 S2, No. pp S139-S144,

Migrenne S; Le Foll C; Levin BE & Magnan C. (2011). Brain lipid sensing and nervous control of energy balance. *Diabetes Metab*, Vol. 37, No. 2, pp 83-88, 1878-1780 (Electronic) 1262-3636 (Linking)

Migrenne S; Marsollier N; Cruciani-Guglielmacci C & Magnan C. (2006). Importance of the gut-brain axis in the control of glucose homeostasis. *Curr Opin Pharmacol*, Vol. 6, No. 6, pp 592-597,

Mullen KL; Pritchard J; Ritchie I; Snook LA; Chabowski A; Bonen A; Wright D & Dyck DJ. (2009). Adiponectin resistance precedes the accumulation of skeletal muscle lipids and insulin resistance in high-fat-fed rats. *Am J Physiol Regul Integr Comp Physiol*, Vol. 296, No. 2, pp R243-251, 0363-6119 (Print) 0363-6119 (Linking)

Newton RU; Taaffe DR; Spry N; Gardiner RA; Levin G; Wall B; Joseph D; Chambers SK & Galvao DA. (2009). A phase III clinical trial of exercise modalities on treatment side-effects in men receiving therapy for prostate cancer. *BMC Cancer*, Vol. 9, No. pp 210, 1471-2407 (Electronic) 1471-2407 (Linking)

Obici S; Feng Z; Arduini A; Conti R & Rossetti L. (2003). Inhibition of hypothalamic carnitine palmitoyltransferase-1 decreases food intake and glucose production. *Nat Med*, Vol. 9, No. 6, pp 756-761,

Obici S; Feng Z; Morgan K; Stein D; Karkanias G & Rossetti L. (2002). Central administration of oleic acid inhibits glucose production and food intake. *Diabetes*, Vol. 51, No. 2, pp 271-275,

Oishi K; Zheng B & Kuo JF. (1990). Inhibition of Na,K-ATPase and sodium pump by protein kinase C regulators sphingosine, lysophosphatidylcholine, and oleic acid. *J Biol Chem*, Vol. 265, No. 1, pp 70-75,

Oomura Y; Nakamura T; Sugimori M & Yamada Y. (1975). Effect of free fatty acid on the rat lateral hypothalamic neurons. *Physiol Behav*, Vol. 14, No. 04, pp 483-486,

Penicaud L; Leloup C; Lorsignol A; Alquier T & Guillod E. (2002). Brain glucose sensing mechanism and glucose homeostasis. *Curr Opin Clin Nutr Metab Care*, Vol. 5, No. 5, pp 539-543,

Peterson HR; Rothschild M; Weinberg CR; Fell RD; McLeish KR & Pfeifer MA. (1988). Body fat and the activity of the autonomic nervous system. *N Engl J Med*, Vol. 318, No. 17, pp 1077-1083,

Proulx K; Cota D; Woods SC & Seeley RJ. (2008). Fatty acid synthase inhibitors modulate energy balance via mammalian target of rapamycin complex 1 signaling in the central nervous system. *Diabetes*, Vol. 57, No. 12, pp 3231-3238,

Proulx K & Seeley RJ. (2005). The regulation of energy balance by the central nervous system. *Psychiatr Clin North Am*, Vol. 28, No. 1, pp 25-38, vii,

Ramos EJ; Romanova IV; Suzuki S; Chen C; Ugrumov MV; Sato T; Goncalves CG & Meguid MM. (2005). Effects of omega-3 fatty acids on orexigenic and anorexigenic modulators at the onset of anorexia. *Brain Res*, Vol. 1046, No. 1-2, pp 157-164,

Randle PJ; Priestman DA; Mistry S & Halsall A. (1994). Mechanisms modifying glucose oxidation in diabetes mellitus. *Diabetologia*, Vol. 37 Suppl 2, No. pp S155-161, 0012-186X (Print) 0012-186X (Linking)

Rapoport SI; Chang MC & Spector AA. (2001). Delivery and turnover of plasma-derived essential PUFAs in mammalian brain. *J Lipid Res*, Vol. 42, No. 5, pp 678-685,

Resh MD. (1999). Fatty acylation of proteins: new insights into membrane targeting of myristoylated and palmitoylated proteins. *Biochim Biophys Acta*, Vol. 1451, No. 1, pp 1-16,

Ronnett GV; Kim EK; Landree LE & Tu Y. (2005). Fatty acid metabolism as a target for obesity treatment. *Physiol Behav*, Vol. 85, No. 1, pp 25-35,

Ross RA; Rossetti L; Lam TK & Schwartz GJ. (2010). Differential effects of hypothalamic long-chain fatty acid infusions on suppression of hepatic glucose production. *Am J Physiol Endocrinol Metab*, Vol. 299, No. 4, pp E633-639, 1522-1555 (Electronic) 0193-1849 (Linking)

Ruge T; Hodson L; Cheeseman J; Dennis AL; Fielding BA; Humphreys SM; Frayn KN & Karpe F. (2009). Fasted to fed trafficking of Fatty acids in human adipose tissue reveals a novel regulatory step for enhanced fat storage. *J Clin Endocrinol Metab*, Vol. 94, No. 5, pp 1781-1788, 1945-7197 (Electronic) 0021-972X (Linking)

Schwinkendorf DR; Tsatsos NG; Gosnell BA & Mashek DG. (2010). Effects of central administration of distinct fatty acids on hypothalamic neuropeptide expression and energy metabolism. *Int J Obes (Lond)*, Vol. No. pp 1476-5497 (Electronic) 0307-0565 (Linking)

Smith QR & Nagura H. (2001). Fatty acid uptake and incorporation in brain: studies with the perfusion model. *J Mol Neurosci*, Vol. 16, No. 2-3, pp 167-172; discussion 215-121,

Tewari KP; Malinowska DH; Sherry AM & Cuppoletti J. (2000). PKA and arachidonic acid activation of human recombinant ClC-2 chloride channels. *Am J Physiol Cell Physiol*, Vol. 279, No. 1, pp C40-50,

Tu Y; Thupari JN; Kim EK; Pinn ML; Moran TH; Ronnett GV & Kuhajda FP. (2005). C75 alters central and peripheral gene expression to reduce food intake and increase energy expenditure. *Endocrinology*, Vol. 146, No. 1, pp 486-493,

Wang R; Cruciani-Guglielmacci C; Migrenne S; Magnan C; Cotero VE & Routh VH. (2006). Effects of oleic acid on distinct populations of neurons in the hypothalamic arcuate nucleus are dependent on extracellular glucose levels. *J Neurophysiol*, Vol. 95, No. 3, pp 1491-1498,

Watkins PA; Hamilton JA; Leaf A; Spector AA; Moore SA; Anderson RE; Moser HW; Noetzel MJ & Katz R. (2001). Brain uptake and utilization of fatty acids: applications to peroxisomal biogenesis diseases. *J Mol Neurosci*, Vol. 16, No. 2-3, pp 87-92; discussion 151-157,

Young JB & Walgren MC. (1994). Differential effects of dietary fats on sympathetic nervous system activity in the rat. *Metabolism*, Vol. 43, No. 1, pp 51-60,

Part 3

Innovative Techniques for the Production of Olive Oil Based Products

10

Meat Fat Replacement with Olive Oil

Basem Mohammed Al-Abdullah[1,*], Khalid M. Al-Ismail[1],
Khaled Al-Mrazeeq[1], Malak Angor[2] and Radwan Ajo[2]
*[1]Department of Nutrition and Food Technology,
Faculty of Agriculture, University of Jordan, Amman,
[2]Al- Huson University Collage, Al- Balqa Applied University, Al-Huson,
Jordan*

1. Introduction

The consumption of convenience foods in the restaurants such as beef or chicken burgers is increasing in Jordan. Burger is a meat product prepared from minced lean meat, with or without addition of other ingredients. The total fat content must not exceed 15% (JS: 1334/2002). In Jordan, burgers are prepared from two main meat sources: beef or chicken. Many efforts have been made to improve the quality and stability of burgers because consumer demand for healthy fast food has rapidly increased in the recent years.

Complete or partial replacement of burger fat with oil rich in monounsaturated fatty acids, such as olive oil may improve the oxidative stability of chicken burger and the nutritional value of beef burger. Another approach that can be followed to improve the quality of beef burgers is the partial replacement of beef meat with chicken meat.

This study aimed at:

1. Studying the effect of partial replacement of beef tallow and chicken fat with olive oil on some chemical and sensory properties of a freshly prepared and stored burger.
2. Studying the effect of partial replacement of beef tallow and meat with chicken meat and fat (50:50) on some chemical and sensory properties of a freshly prepared and stored burger.
3. Studying the effect of grilling on some chemical and sensory properties of a freshly prepared and stored burger formulations.

Five burger formulations were prepared and studied during storage and after grilling at 75°C for 20 minutes. These formulations were: beef, chicken, mixed beef and chicken (50:50), beef with olive oil and chicken with olive oil.

The effect of storage and grilling was evaluated by determining cooking loss by using weight differences between raw and cooked burgers, thiobarbituric acid reactive substances (TBARS) (Faustman, *et al.*, 1992), fatty acid profile using GLC analysis; fatty acid methyl esters (FAMEs) of the burger samples were prepared according to Chritopherson and Giass (1969) method, cholesterol and 7-ketocholesterol; cold saponification and extraction was

---

* Corresponding Author

carried out according to the method used by Sander, *et al.* (1988) and the trimethylsilyl derivatives (TMS) of cholesterol and cholesterol oxides were carried out according to the method used by Pie, *et al.* (1990).

2. Moisture, fat and protein content

The moisture, fat and protein contents for both beef burger treatments before grilling, were about 65.50%, 15.11%, and 18.20%, respectively. The moisture, fat and protein contents of both chicken burger samples before grilling were about 66.50%, 15%, and 17.50%, respectively.

The moisture loss percentage of the freshly prepared treatments due to grilling was between 20.37-25.62%, and fat loss was between 18.85-21.51%. On the other hand, the increase in protein contents was (96-116%) of the burger samples. Moisture and fat contents of the grilled samples were lower than those of raw samples, while protein content was higher. This is mainly due to the loss of water and fat.

3. Oxidative rancidity measured by TBARS test

The initial TBARS values of the beef sample expressed as mg malondialdehyde/kg meat, were about two times greater than those of chicken sample. These results reflect the quality of the raw materials, which in the case of beef, it already had a high initial degree of peroxidation. Inappropriate storage conditions of meat, together with the action of light, oxygen and the presence of myoglobine probably accelerated oxidation.

Characterisic	Time of storage (month)	*Treatment**				
		Beef	Chicken	Mixed	Beef with olive oil	Chicken with olive oil
Raw	0	$_b2.26^a$	$_b1.21^c$	$_b2.09^b$	$_b2.27^a$	$_b0.74^d$
	1	$_a2.71^d$	$_a3.00^c$	$_a2.59^e$	$_a5.07^b$	$_a5.23^a$
	3	$_a2.62^a$	$_b1.20^c$	$_b2.13^b$	$_b2.59^a$	$_b0.77^d$
Grilled	0	$_b0.57^d$	$_b5.09^a$	$_b2.10^c$	$_c0.38^e$	$_b3.76^b$
	1	$_a1.09^d$	$_a7.04^b$	$_a3.68^c$	$_a0.91^e$	$_a7.99^a$
	3	$_a1.03^d$	$_c4.53^a$	$_c1.53^c$	$_b0.79^e$	$_c3.12^b$

Each value is the mean of three replicates.
* Values within the same column with different subscripts denote significant differences ($p \leq 0.05$) between storage times according to LSD.
** Values within the same row with different superscripts denote significant differences ($p \leq 0.05$) between the treatments according to LSD.

Table 1. Thiobarbituric acid reactive substances values (TBARS) expressed as mg malondialdehyde /kg meat for the raw and grilled burger samples during storage time.

It can be observed that TBARS values increased during the first month.The increase was higher in the chicken sample and those with olive oil than those of beef sample.These results might be explained by the fact that the fatty acids of these samples have higher degree of unsaturation when compared with those of beef.

Melton (1985) reported that oxidized flavors were detectable at TBARS numbers of 0.3-1.0 in beef or pork, 1or 2 in chicken, and higher than 3 in turkey. The TBARS values obtained in this study, remarkably exceeded these ranges. So it can be assumed that these high values of TBARS could be attributed to oxidation as well as other interferences.

On the other hand, decrease in TBARS values noticed at the end of storage period were 85, 60, 47, 18 and 3% for chicken with olive oil, chicken, beef with olive oil, mixed and beef treatments, respectively. This behavior may be ascribed to the combination of aldehydes with other compounds and to the loss of volatile aldehydes (Severini, et al., 2003).

Different trends were observed on the effect of grilling on TBARS values, since TBARS values decreased in both beef samples, whereas they increased in both chicken samples. This finding may be attributed to the fact that chicken fat contains higher levels of PUFA, which are prone to higher level of oxidation.

4. Cholesterol and cholesterol oxides

It was evident that cholesterol content of the raw and grilled chicken sample was about 39% higher than those of beef sample. This is due to the use of chicken skin which contains high level of cholesterol in chicken burger. Mixed meat samples had cholesterol content which was about 15% lower than chicken and 18% higher than those of beef.

Substitution of the added beef and chicken fat with olive oil resulted in a considerable decrease in cholesterol contents. The reduction in beef and chicken samples was about 53% and 58%, respectively

Characteristic	Time of storage (month)	*Treatment				
		Beef	Chicken	Mixed	Beef with olive oil	Chicken with olive oil
Raw	0	$_a$333.87	$_a$462.10	$_a$391.67	$_a$157.70	$_a$193.43
	1	$_a$331.27	$_a$461.67	$_a$390.66	$_a$156.61	$_a$193.03
	3	$_a$331.30	$_a$460.27	$_a$390.47	$_a$155.73	$_a$192.00
	**Means	^{c}332.15$_a$	^{a}461.35$_a$	^{b}390.93$_a$	^{e}156.68$_a$	^{d}192.82$_a$
Grilled	0	$_a$331.73	$_a$460.13	$_a$390.27	$_a$156.47	$_a$191.13
	1	$_a$330.23	$_a$459.37	$_a$389.30	$_a$154.92	$_a$191.07
	3	$_a$330.93	$_a$459.11	$_a$389.13	$_a$154.67	$_a$190.28
	**Means	^{c}330.96$_a$	^{a}459.54$_a$	^{b}389.57$_a$	^{e}155.35$_a$	^{d}190.83$_a$

Each value is the mean of three replicates.
* Values within the same column with same subscripts are not significantly (p> 0.05) different according to LSD.
** Values within the same row with different superscripts denote significant differences (p≤ 0.05) between treatments according to LSD, whereas values within the same column with same subscripts denote no significant (p> 0.05) differences among raw and grilled samples according to LSD.

Table 2. Cholesterol content (mg/100 g fat) for the raw and grilled burger samples during storage.

Characteristic	*Treatment**				
	Beef	Chicken	Mixed	Beef with olive oil	Chicken with olive oil
Raw	$_c50.12^a$	$_a70.82^a$	$_b59^a$	$_e23.78^a$	$_d29^a$
Grilled	$_c38.76^b$	$_a56^b$	$_b45.42^b$	$_e17.77^b$	$_d23^b$

Each value is the mean of three replicates.
* Means in the same row with the different subscripts denote significant differences among treatments of burger ($p \leq 0.05$) according to LSD.
** Means in the same column with different superscripts denote significant differences among raw and grilled burger samples ($p \leq 0.05$) according to LSD.

Table 3. Cholesterol values (mg/100g burger) for the raw and grilled burger samples.

Storage time and grilling did not affect cholesterol contents of all treatments, calculated on the fat basis (mg cholesterol/100g fat).

However, cholesterol content calculated on the burgers basis (mg cholesterol/100g burger) showed lower cholesterol in grilled samples compared to the raw one. The reduction was about 23, 21, 23, 25 and 21% for beef, chicken, mixed, beef with olive oil and chicken with olive oil samples, respectively. This reduction might be due to the loss of fat during cooking.

7-ketocholesterol was used in this study as a tracer of the degree of cholesterol oxidation, because of its fast and continuous formation at levels relatively high with respect to the other oxidation products (Park and Addis, 1985). moreover, the chromatographic peak of 7-ketocholesterol does not overlap with other peaks of cholesterol oxides products and components of food matrices (Rodriguez-Estrada, et al., 1997).

In this study, there was no detectable amount of 7-ketocholesterol in all raw and grilled samples, indicating that storage and grilling did not affect the stability of cholesterol against oxidation. This could be explained by the fact that grilling conditions were not severe, since the maximum temperature of grilling was about 75°C and the time of grilling did not exceed 20 minutes. Cholesterol shows high oxidation stability at temperature below 100°C (Kyoichi, et al., 1993). Furthermore, the grilling machine permitted low oxygen level to be in contact with burger during grilling because the upper part of the grill was closed and directly came into contact with the burgers.

5. Fatty acids profile

The effect of formulation, grilling and storage period on SFA, MUFA and PUFA contents of the burgers was observed. As expected, fatty acid composition of burgers reflected the fatty acid composition of the tissues and the fat used for their manufacturing.

It is well known that SFA are considered as a primary cause of hypercholesterolemia, and MUFA provide the body of essential fatty acids and decrease LDL cholesterol in the body (Mattson and Grundy, 1985). On the other hand, the addition of beef meat and fat to chicken burger enhanced its oxidative stability by increasing SFA by 32% and decreasing PUFA content by 34%, approximately. PUFA are easily prone to oxidation generating short chain compounds that deteriorate the sensory properties of the meat products.

Fatty acid	Treatment									
	Beef		Chicken		Mixed		Beef with olive oil		Chicken with olive oil	
	Raw	Grilled	Raw	Grilled	Raw	Grilled	Raw	Grilled	Raw	Grilled
Myristic C14:0	1.36	1.34	0.58	0.53	0.88	0.76	0.25	0.24	0.29	0.22
Palmitic C16:0	34.78	31.79	26.71	26.58	30.69	28.87	16.71	15.42	17.25	14.45
Palmitoleic C16:1	1.01	1.48	4.72	4.62	3.02	3.73	0.81	1.11	2.68	3.34
Stearic C18:0	22.36	20.57	6.13	6.00	12.61	10.48	10.21	8.93	5.86	4.81
Oleic C18:1	37.72	39.65	42.84	42.88	39.93	42.88	58.84	63.32	59.92	65.37
Linoleic C18:2	1.81	3.1	17.82	17.81	11.60	12.03	8.63	8.94	11.97	11.94
Linolenic C18:3	0.35	0.96	1.10	0.88	0.79	0.83	0.86	0.89	1.26	0.93
Arachidic C20:0	0.04	traces	0.02	traces	0.02	traces	0.03	traces	0.01	traces

Each value is the mean of three readings of fatty acids after samples formulation.

Table 4. Means values of fatty acids profile (g/100g fat) for the raw and grilled burger samples after formulation.

Another strategy for changing fatty acid profile of meat products rather than meat mixing is the replacement of animal fats by vegetable oils. Olive oil is a vegetable oil whose MUFA content is high. The MUFA, PUFA and SFA contents were about 72%, 10% and 13%, respectively. The addition of olive oil in place of beef and chicken fat changed the fatty acids composition of the beef and chicken burgers. The decrease in SFA of beef sample was about 54%, whereas the increase in MUFA and PUFA contents was about 54% and 33.9%, respectively, of their original contents in beef fat. On the other hand, the increase in MUFA was about 32%, whereas the decrease in SFA and PUFA contents was about 30% of their original contents in chicken fat. The decrease in SFA contents in these burger samples was due to the decrease in myristic, palmitic and stearic acid contents, while the increase in MUFA was due mainly to oleic acid, since the addition of olive oil decreased the palmitoleic acid contents. The increase in PUFA content of beef sample was mainly due to the increase in linoleic and to a less extent to the increase in linolenic content.

MUFA and PUFA contents showed gradual and significant decrease for all treatments during storage period, especially at the end of storage. This may be due to the oxidation of unsaturated fatty acids.

In the case of PUFA, the decrease in their contents of beef with olive oil was lower than in beef with tallow ($\approx$ 47%), while chicken samples showed reverse trend, since the decline in PUFA contents of chicken was about 8% compared to 22% in chicken with olive oil.

Charac-teristic	Time of storage (month)	·Treatment**									
		Beef		Chicken		Mixed		Beef with olive oil		Chicken with olive oil	
		Raw	Grilled	Raw	Grilled	Raw	Grilled	Raw	Grilled	Raw	Grilled
SFA	0	a58.54a	a53.70b	a33.42a	a33.11b	a44.20a	a40.11b	a27.20a	a24.61b	a23.40a	a19.48b
	1	a58.28a	a53.75b	a33.50a	a33.25b	a44.57a	a40.24b	a26.93a	a24.90b	a23.61a	a19.50b
	3	a58.89a	a53.63b	a33.76a	a33.57b	a44.61a	a40.49b	a27.13a	a24.92b	a23.72a	a19.73b
MUFA	0	a38.73b	a41.13a	a47.56a	a47.50a	a42.95b	a46.69a	a59.65b	a64.43a	a62.60b	a68.71a
	1	b38.11b	b39.99a	b45.87a	b45.76a	b41.70b	b46.02a	b56.90b	b63.46a	b60.19b	b67.72a
	3	c35.35b	c36.50a	c35.17a	c40.60a	c36.58b	c38.72a	c47.22b	c54.50a	c50.59b	c57.52a
PUFA	0	a2.16b	a4.06a	a18.92a	a18.69a	a12.39b	a12.86a	a9.49b	a9.82a	a13.26a	a12.87a
	1	b1.53b	b2.47a	b17.82b	b18.55a	b11.93a	b11.42b	b7.29a	b7.93a	b12.59b	a12.77a
	3	c1.15b	b 2.69a	b17.64b	a18.42a	c11.29a	c10.79b	c5.94b	b7.77a	c10.34b	b12.02a

Each value is the mean of three replicates.
* Values within the same column with different subscripts are significantly ($p \leq 0.05$) different according to LSD.
** Values within the same row with different superscripts denote significance different ($p \leq 0.05$) among raw and grilled sample according to LSD

Table 5. Effect of formulation, storage time and grilling on fatty acids profile (g/100g fat) of the burger samples.

Grilling significantly decreased SFA, and increased MUFA contents of all samples, except for MUFA contents of chicken sample which remained constant. PUFA contents, in general, increased in most samples, but in some cases there was no clear trend.

6. Cooking loss

Chicken sample with olive oil showed lower cooking loss in weight due to grilling when compared to the corresponding samples without olive oil. This result showed the ability of protein matrix to bind monounsaturated fat. Chicken samples with olive oil had lower cooking loss in weight when compared to beef samples which was due to the highest water holding capacity, lipid capacity and lipid stability of chicken meat rather than beef meat.

Characteristic	Time of storage (month)	·Treatment**				
		Beef	Chicken	Mixed	Beef with olive oil	Chicken with olive oil
Cooking loss%	0	b49.69c	b50.22b	b51.30a	b50.26b	b43.28d
	1	b49.86c	b50.48b	b51.53a	b50.21b	b43.02d
	3	a51.70c	a52.63b	a53.17a	a52.78b	a47.39d

Each value is the mean of three replicates.
* Values within the same column with different subscripts are significantly ($p \leq 0.05$) different according to LSD.
** Values within the same row with different superscripts denote significant differences ($p \leq 0.05$) according to LSD.

Table 6. Percentage cooking loss in weight of burger samples during storage period.

In the mixed treatment we expected that cooking loss value will be between beef and chicken sample values, but unexpected result was obtained, the outcome showed that mixed treatment had the highest cooking loss in weight. More investigation is needed to explain the results.

The highest cooking loss was found after three months of storage which might be due to the weakness of protein matrix to entrap moisture and fat during storage, moreover, this weakness of protein matrix results in decrease of water and lipid holding capacity and stability, which might be due to denaturation of protein during frozen storage.

7. Sensory evaluation

Cooked burgers from each treatment were evaluated by 18 panelists from the sensory evaluation team at the Department of Nutrition and Food Technology. The panelists were both male and female, and were of different ages; they were requested to taste each sample separately without comparing it with other samples. Panelists were familiarized with the questionnaire form used. The samples were evaluated for desirability in appearance, color, tenderness, flavor, juiciness and overall acceptability using a 9-hedonic scale test as described by LARMOND (1991), varying from 9 (like extremely) to 1 (dislike extremely). Pieces of bread and water were used to neutralize the taste between samples.

The sensory evaluation results showed that all the sensory characteristics did not exceed the range like moderately, or fell to dislike slightly. This low score given by the panelists for all samples might be attributed to the fact that the prepared burgers were free of any added ingredients or additives that are usually added to these type of products such as spices, salt, protein derivatives of vegetable origin, dietary fibers, antioxidants, flavor enhancers and other additives which result in enhancing the sensory characteristics and the stability of the meat products.

Since the fat content of all burger treatments was about 15%, these products might contain up to 20-30% of fat to give the desirable succulence and texture.

Mixing of chicken with beef meat enhanced the sensory characteristics of the beef. In general, mixed sample had sensory scores higher than beef sample, and were close to the chicken sample. Mixed formulation was the most stable with respect to the sensory characteristics during the storage period. Freshly prepared mixed formulation samples had appearance and color scores (6.94 and 6.89, respectively) higher than those of the beef and chicken samples.(6.11 and 6.83, respectively for appearance) and (5.67 and 6.61, respectively for color). This may be due to the dilution of the redness color of beef meat as well as the dilution of the yellowness of the chicken meat which resulted in moderate appearance and color between beef and chicken meats (between redness and yellowness), since beef meat contains more myoglobin than chicken.

Appearance and color are related sensory qualities, so this modification in color of the mixed treatment affected the appearance, which in role affected the panelist's evaluation.

Tenderness evaluation of meat and meat products by panelists is correlated mainly with juiciness. Therefore, close scores of tenderness and juiciness of beef chicken and mixed treatments were observed. Tenderness and juiciness scores of the mixed formulations were significantly higher than those of beef, and very close to those of chicken. This indicated that tenderness and juiciness are strongly related to the type of meat more than to other factors.

Characteristic	Time of storage (month)	Treatment				
		Beef	Chicken	Mixed	Beef with olive oil	Chicken with olive oil
Appearance	0	${}_a6.11^a$	${}_a6.83^a$	${}_a6.94^a$	${}_a6.00^a$	${}_a6.56^a$
	1	${}_a6.06^{ab}$	${}_{ab}6.22^{ab}$	${}_a7.00^a$	${}_a5.33^b$	${}_a5.61^b$
	3	${}_a5.94^{ab}$	${}_b5.50^b$	${}_a6.83^a$	${}_b4.16^c$	${}_a5.72^b$
Color	0	${}_a5.67^b$	${}_a6.61^{ab}$	${}_a6.89^a$	${}_a5.88^b$	${}_a6.67^{ab}$
	1	${}_a5.56^{bc}$	${}_{ab}6.00^{ab}$	${}_a7.00^a$	${}_a5.39^c$	${}_a5.61^{bc}$
	3	${}_a6.33^a$	${}_b5.11^b$	${}_a6.44^a$	${}_b4.00^c$	${}_a5.61^{ab}$
Tenderness	0	${}_a4.44^b$	${}_a6.56^a$	${}_a6.10^a$	${}_a4.27^b$	${}_a6.44^a$
	1	${}_a4.55^b$	${}_a6.33^a$	${}_a6.72^a$	${}_a4.50^b$	${}_a6.50^a$
	3	${}_a4.60^b$	${}_a6.22^a$	${}_a6.67^a$	${}_a4.33^b$	${}_a6.17^a$
Flavor	0	${}_a4.94^b$	${}_a6.33^a$	${}_a5.78^{ab}$	${}_a5.06^b$	${}_a6.06^a$
	1	${}_a5.06^{bc}$	${}_a5.88^{ab}$	${}_a6.28^a$	${}_{ab}4.72^c$	${}_a5.12^{bc}$
	3	${}_a4.83^{bc}$	${}_a5.50^{ab}$	${}_a5.94^a$	${}_b3.94^c$	${}_a5.63^{ab}$
Juiciness	0	${}_a4.17^b$	${}_a6.44^a$	${}_a6.11^a$	${}_a4.44^b$	${}_a6.17^a$
	1	${}_a4.27^b$	${}_a5.61^a$	${}_a5.67^a$	${}_a4.22^b$	${}_a5.44^a$
	3	${}_a4.44^b$	${}_a5.61^a$	${}_a5.83^a$	${}_a3.56^c$	${}_a5.50^{ab}$
Overall acceptability	0	${}_a5.38^b$	${}_a6.52^a$	${}_a6.39^{ab}$	${}_a5.39^b$	${}_a6.10^{ab}$
	1	${}_a5.00^c$	${}_{ab}6.06^{ab}$	${}_a6.50^a$	${}_{ab}4.44^c$	${}_a5.17^{bc}$
	3	${}_a5.22^a$	${}_b5.51^a$	${}_a6.00^a$	${}_b3.72^b$	${}_a5.83^a$

Means in the same column with the same subscripts denote no significant differences among treatments of burger (p> 0.05) according to LSD.
Means in the same row with different superscripts denote significant differences among treatments of burger (p≤ 0.05) according to LSD.
Means are the average of 18 reading.

Table 7. Effect of formulation and storage time on sensory evaluation scores for the burger samples.

Sensory scores	Appearance	Color	Tenderness	Flavor	Juiciness	Overall acceptability
Appearance	1.00	0.95*	0.60*	0.79*	0.72*	0.90*
Color	0.95*	1.00	0.59*	0.71*	0.70*	0.85*
Tenderness	0.60*	0.59*	1.00	0.78*	0.96*	0.79*
Flavor	0.79*	0.71*	0.78*	1.00	0.84*	0.92*
Juiciness	0.72*	0.70*	0.96*	0.84*	1.00	0.88*
Overall acceptability	0.90*	0.85*	0.79*	0.92*	0.88*	1.00

* Correlation is significant at the 0.05 level

Table 8. Pearson's correlation coefficients between the sensory scores for the burger formulations.

Substitution of meat fat in beef and chicken samples with olive oil, in general, did not affect the sensory characteristics, since no significant differences were found between the sensory scores of the samples with and without olive oil. Beef with olive oil showed lower sensory scores after three months of storage compared to the beef sample with tallow, whereas the sensory characteristics of the chicken with olive oil remained stable during the storage period.

Although chicken with olive oil treatment showed lower cooking loss compared with the chicken treatment The tenderness and juiciness scores of these two treatments were not significantly different.

Storage time did not significantly affect the sensory evaluation scores of each treatment, except for chicken in which the appearance, color and overall acceptability at the end of storage were lower than the initial values. Appearance, color, flavor and overall acceptability of beef with olive oil also were affected by storage time. This decline in sensory parameters of these samples should be attributed to oxidation

In conclusion, it could be observed that the addition of olive oil did not affect the sensory properties of chicken burger, but it had a slight negative effect on these properties of beef burger, and addition of chicken meat to beef burger improved their sensory properties, which was very close to those of chicken sample. In addition, although, the fatty acid oxidation measured by TBARS of all treatments during storage and by grilling was relatively high, but it didn't affect significantly the sensory properties of their samples.

As a result of this research, it is recommended to introduce olive oil in burgers and other potential meat products to improve their nutritional value and to reduce their cholesterol content, and also to produce burger by mixing chicken and beef meat to enhance the sensory properties of the beef and to improve the oxidative stability of the chicken. However, Further studies are needed to determine the most suitable ratio of chicken/beef meat and fat to be used in burger formulas which give the best chemical and sensory properties.

8. References

Chritopherson, S. and Giass, R. (1969). Preparation of milk fat methyl esters by alcoholysis in an essentially nonalcoholic solution. *Journal of Dairy Science,* 52, 1289-1290.

Fernández, J., Pérez-Alvarez, J. and Fernández-López, J. (1997). Thiobarbituric acid test for monitoring lipid oxidation in meat. *Food Chemistry,* 59(3), 345-353.

Faustman, C., Yin, M. and Nadeau, D. (1992). Color stability, lipid stability, and nutrient composition of red and white veal. *Journal of Food Science,* 57, 302-304.

Jordanian Institute for Standards and Metrology, The Hashemite Kingdom of Jordan. (2002). Standard No. JS: 1334/2002, (2nd ed). Meat and Meat Products: *Poultry-Chilled and/or Frozen Chicken /or Burger.* Amman, Jordan. (In Arabic).

Kyoichi. O., Takehiro, K., Koji, Y. and Michihiro, S. (1993). Oxidation of cholesterol by heating. *Journal of the American Chemists' Society,* 41, 1198-1202.

Larmond, E. (1991). *Laboratory Methods for Sensory Evaluation of Food,* (2nd ed). Ottawa: Canadian Department of Agriculture Publication.

Mattson, F. H. and Grundy, S. M. (1985). Comparison of dietary saturated, monounsaturated and polyunsaturated fatty acids on plasma lipids and lipoproteins in man. *Journal of Lipid Research,* 26, 194-203.

Melton, S. L. (1985). Methodology for following lipid oxidation in muscle foods. *Journal of Food Technology*, 38, 105-111.

Papadina, S. N. and Bloukas, J. G. (1999). Effect of fat level and storage conditions on quality characteristics of traditional Greek sausage. *Meat Science*, 51, 103-113.

Park, S. W. and Addis, P. B. (1985). HPLC determination of C7 oxidized cholesterol derivatives in foods. *Journal of Food Science*, 50, 1437-1444.

Pie, J. E., Spaphis, K. and Seillan, C. (1990). Evaluation of oxidation degradation of cholesterol in food and food ingredients: identification and quantification of cholesterol oxides. *Journal of Agricultural and Food Chemistry*, 38, 973-979.

Rodriguez-Estrada, M. T., Penazzi, G., Caboni, M. F., Bertacco, G. and Lercker, G. (1997). Effect of different cooking methods on some lipid and protein components of hamburgers. *Meat Science*, 45, 365-375.

Sander, B. D., Smith, D. E. and Addis, P. B. (1988). Effects of processing stage and storage conditions on cholesterol oxidation products in butter and cheddar cheese. *Journal of Dairy Science*, 71, 3173-3178.

Severini, C., De Pilli, T. and Baiano, A. (2003). Partial substitution of pork backfat with extra-virgin olive oil in 'salami' products: effects on chemical, physical and sensorial quality. *Meat Science*, 64, 323-331.

Wilson, N., Dyett, E., Hughes, R. and Jones, C. (1981). *Meat and Meat Products: Factors Affecting Quality Control*. London: Applied Science Publishers.

11

Meat Products Manufactured with Olive Oil

S.S. Moon[1], C. Jo[2], D.U. Ahn[3], S.N. Kang[4], Y.T.Kim[1] and I.S. Kim[4]

[1]Sunjin Meat Research Center
[2]Chungnam National University
[3]Iowa State University
[4]Gyeongnam National University of Science and Technology
[1,2,4]Korea
[3]USA

1. Introduction

Consumer perception of processed meat products is a critical issue for the meat industry. In recent years consumers are increasingly conscious about healthy diet. However, most of the processed meat products contain high amounts of fat, which are related to chronic diseases such as obesity and cardiovascular heart diseases. Health organizations have suggested to reduce the intake of total dietary fat, particularly saturated fatty acids and cholesterol, as a mean to prevent cardiovascular heart diseases (NCEP, 1988). Consumers now want low or reduced-animal fat products with high palatability and nutritional quality (Pietrasik & Duda, 2000).

Animal fat is a major factor that determines the eating quality of meat products including texture, flavor and mouth-feel (Keeton, 1994). Therefore, reducing fat levels in meat products is not as simple as using less amounts of fat in the formulation. Twenty percent or higher reduction of fat content in meat products can lead to an unacceptable product texture, flavor and appearance (Miles, 1996). Total substitution of fat with water produces unacceptably soft and rubbery product with an increased moisture loss during processing (Claus & Hunt, 1991).

The problems caused by fat reduction in processed meat products can be minimized by replacing animal fat with fat replacers (Colmenero, 1996). Several studies have demonstrated that replacing animal fat with soy products or carbohydrate is successful in textural and sensory properties of low-fat products (Decker et al., 1986; Berry & Wergin, 1993; Yusof & Babji, 1996). Isolated soy proteins (ISP) were successfully incorporated into meat products to reduce fat, improve yields, and enhance emulsion stability. Carageenan increases yield, consistency, sliceability, and cohesiveness, while decreasing purge in low-fat products (Foegeding & Ramsey, 1986; Xiong et al., 1999; Lin & Mei, 2000). Maltodextrin, which is a hydrolysis by-product of starch, is widely used in foods as a funcitonal biopolymer that provides desirable texture, stability, appearance, and flavor (Wang & Wang, 2000).

Olive oil is a vegetable oil with the highest level of monounsaturated fatty acids (MUFA) and has attracted attention as a replcacer for animal fat in processed meat products. Olive oil has a high biological value due to a favorable mix of predominantly MUFA and naturally occurring antioxidants including vitamin E, vitamin K, carotenoids and polyphenols such as hydroxytyrosol, tyrosol and oleuropein. Oleic acid makes up 92% of the MUFA in foods, and 60-80% of the oleic acid comes from olive oil (Pérez-Jiménez et al., 2007). Olive oil contains 56-87% monosaturated, 8-25% saturated and 3.6-21.5% polyunsaturated fatty acids (IOOC, 1984). The potential health benefits of olive oil include an improvement in lipoprotein profile, blood pressure, glucose metabolism and antithrombotic profile. It is also believed that olive oil has a positive influence in reducing inflammation and oxidative stress. Thus, intake of MUFA may protect against age-related cognitive decline and Alzheimer's disease. Olive oil is also reported to help prevent breast and colon cancer (Pérez-Jiménez et al., 2007, Waterman & Lockwood, 2007).

This chapter discusses the effect of olive oil on the quality of emulsion-type sausage (Moon et al., 2008) and pork patty (Hur et al., 2008) when used as an animal fat replacer in the products. The grade of olive oil used were extra virgin olive oil(defined by the European Union Commission reg. No. 1513/2001).

2. Fat replacers in processed meat products

Most efforts in developing low-fat meat products to satisfy concerned consumers have been focused on reducing fat and/or substituting animal fats in the formula with plant oils. Fat is an important determinant for the sensory properties of meat and meat products, and thus a simple reduction of animal fat content in the formulation can lead to a product with poor sensory quality. Therefore, strategies to reduce animal fat while retaining traditional flavor and texture of meat products.

Juiciness and mouthfeel are very closely related to the fat content in meat products. To a large extent these sensory quality can be retained by using binders in low-fat and/or healthy meat products. Binders have been added to meat products for many years for both technological reasons and cost savings. Many binders with a number of different properties are available, but all those used in value-added meat products are to improve water binding capacity. Among the binders, carrageenan is the most widely used in meat industry. According to Varnam & Sutherland(1995), iota-carrageenan with calcium ions forms a syneresis-free, clear plastic gel with good resetting properties after shear. It is particularly recommended for use in low-fat products. Iota-carrageenan has very good water retention properties, and enhance cold solubility and freeze-thaw characteristics of processed products. The presence of NaCl in solution inhibits swelling of carrageenan but this difficulty can be solved by using NaCl encapsulated with partially hydrogenated vegetable oil such as olive oil, soya oil, corn oil and palm oil. Hydrogenated corn oils or palm oils are particularly effective in replacing beef fat. Soya oil emulsion is also effective at levels up to 25%, especially when used in conjunction with isolated soya proteins (Varnam & Sutherland, 1995).

Olive oil can be used in processed meat products an an oil-in-water emulsion form (Hoogencamp, 1989). Briefly, water is heated to 60-65°C. This water is homogenized with the isolated soy protein (42.15%, w/w) and the mixture is cooled to 5°C and then placed in a chilled cutter. After homogenizing for 1 min, olive oil is added while homogenization is continued. Finally, the mixture is homogenized for additional 3 min and then used for manufacturing sausages and patties.

The incorporation of olive oil has been studied in fermented sausages (Bloukas et al., 1997; Kayaardi & Gök, 2003; Koutsopoulos et al., 2008) and beef patties (Hur et al, 2008). Partial replacement of animal fats with olive oil has also been tested (ranging between 3–10 g of olive oil per 100 g of product) in frankfurter sausages and low-fat products. Previous studies (Jiménez-Colmenero, 2007; López-López et al., 2009b) indicated that partial replacement of pork backfat with olive oil increased MUFA contents without significantly altering the n-6/n-3 ratio.

3. Incorporation of olive oil in meat products

To develop healthier meat products, various technological options of replacing animal fat have been studied (Jiménez-Colmenero, 2007). Olive oil has been incorporated in meat emulsion systems such as frankfurters in liquid (Lurueña-Martinez et al., 2004; López-López et al., 2009a, 2009b) or interesterified form (Vural et al., 2004). However, oil-in-water emulsion is the most suitable technological option for stabilizing the non-meat fats added to meat derivatives as ingredients due to physicochemical properties (Bishop et al., 1993; Djordjevic et al., 2004). There are a number of procedures that can be used to produce a plant or marine oil-in-water emulsions (with an emulsifier, typically a protein of non-meat origin) for meat products (Jiménez-Colmenero, 2007), but only sodium caseinate has been used to stabilize olive oil for incorporation in frankfurter-type products (Paneras & Bloukas, 1994; Ambrosiadis et al., 1996; Paneras et al., 1998; Pappa et al., 2000; Choi et al., 2009).

Tables 1 and 2 are examples of fmomulas that use olive oil and different fat replacers in producing an emulsion-type sausage and pork patty.

Ingredients (%)		Control	ICM [1]	ICMO [2]
Pork ham		68.95	73.24	71.57
Pork backfat		19.25	-	-
Ice/water		9.75	7.71	9.38
Fat replacer	ICM[1]	-	17.00	12.00
	Olive Oil	-	-	5.00
NPS [3]		1.30	1.30	1.30
Phosphate		0.20	0.20	0.20
Sugar		0.50	0.50	0.50
Monosodium glutamate		0.05	0.05	0.05
Total		100	100	100

[1] Isolated soy protein: carrageenan: maltodextrin: water = 2:1:1:20.

[2] ICM+Olive Oil.

[3] NaCl: NaNO2 = 99:1.

Table 1. Formulation of emulsion-type low-fat sausages manufactured with and without fat replacers.

	C	T 1	T 2	T 3
Lean pork	83.5	81.0	80.5	80.0
Pork back fat	10.0	5.0	5.0	5.0
Olive oil	-	5.0	5.0	5.0
ISP	-	0.5	0.5	0.5
Carageenan	-	-	0.5	0.5
Maltodextrin	-	-	-	0.5
Salt	1.2	1.2	1.2	1.2
Black pepper	0.3	0.3	0.3	0.3
Water	5.0	7.0	7.0	7.0
Total	100	100	100	100

[1]C, 10 % backfat; T1, 5 % backfat + 5% olive oil + 0.5 % isolated soy protein; T2, 5% backfat + 5% olive oil + 0.5% isolated soy protein + 0.5% carageenan (T2). T3, 5% backfat + 5% olive oil + 0.5% isolated soy protein + 0.5% carageenan + 0.5% martodextrin.

Table 2. Formulation of pork patty with fat replacers

3.1 Chemical composition and nutritional value of meat products manufactured with olive oil

The chemical composition of emulsion-type sausages indicated that fat content was reduced by replacing the pork backfat with ICM, but increased with added olive oil (Table 3). Replacing backfat with fat replacers resulted in increased fat content at day 30 for ICM and day 15 and 30 for ICMO; however, the control was not differ. These results could be due to increased moisture loss (%) with longer storage time. ICM and ICMO had higher moisture content than control. When pork backfat is fully replaced by oil-in-water emulsion, which contains 52% olive oil, the sausage contains approximately 13 g of olive oil per 100 g of product. This means a considerable increase in the proportion of MUFA. Olive oil can make up almost 70% of the total fat content of the sausage. The caloric content of sausages was 225-245 kcal/100 g, and 70% of which were from fat. In traditional sausages, all are supplied by animal fat, whereas, in the sausage replaced with olive oil, the animal fat supplied only 20%. The other 50% is from the olive oil. It was suggested that meat products, strategically or naturally enriched with healthier fatty acids, can be used to achieve desired biochemical effects without dietary supplements or changing dietary habits (Jiménez-Colmenero et al., 2010).

Up to 7 – 13 g of olive oil could be added per 100 g sausages as an animal fat replacer. However, the purpose of replacing animal fat with olive oil is to produce low-fat products, and consequently such high proportion of olive oil is not desirable (Jiménez-Colmenero et al., 2007). One of the fundamental strategies in developing a healthier lipid formula is concentrating active components in target food products to enable the cosumption of recommended intake levels with normal portion sizes. Dietary models provided by the World Health Organization (2003) suggested that MUFA should be the major dietary fatty acids. If MUFAs are the predominant fatty acids in a product, the total fat intake would not be substantial (Pérez-Jiménez et al., 2007).

Protein content of the sausage (ICMO) containing ICM and olive oil was higher than that of the control. This could be attributed to higher lean content and ISP in the formulation of

ICMO. Therefore, the replacement of animal fat with olive oil may produce products with healthier lipid composition (higher MUFAs, mainly oleic acid) without substantial deterioration in nutritional quality.

In pork patty study, moisture content was significantly higher in the products with olive oil+ISP+carageenan (T2) and T2 with maltodextrin (T3) when compared with control and that with olive oil+ISP (T1) (Table 4). In contrast, control and T1 had significantly higher crude protein than T2 and T3. Crude fat content was higher in T1 and T2. The pork patty with olive oil treatment had higher ash content than control. Pietrasik and Duda (2000) reported that the increased weight losses when the reduction of fat is accompanied by an increase in the proportion of moisture, and protein levels remain essentially the same. However, substitution of backfat with olive oil produced pork patty not only with higher in moisture but also higher fat content than control in this study. Thus, it can be assumed that olive oil substitution for backfat may not induce weight loss of pork patty. These results agreed with Pappa et al. (2000) who reported no significant difference in yield when olive oil was replacing pork fat in low-fat frankfurters.

Treatment	Fat (%)	Protein (%)	Moisture (%)
Control			
1 day	19.72±1.56[a]	15.06±0.71[d]	61.96±1.78[d]
15 days	19.36±1.34[a]	15.16±0.49[d]	61.34±1.40[d]
30 days	19.62±1.44[a]	15.34±0.70[d]	61.15±1.46[d]
ICM[1)]			
1 day	3.34±0.63[e]	18.38±0.96[a]	74.58±1.15[a]
15 days	3.21±0.59[e]	18.23±0.84[a]	74.24±1.06[a]
30 days	4.63±0.46[d]	17.79±0.52[ab]	72.77±0.58[b]
ICM O[1)]			
1 day	7.35±0.19[c]	16.70±0.75[bc]	73.24±0.75[ab]
15 days	8.65±0.29[b]	17.27±0.50[ab]	71.08±0.95[c]
30 days	8.58±0.42[b]	16.60±0.49[bc]	71.12±1.06[c]

[1)] See Table 1.
[a-e] Means ± S.E. with different letters in the same column indicate significant differences ($p<0.05$).

Table 3. Chemical composition of emulsion-type low-fat sausages with or without fat replacers

	C	T1	T2	T3
Moisture	60.42±0.65[B2)]	60.32±1.05[B]	62.15±0.22[A]	61.63±0.37[AB]
Crude Protein	23.37±0.44[A]	20.28±0.62[BC]	19.54±0.76[C]	21.30±1.84[B]
Crude fat	14.93±0.90[B]	17.34±0.41[A]	16.29±1.05[A]	14.88±0.85[B]
Ash	1.28±0.02[B]	2.06±0.13[A]	2.02±0.03[A]	2.19±0.11[A]

[1)] See Table 2.
[2) A-C] Means ± SD with different superscripts in the same row significantly differ at p<0.05.

Table 4. Proximate compositions in pork patty made by substituted olive oil for backfat

3.2 Physicochemical properties of meat products manufactured with olive oil

The water holding capacity (WHC) of meat products provide succulent texture and mouthfeel to consumers. A number of studies have proved that there are an inverse relationship between fat content and the amount of water released (Hughes et al., 1997). In Table 5, ICMO was not difference in WHC when compared with the control. It means that olive oil can be combined with other fat replacers such as ISP and carrageenan to improve WHC in meat products. In the case of ICMO, which was emulsified with ISP and carrageenan, the release of water seemed to be protected during storage days.

Cooking loss of meat products is usually influenced by fat content. The products with higher fat content lose less water after cooking ((Jiménez-Colmenero et al., 2007) because high-fat products contain less water. The cook losses of the low-fat sausages manufactured with olive oil and fat replacers (ICM and ICMO) were lower than those of the control (Table 5). However, when the reduction of fat contents in the sausages was considered, the increase of cook loss is not significant. Some fat replacers such as whey protein, carrageenan and tapioca starch could reduce the cook loss of low-fat sausages due to water retainability (Lyons et al., 1999).

Treatment		WHC (%)	Cook loss (%)
Control			
	1 day	71.02±1.17[a]	13.30±0.37[cd]
	15 days	69.52±0.89[ab]	13.18±0.53[d]
	30 days	68.33±0.93[b]	13.86±0.52[bcd]
ICM[1)]			
	1 day	68.32±0.59[b]	14.37±0.82[bc]
	15 days	67.95±0.95[bc]	14.78±0.48[a]
	30 days	66.77±0.59[c]	14.90±0.40[a]
ICMO[1)]			
	1 day	69.79±0.43[ab]	13.13±0.54[d]
	15 days	69.12±1.18[ab]	14.01±0.34[bc]
	30 days	68.28±0.82[b]	14.61±0.52[ab]

[1)] See Table 1., [a-d] Means ± S.E. with different letters in the same column indicate significant differences (*p*<0.05).

Table 5. Water holding capacity (WHC, %) and cook loss (%) of low-fat sausages with or without fat replacers

	C	T1	T2	T3
pH	5.82±0.03[A]	5.75±0.02[B]	5.78±0.01[B]	5.78±0.02[B]
WHC (%)	79.05±2.22[A2)]	72.05±1.12[B]	80.39±14.58[B]	83.99±12.65[A]
Fat retention (%)	79.31±0.02[C]	83.97±0.01 [B]	84.64±1.06 [B]	86.61±1.28 [A]
Cooking loss (%)	28.05±0.70	27.30±0.69	27.72±1.10	26.95±1.61

[1)] See Table 2., [2)] [A-C] Means ± SD with different superscripts in the same row significantly differ at p<0.05.

Table 6. Changes of physical characteristics in pork patty made by substituted olive oil for backfat

On other hand, WHC of pork patty was significantly higher in control and T3 than T1 and T2. Control had higher pH than olive oil-added pork patties, but no significant differences were found among the samples with 50% olive oil substitution for backfat. Fat retention was higher in the olive oil-substuted samples than control. Especially T3, the patty with olive oil+ISP+carageenan, showed the highest fat retention. However, cooking loss was not different among the treatments. In this present study, WHC was steadily decreased as olive oil substitution level increased. However, this does not mean that the quality of pork patty decreased, because fat retention was higher in olive oil-added pork patties, and cooking loss was not significantly different.

In other meat product studies, Kayaardi and Gok (2003) reported that replacing beef fat with olive oil had no effect on the pH value of the Soudjouks samples. Luruena-Martinez et al. (2004) and Muguerza et al. (2002) reported that the addition of olive oil did not produce significant differences in cooking losses of sausage but made the sausage lighter in color and more yellow (Muguerza et al., 2002). In contrast, Bloukas et al. (1997) reported that the higher the olive oil content, the higher the weight loss, probably due to higher amounts of water added. Hur et al. (2008) repoprted that WHC was decreased but fat retention was increased by olive oil substitution.

3.3 Color and lipid oxidation of meat products manufactured with olive oil

Color of meat products is an important quality parameter for purchase decision by consumers. The most common cause for changing color is the formation of metmyoglobin by oxygen-dependent meat enzymes. Aerobic micro-organisms are successfully competing with meat pigments for oxygen. Formation of metmyoglobin can vary, and occasionally discolored areas are present adjacent to and fully demarcated areas where coloration is bright pink. Use of low-quality fat containing high levels of peroxides can cause oxidation of meat pigments (Varnam & Sutherland, 1995).

Varnam & Sutherland(1995) reported that sausages can have a number of specific quality issues: 'Pressure marks' are the result of oxygen deficiency where packed sausages are in close contact to each other. Pigment is initially converted to reduced myoglobin and subsequently, as some diffusion of oxygen occurs, to metmyoglobin. 'White spot' appears to be an oxidative defect, which involves formation of circular grey or white areas that increase in size with continuing storage. It could be associated with low SO_2 levels and use of fats with a high peroxide content.

The sausage incorporated with ICM and olive oil as fat replacers showed higher yellowness and redness (Table 7). Yellower color could be from the original color of olive oil and redder color from higher lean ratio, which includes higher myoglobin content, compared to traditional sausages (control).

Olive oil and ISP are known to have antioxidant properties. The sausages emulsified with ISP and olive oil (ICMO) inhibited lipid oxidation (Table 7). The progress of lipid oxidation can cause changes of meat quality including color, flavor, odor, texture and even the nutritional value in meat products (Fernandez et al., 1997). The stability of fat often limits the shelf life of meat products. The incorporation of olive oil and ISP into meat products may improve the shelf life of the products due to their antioxidant properties. In our study, TBARS values of ICMO were lower than those of the control on days 15 and 30. The TBARS of ICMO sample remained constant throughout the 30 days of storagebut those of the control and ICM increased ($p < 0.05$) from days 15 to 30. The higher TBARS value for the control on each storage day might be due to high fat content in control sausages.

Treatment	Lightness (L^*)	Redness (a^*)	Yellownes s (b^*)	TBARS (mg malonaldehyde/kg sample)
Control				
1 day	78.39±0.37[a]	11.06±0.21[b]	3.45±0.17[b]	0.16±0.03[c]
15 day	77.25±0.64[ab]	10.41±0.19[b]	2.34±0.24[c]	0.22±0.03[b]
30 day	76.41±0.88[b]	10.22±0.09[b]	2.49±0.61[bc]	0.32±0.05[a]
ICM[1)]				
1 day	74.95±0.69[c]	12.13±0.40[a]	3.30±0.16[b]	0.16±0.02[c]
15 day	73.48±0.98[cde]	10.42±0.07[b]	2.42±0.24[bc]	0.14±0.04[cd]
30 day	71.69±1.31[e]	10.29±0.13[b]	2.20±0.05[c]	0.24±0.02[b]
ICMO[1)]				
1 day	73.45±0.18[de]	11.80±0.64[ab]	4.04±0.13[a]	0.17±0.02[c]
15 day	72.49±0.17[e]	10.46±0.25[b]	2.44±0.15[bc]	0.15±0.04[cd]
30 day	72.01±0.65[e]	10.31±0.06[b]	2.79±0.13[bc]	0.20±0.03[bc]

[1)] See Table 1., [a-e] Means ± S.E. with different letters in the same column indicate significant differences ($p<0.05$).

Table 7. Color and lipid oxidation of low-fat sausages with or without fat replacers

L*-value of raw pork patty was higher in control and T1 than other samples, but no significant difference were found after cooking (Table 8). a*-value was significantly higher in control than the samples with olive oil-added products in both raw and cooked states. It can be assumed that redness may be higher in control than olive oil-added pork patties, but lightness and yellowness may not be much different. Paneras et al. (1998) also reported differences in color when low fat frankfurters were produced with different levels of vegetable oils. Low-fat frankfurters were darker, redder and more yellow than high fat frankfurters. However, Marquez et al. (1989) found no differences in color parameters by oil treatments in beef frankfurters. These studies indicated that the change of meat color by oil treatment can vary depending upon the meat products.

	Color	C	T1	T2	T3
Raw sample	L*	55.89±1.46[A2)]	55.31±0.96[A]	52.00±0.62[B]	52.58±1.32[B]
	a*	13.86±0.35[A]	11.75±0.63[B]	11.75±0.45[B]	11.84±0.52[B]
	b*	9.46±0.09	9.77±0.48	9.04±0.70	9.48±0.49
Cooked sample	L*	62.11±5.90	63.98±3.58	66.71±0.40	66.26±1.94
	a*	7.60±0.30[A]	7.02±0.33[B]	6.67±0.13[BC]	6.07±0.24[C]
	b*	9.37±0.73[B]	11.06±0.08[A]	8.57±0.56[C]	9.80±0.93[AB]

[1)] See Table 2.,
[2)] [A-C] Means ± SD with different superscripts in the same row significantly differ at p<0.05.

Table 8. Changes of meat color in pork patty by substituting backfat with olive oil

Chin et al. (1999) and Claus et al. (1990) found that redness and lightness values were more affected by fat/lean ratio and myoglobin concentration of the lean part. Muguerzaet al. (2002) and Bloukas et al. (1997) also found that replacing, in part, backfat with olive oil produced yellower sausages than controls. Muguerza et al. (2002) reported that antioxidant present in olive oil and ISP helped maintaining color by minimizing color oxidation. The present study is in agreement with the findings of other researchers (Kayaardi & Gök, 2003; Ansorena & Astiasarán, 2004; Bloukas et al., 1997) who reported increase of lipid oxidation in meat products during fermentation and ripening period. They found that replacing animal fat with olive oil was effective for inhibiting the lipid oxidation during storage. Our previous and present results indicated that replacing animal fat with olive oil can be effective in inhibiting lipid oxidation in meat products during storage.

3.4 Texture and sensory properties of meat products manufactured with olive oil

Textural properties of the emulsion-type sausages are affected by the replacement of backfat with olive oil emulsion (Table 9). In general, frankfurters made with oil-in-water emulsions presented higher hardness, cohesiveness and chewiness and lower adhesiveness than traditional frankfurters. The textural properties of frankfurters manufactured with olive oil are influenced by the characteristics of oil-in-water emulsion and its role in the meat protein matrix. Frankfurters with olive oil emulsion containing caseinate or soy protein presented similar hardness and chewiness to control, but those with soy protein presents higher springiness and cohesiveness (Jiménez-Colmenero et al. 2010) (Table 9).

The frankfurters containing olive oil emulsion with caseinate or soy protein had higher hardness, cohesiveness, gumminess and chewiness values than the traditional sausages. The result of texture might be due to the reduced fat content sausages. In high fat frankfurters, in which pork backfat is replaced by olive oil, generally have less flavor intensity and are harder and less juicy (Jiménez-Colmenero et al., 2010). However, these differences are marginal, and the frankfurters received similar scores for general appearance and acceptability (Jiménez-Colmenero et al., 2010). Partial substitution of animal fat with olive oil reduced juiciness scores.

Parameter	Control	ICM[1]	ICMO[1]
Hardness (kg)	0.33±0.04b	0.42±0.02a	0.40±0.03a
Cohesiveness	60.85±1.52b	66.47±0.90a	66.09±0.54a
Springiness	13.11±0.27	13.53±0.04	13.23±0.24
Gumminess (g)	19.26±0.88b	22.09±0.65a	21.74±0.30a
Chewiness (g)	228.70±6.02b	271.28±6.30a	268.11±8.55a

[1] See Table 1.
a-b Means ± S.E. with different letters in the same row indicate significant differences ($p<0.05$).¶(9pt)

Table 9. Textural attributes of low-fat sausages with or without fat replacers

The textural properties of pork patties are presented in Table 10. Brittleness and hardness were significantly higher in the patties with olive oil than control, whereas springiness was the lowest in T1. Cohesiveness, gumminess and chewiness were significantly higher in T2 and T3 than control and T1. Chin et al. (1999) found higher hardness values when animal fat was replaced with a mixture of ISP and carrageenan in 30% fat bologna sausages. These results are similar to the findings of Crehan et al. (2000), who reported that added maltodextrin treatment as a fat replacer had higher hardness, gumminess and chewiness than control in 12% fat sausages. The present study was also supported by the findings of Pietrasik and Duda (2000) who reported that replacing backfat with the mixture of carrageenan and ISP was positively correlated with hardness, cohesiveness, gumminess and chewiness. Bloukas et al. (1997) found that fermented sausages with direct incorporation of olive oil in liquid form were softer than control sausages. Luruena-Martinez et al. (2004) also reported that olive oil addition together with fat reduction caused a significant decrease in hardness and the related parameters such as chewiness and gumminess due to high monounsaturated fat in the product. In contrast, we found that pork patties made with olive oil were not only harder but also higher in other mastication power compared with control. Usually, a decrease in textural properties with the increase in olive oil are expected because a solid fat is replaced with a liquid oil. (, the changes of mechanical texture should be influenced by other ingredients such as a carageenan and maltodextrin used in this study.

	C	T1	T2	T3
Brittleness (g)	0.42±0.11[B2]	0.72±0.17[A]	0.72±0.03[A]	0.60±0.17[AB]
Hardness (g)	470±40.0[B]	720±16.0[A]	730±40.0[A]	600±17.0[A]
Cohesiveness (%)	49.44±6.49[AB]	37.53±10.17[B]	52.04±1.74[A]	54.09±6.34[A]
Springiness (%)	13.64±0.08[A]	11.83±1.67[B]	13.66±0.31[A]	13.69±0.15[A]
Gumminess (g)	23.08±2.09[B]	27.58±12.44[AB]	37.84±2.74[A]	31.75±5.72[AB]
Chewiness (g)	314.87±27.14[B]	312.43±90.27[B]	517.31±47.06[A]	434.42±75.27[A]

[1] See Table 2.
[2] A-B Means ± SD with different superscripts in the same row significantly differ at $p < 0.05$.

Table 10. Changes in the textural properties of pork patties by substituting backfat with olive

In sensory evaluation, ICMO was rated the lowest for color and overall acceptability when compared with the control, traditional sausages (Table 9). Muguerza et al. (2002) reported that sausages, which replaced 30 or 20% backfat with 20% olive oil, were rated worse for color, odor and taste than without added olive oil. However, panels did not recognize the differences in flavor and juiciness between ICMO and traditional sausages in the present study. Bloukas and Paneras (1993) found that low-fat frankfurters (11% fat content) with olive oil had similar flavor but were less palatable than the traditional frankfurters (28% fat content). Lyons et al. (1999) also found that the combination of whey protein concentrate, carrageenan and starch resulted in a low-fat sausage with similar mechanical and sensory characteristics to 20% full-fat sausages. High fat sausages (26%) are less firm and juicy than low-fat sausages (10%) made with a combination of olive, cottonseed and soybean oils but it is difficult to realize the differences in overall acceptability (Jiménez-Colmenero et al., 2010).

Sensory attributes	Control	ICM[1]	ICMO[1]
Color	6.10±0.88[a]	6.50±0.97[a]	4.60±0.70[b]
Aroma	5.60±0.70	5.90±0.48	5.50±0.53
Flavor	5.90±0.88	6.10±0.74	5.50±1.08
Tenderness	5.36±0.42[b]	6.10±0.37[a]	5.87±0.64[ab]
Juiciness	5.90±0.74	6.00±0.94	6.00±1.05
Overall acceptability	6.10±0.74[ab]	6.25±0.79[a]	5.50±0.85[b]

[1] See Table 1.

[a-b] Means ± S.E. with different letters in the same row indicate significant differences ($p<0.05$).

Table 11. Sensory attributes of low-fat sausages with or without fat replacers

The sensory evaluation of pork patties (Table 12) indicated that color, aroma and flavor of control were higher than those of the olive oil-added ones, whereas tenderness was higher in olive oil-added samples.

	C	T1	T2	T3
Color	6.90±0.32[A]	6.40±0.52[AB]	6.30±0.67[B]	6.50±0.53[AB]
Aroma	6.90±0.88[A]	5.70±0.48[B]	5.70±0.48[B]	5.40±0.52[B]
Flavor	6.40±0.52[A]	5.60±0.70[B]	5.60±0.52[B]	5.60±0.70[B]
Tenderness	5.20±0.42[B]	5.70±0.67[AB]	5.50±0.53[AB]	5.90±0.74[A]
Juiciness	5.00±0.82	4.70±0.67	4.80±0.63	4.90±0.74
Overall acceptability	7.20±0.42[A]	6.40±0.84[B]	6.50±0.71[B]	6.80±0.63[AB]

[1] See Table 2.,

[2] [A-B] Means ± SD with different superscripts in the same row significantly differ at $p<0.05$.

Table 12. Changes of sensory evaluation value in pork patty made by substituted olive oil for backfat

Control was significantly higher in overall acceptability than olive oil-added pork patties. The substitution of pork backfat with olive oil is limited as it may affect the taste of the pork patty. Pappa et al. (2000) reported that the replacing pork backfat with olive oil positively affected the overall acceptability of the low-fat frankfurters. In contrast, Bloukas and Paneras (1993) reported that low-fat frankfurters produced by total replacement of pork backfat with olive oil had lower overall palatability than high-fat frankfurters produced with pork backfat. The ingredients used or the amount of olive oil added in the formula could have influenced this difference in sensory scores. Also, the effect of olive oil substitution of backfat on quality can vary depending upon meat products. The patties with olive oil had lower sensory evaluation scores. Meanwhile, tenderness was higher in the sample with olive oil than the control. Paneras et al. (1998) reported that low-fat frankfurters produced with vegetable oils were firmer and less juicy than high-fat controls. A possibility of reducing the negative effects due to the high fat content of these products is partially substituting pork backfat with other ingredients (Muguerza et al., 2001). Fat is very important for the rheological and structural properties of meat products and the formation of a stable emulsion (Luruena-Martinez et al., 2004). The tenderness of olive oil-added pork patties were higher than control because olive oil is more fluid than backfat in sensory evaluation.

4. Conclusion

The addition of olive oil to a mixture of fat replacer resulted in somewhat undesirable color and overall acceptability, but lipid oxidation was inhibited. Soem quality problems including color of sausages can be minimized by combining carrageenan, maltodextrin and isolated soy protein with olive oil. The physical properties of pork patties made with olive oil emulsions were stable when compared with commercial pork patties, but they were significantly influenced by other ingredients in the oil emulsions. In conclusion, the use of olive oil in meat products to replace backfat may have a beneficial effect to human health. However, sensory quality of the products needs further improvment so that the product is compatible to conventional products

5. Acknowledgment

This work was supported by Priority Research Centers Program through the National Research Foundation of Korea (NRF) funded by the Ministry of Education, Science and Technology (2009–0093813).

6. References

Ambrosiadis, J.; Vareltzis, K. P. & Georgakis, S. A. (1996), Physical, chemical and sensory characteristics of cooked meat emulsion style products containing vegetable oils. *International Journal of Food Science and Technology*, Vol.31, No.2, (April 1996), pp. 189–194.

Ansorena, D. & Astiasarán, I. (2004), Effect of storage and packaging on fatty acid composition and oxidation in dry fermented sausages made with added olive oil and antioxidants. *Meat Science*, Vol.67, No.2, (June 2004), pp. 237-244.

Bishop, D. J.; Olson, D. G. & Knipe, C. L. (1993), Pre-emulsified corn oil, pork fat, or added mositure affect quality of reduced fat bologna quality. *Journal of Food Science*, Vol.58, No.3, (May 1993), pp. 484–487.

Bloukas, J. G. & Paneras, E. D. (1993). Substituting olive oil for pork backfat affects quality of low-fat frankfurters. *Journal of Food Science*. Vol.58, No.4, (July 1993), pp. 705-709.

Bloukas, J. G.; Paneras, E. D. & Fournitzis, G. D. (1997), Effect of replacing pork backfat with olive oil on processing and quality characteristics of fermented sausages. *Meat Science*, Vol.45, No.2, (February 1997), pp. 133-144.

Chin, K. B.; Keeton, J. T.; Longnecker, M. T. & Lamkey, J. W. (1999), Utilization of soy protein isolate and konjac blends in a low gat bologna(model system). *Meat Science*, Vol.53, No.1, (September 1999), pp. 45-57.

Choi, Y. S.; Choi, J. H.; Han, D. J.; Kim, H. Y.; Lee, M. A.; Kim, H. W.; Jeong, J. Y. & Kim, C. J. (2009). Characteristics of low-fat emulsion systems with pork fat replaced by vegetable oils and rice bran fiber. *Meat Science*, Vol.82, No.2, (June 2009), pp. 266–271.

Claus JR,; Hunt MC. & Kastner CL. (1990), Effects of substituting added water for fat on the textural, sensory, and processing characteristics of bologna. *J. Muscle Foods*, Vol.1, No.1, (January 1990), pp. 1-21.

Claus, J. R. & Hunt, M. C. (1991). Low-fat, high-added water bologna formulated with texture-modifying ingredients. J. Food Sci. Vol.56, No.3, (May 1991), pp. 643-647.

Colmenero, F. J. (1996), Technologies for developing low-fat meat products. *Trends in Food Sci Technol.*, Vol.7, (1996), pp. 41-48.

Crehan, C. M.; Hughes, E.; Troy, D. J. & Buckley, D. J. (2000), Effects of fat level and maltodextrin on the functional properties of frankfurters formulated with 5, 12 and 30% fat. *Meat Science*, Vol.55, No.4, (August 2000), pp. 463-469.

Decker, C. D.; Conley, C. C. & Richert, S. H. (1986), *Use of isolated soy protein in the development of frankfurters with reduced level of fat, calories, and cholesterol*. Proceedings of the European Meeting of Meat Research Workers. Food Science and Technology. Vol.7, No.32, (1986), pp. 333-336.

Djordjevic, D.; McClements, D. J. & Decker, E. A. (2004), Oxidative stability of whey protein-stabilized oil-in-water emulsions at pH 3: Potential omega-3 fatty acid delivery systems (Part B). *Journal of Food Science*, Vol.69, No.5,(June 2004), pp. C356–C362.

Fernandez, J.; Perez-Alvarez, A. & Fernandez-Lopez, J. A. (1997), Thiobarbituric acid test for monitoring lipid oxidation in meat. *Food Chem.*, Vol.59, No.3, (July 1997), pp. 345-353.

Foegeding, E. A. & Ramsey, S. R. (1986). Effects of gums on low-fat meat batters. *J. Food Sci.* Vol.51,No. 1, (January 1986), pp. 33-36.

Honikel, K. O. (1987), The water binding of meat. *Fleischwirtschaft*. Vol.67, (1987), pp. 1098-1102.

Hoogenkamp, H. W. (1989), Low-fat and low-cholesterol sausages. *Fleischwirtschaft*. Vol.40, (1989), pp. 3-4.

Hughes, E.; Cofrades, S. & Troy, D. J. (1997), Effects of fat level, oat fibre and carrageenan on frankfurters formulated with 5, 12 and 30% fat. *Meat Science*, Vol.45, No.3, (March 1997), pp. 273-281.

Hur, S. J.; Jin, S. K., & Kim, I. S. (2008), Effect of extra virgin olive oil substitution for fat onquality of pork patty. *Journal of the Science of Food and Agriculture*, Vol.88, No.7, (March 2008), pp. 1231–1237.

IOOC (International Olive Oil Council). (1984), *International trade standards applying to olive oil and olive residue oils.* COI/T.

Jiménez-Colmenero, F. (2007), Healthier lipid formulation approaches in meat-based functional foods. Technological options for replacement of meat fats by non-meat fats. *Trends in Food Science and Technology,* Vol.8, No.11, (2007), pp. 567–578.

Jiménez-Colmenero, F.; Herrero, A.; Pintado, T.; Solas, M. T. & Ruiz-Capillas, C. (2010), Influence of emulsified olive oil stabilizing system used for pork backfat replacement in frankfurters. *Food Research International.* Vol.43, No.8, (October 2010), pp. 2068-2076.

Kayaardi, S. & Gök, V. (2003), Effect of replacing beef fat with olive oil on quality characteristics of Turkish soudjouk (sucuk). *Meat Science,* Vol.66, No.1, (January 2004), pp. 249–257.

Keeton J. T. (1994), Low-fat meat products-technological problems with processing. *Meat Science,* Vol.36, No. 1-2, (1994), pp. 261-276.

Koutsopoulos, D. A., Koutsimanis, G. E., & Bloukas, J. G. (2008), Effect of carrageenan level and packaging during ripening on processing and quality characteristics of low-fat fermented sausages produced with olive oil. *Meat Sci,.* Vol.79, No.1, (May 20008), pp. 188–197.

Lin, K. W. & Mei, M. Y. (2000), Influences of gums, soy protein isolate, and heating temperature on reduced-fat meat batters in a model system. *J. Food Sci.* Vol.65, No.1, (January 2000), pp. 48-52.

López-López, I.; Bastida, S.; Ruiz-Capillas, C.; Bravo, L.; Larrea, T.; Sanchez-Muniz, F.; Cofradesa, S. & Jiménez-Colmenero, F. (2009a), Composition and antioxidant capacity of low-salt meat emulsion model systems containing edible seaweeds. *Meat Science,* Vol.83, No.3, (November 2009), pp. 255–262.

López-López, I.; Cofrades, S. & Jiménez-Colmenero, F. (2009b), Low-fat frankfurters enriched with n–3 PUFA and edible seaweed: Effects of olive oil and chilled storage on physicochemical, sensory and microbial characteristics. *Meat Science,* Vol.83, No.1, (May 2009), pp. 148–154.

Lurueña-Martínez, M. A.; Vivar-Quintana, A. M. & Revilla, I. (2004), Effect of locust bean/xanthan gum addition and replacement of pork fat with olive oil on the quality characteristics of low-fat frankfurters. *Meat Science,* Vol.68, No.3, (Nobember 2004), pp. 383–389.

Lyons, P. H.; Kerry, J. F.; Morrissey, P. A. & Buckley, D. J. (1999), The influence of added whey protein/carrageenan gels and tapioca starch on the textural properties of low fat pork sausages. *Meat Science,* Vol.51, No.1, (January 1999), pp. 43-52.

Marquez, E. J.; Ahmed, E. M.; West, R. L. & Johnson, D. D. (1989), Emulsion stability and sensory quality of beef frankfurters produced at different fat or peanut oil levels. *Journal of Food Science.* Vol.54, No.4, (July 1989), pp. 867-873.

Miles R. S. (1996), *Processing of low fat meat products.* Proceedings of 49th Reciprocal Meat Conference, American Meat Science Association, Chicago, IL, (June 1996).

Moon, S. S.; Jin, S. K.; Hah, K. H. & Kim, I. S. (2008), Effects of replacing backfat with fat replacers and olive oil on the quality characteristics and lipid oxidation of low-fat sausgae during storage. *Food Sci. Biotechnol.* Vo.17, No.2, (April 2008), pp. 396-401.

Muguerza, E., Gimeno, O.; Ansorena; D., Bloukas, J. G. & Astiasaran, I. (2001), Effect of replacing pork backfat with pre-emulsified olive oil on lipid fraction and sensory quality of Chorizo de Pamplona – a traditional Spanish fermented sausage. *Meat Science*, Vol.59, No.4, (November 2001), pp. 251-258.

Muguerza, E.; Fista, G.; Ansorena, D.; Astiasarán, I. & Bloukas, J. G. (2002), Effect of fat level and partial replacement of pork backfat with olive oil on processing and quality characteristics of fermented sausages. *Meat Science,* Vol.61, No.4, (August 2002), pp. 397-404.

NCEP (National Cholesterol Education Program). (1988), The effect of diet on plasma lipids, lipoproteins and coronary heart disease. *J. American Diet Association,* No.88, (1988), pp. 1373-1400.

Paneras, E. D. & Bloukas, J. G. (1994), Vegetable-oils replace pork backfat for low-fat frankfurters. *Journal of Food Science*, Vol.59, No.4, (July 1994), pp. 725–733.

Paneras, E. D., Bloukas, J. G. and Filis, D. G. (1998), Production of low-fat frankfurters with vegetable oils following the dietary guide lines for fatty acids. *Journal of Muscle Foods*, Vol.9, No. 2, (April 1998), pp. 111-126.

Pappa, I. C.; Bloukas, J. G. & Arvanitoyannis, I. S. (2000), Optimization of salt, olive oil and pectin level for low-fat frankfurters produced by replacing pork backfat with olive oil. *Meat Science*, Vol.56, No.1, (July 2000), pp. 81–88.

Pérez-Jiménez, F.; Ruano, J.; Perez-Martinez, P.; Lopez-Segura, F. & Lopez-Miranda J. (2007), The influence of olive oil on human health: not a question of fat alone. *Mol Nutr Food Res.*, Vol.51, No.10, (Octover 2007), pp. 1199-1208.

Pietrasik Z. & Duda Z. (2000), Effect of fat content and soy protein/carrageenan mix on the quality characteristics of comminuted, scalded sausages. *Meat Science*, Vol.56, No.2, (October 2000), pp. 181-188.

Varnam, A. H. & Sutherland, J. P. (1995), Meat and meat products. Chapman and Hall, ISBN 0-412-49560-0, London, UK.

Vural, H.; Javidipour, I. & Ozbas, O. O. (2004), Effects of interesterified vegetable oils and sugarbeet fiber on the quality of frankfurters. *Meat Science*, Vol.67, No.1, (May 2004), pp. 65–72.

Wang, Y. J. & Wang, L. (2000), Structures and properties of commercial maltodextrins from corn, potato, and rice starches. *Starch*, Vol.52, No. 8-9, (September 2000), pp. 296-304.

Waterman, E. & Lockwood, B. (2007), Active components and clinical applications of olive oil. *Alternative Medicine Review*, Vol.12, No.4, (December 2007), pp. 331–342.

WHO. (2003), *Preventing and Combating Cardiovascular Diseases in the Community (Technical Report Series 732) WHO Expert Committee Report*, World Health Organization, Geneva, Switzerland.

Xiong, Y. L.; Noel, D. C. & Moody, W. G. (1999), Textural and sensory properties of low-salt beef sausage with added water and polysaccharides affected by pH and salt. *J. Food Sci.* Vol.64, No.3, (May 1999), pp. 550-554.

Yusof S. C. & Babji, A. S. (1996), Effect of non-meat proteins, soy protein isolate and sodium caseinate, on the textural properties of chicken bologna. *Inter J. Food Sci Nutr.*, Vol.47, No.4, (July 1996), pp. 323-329.

12

Olive Oil as Inductor of Microbial Lipase

Marie Zarevúcka
Institute of Organic Chemistry and Biochemistry,
Czech Republic

1. Introduction

Increasing interest in lipases has been observed at the end of the last century, due to their potential application, in (bio)degradation as well as in (bio)synthesis of glycerides. The advantages of the enzymatic hydrolysis over the chemical process consist of less energy requirements and higher quality of the obtained products. Beside this, lipases are also efficient in various reactions such as esterification, transesterification and aminolysis in organic solvents. Examples in the literature are numerous. Lipases are used in different fields such as resolution of racemic mixtures, synthesis of new surfactants and pharmaceuticals, bioconversion of oils and fats and detergency applications.

Lipase activity has been found in different moulds, yeasts and bacteria. Numerous papers have been published on selection of lipase producers and on fermentation process. This kind of information is important in order to identify optimal operation conditions for enzyme production. Previous studies on the physiology of lipase production showed that the mechanisms regulating biosynthesis vary widely in different microorganisms. Obtained results showed that lipase production seems to be constitutive and independent of the addition of lipid substrates to the culture medium. However, their presence can enhance the level of produced lipase activity. On the other hand, it is well known that, in other microorganisms, lipid substrates are necessary for lipase production. These enzymes are generally produced in the presence of a lipid such as oil or triacylglycerols or any other inductor, such as fatty acids. Lipidic carbon sources seem to be essential for obtaining a high lipase yield. The review is focused on the olive oil as lipase inductor.

1.1 Lipases

Lipases, (triacylglycerol acylhydrolases; EC 3.1.1.3.) are one of the most important classes of hydrolytic enzymes that catalyse both hydrolysis and synthesis of esters. Hydrolysis of a triacylglycerols by lipases can yield di- and monoacylglycerols, glycerol and free fatty acids. Lipases are valuable biocatalysts with diverse applications. Although lipases share only 5% of the industrial enzyme market, they have gained focus as biotechnologically valuable enzymes. They play vital roles in food, detergent and pharmaceutical industries.

Commercial microbial lipases are produced from bacteria, fungi and actinomycetes (Babu & Rao, 2007). Their industrial importance arises from the fact that they act on a variety of substrates promoting a broad range of biocatalytic reactions. Lipases from different sources show different substrate specificities and they are widely used in industrial applications for biosynthesis (Jaeger & Eggert, 2002).

Most of the lipases, which are used in laboratory investigations and/or in industrial production, are substrate tolerant enzymes, which accept a large variety of natural and synthetic substrates for biotransformation. Microbial lipases are mostly inducible extracellular enzymes, synthesized within the cell and exported to its external surface or environment. Lipases are ubiquitous enzymes which are widely distributed in plants, microbes and higher animals. Microbial sources are superiour to plants and animals for enzyme production and this can be attributed to easy cultivation and genetic manipulation (Hasan et al., 2006). Each microorganism requires a different carbon source to produce lipase at its maximum level.

Microbial lipases are mostly extracellular and their production is greatly influenced by medium composition besides physicochemical factors such as temperature, pH, and dissolved oxygen. The major factor for the expression of lipase activity has always been reported as the carbon source, since lipases are inducible enzymes. These enzymes are generally produced in the presence of a lipid such as oil or triacylglycerol or any other inductor, such as fatty acids, hydrolysable esters, Tweens, bile salts, and glycerol. Lipidic carbon sources seem to be essential for obtaining a high lipase yield. However, nitrogen sources and essential micronutrients should also be carefully considered for growth and production optimization. These nutritional requirements for microbial growth are fulfilled by several alternative media as those based on defined compounds like sugars, oils, and complex components such as peptone, yeast extract, malt extract media, and also agroindustrial residues containing all the components necessary for microorganism development. A mix of these two kinds of media can also be used for the purpose of lipase production. The main studies available in the literature since 2000 covering these subjects are presented below, divided by the kind of microorganisms used (Fernandes et al., 2007; Li et al., 2004; Tan et al., 2003).

1.2 Lipase catalytic properties

Lipase hydrolysis of water-insoluble substrates results from adsorption of the enzyme to the substrate-water interface, which can induce a conformational change in the enzyme structure, causing reaction rates to be influenced by both this adsorptive interaction as well as interaction with substrates. When lipases are active in organic solvents in which substrates are soluble, reactions follow normal enzyme kinetic models (Martinelle & Hult, 1995).

Reactions in which lipases may be involved, both in nature and in laboratory or industrial application, are: (a) enzyme-catalyzed hydrolysis, (b) enzyme-catalyzed esterification, (c) enzyme-catalyzed transesterification by acidolysis, (d) enzyme-catalyzed transesterification by alcoholysis, (e) enzyme-catalyzed interesterification and (f) enzyme-catalyzed aminolysis (Scheme 1).

1.3 Plant oils

The major components of fats and vegetable oils (98%) are triacylglycerols, which consist of glycerol molecules esterified with three long-chain fatty acids. The remainder of the oil, although only a small part in proportion to triacylglycerols, includes a very large number of minor compounds, including the phenolics and the sterols. These compounds give olive oil its unique flavour and contribute greatly to the nutritional benefits.

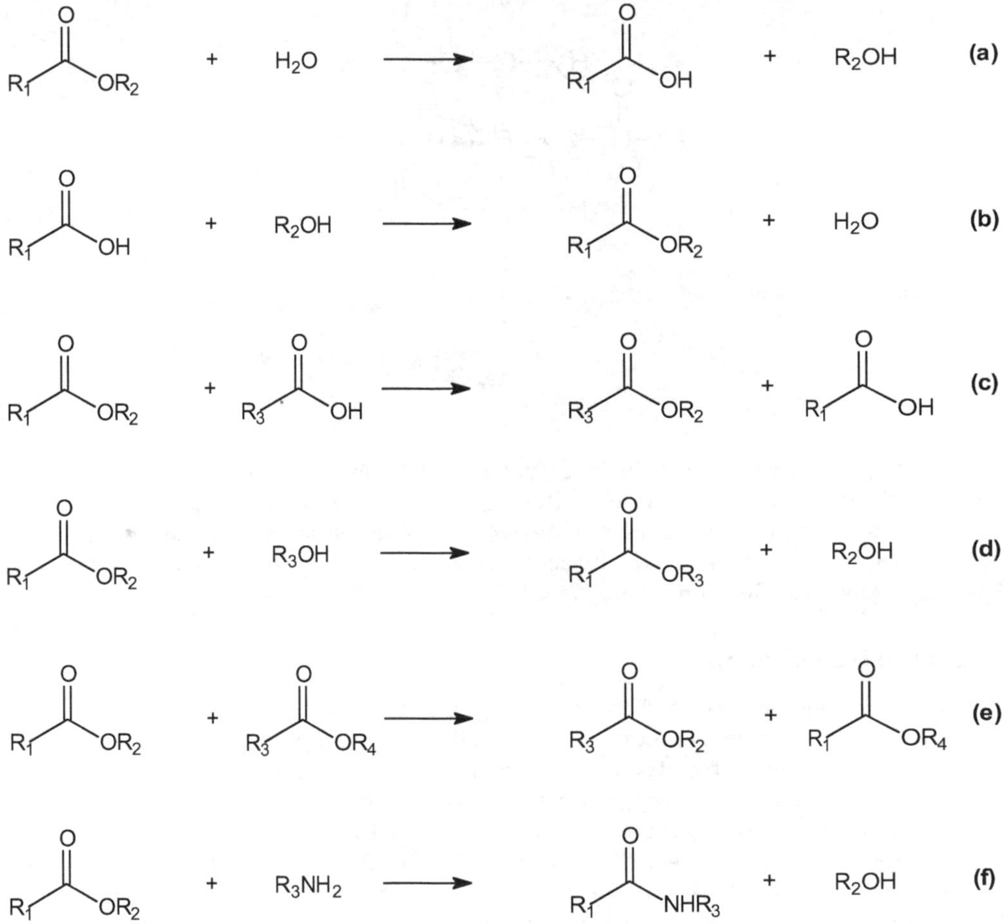

Scheme 1. Processes catalyzed by lipases: (a) enzyme-catalyzed hydrolysis, (b) enzyme-catalyzed esterification, (c) enzyme-catalyzed transesterification by acidolysis, (d) enzyme-catalyzed transesterification by alcoholysis, (e) enzyme-catalyzed interesterification and (f) enzyme-catalyzed aminolysis.

The structure of triacylglycerol molecule is depicted in Fig. 1. The seed triacylglycerols are usually characterized by predominance of C_{18}-unsaturated and polyunsaturated fatty acids, and this distinguishes them from animals fats, which are generally of a more saturated nature. The C_{18}-unsaturated fatty acids (oleic, linoleic, and linolenic) are particularly important and govern, to a large degree, the physical properties of the oil and hence its use and commercial value.

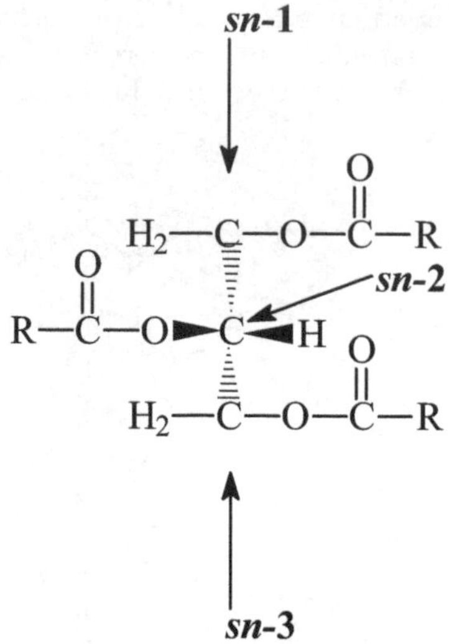

$$sn\text{-}1$$

$$\downarrow$$

$$\uparrow$$

$$sn\text{-}3$$

Fig. 1. The particular fatty acids in the plant triacylglycerols are not distributed randomly between the different sn-carbon atoms. It is a general rule that saturated species of fatty acids are confined to the positions sn-1 and sn-3 with some enrichment in the first position, and that the polyunsaturated C_{18} fatty acids are located mainly at position sn-2 (Gunstone & Ilyas-Qureshi, 1965; Gunstone et al., 1965)

2. Lipase biosynthesis

Lipase production requires carbon and nitrogen sources as required by any fermentation process. Most of the lipases production studies do not use simple sugars as carbon sources. They rather use lipid substrates as sole carbon sources (Zhang et al., 2009a; Zhang et al., 2009b; Hun et al., 2003). Lipase production is rarely constitutive and the quantity of the extracellular lipase produced is meagre (Lee et al., 2001). Hence inductors like vegetables oils (Kumar et al., 2005), Tween 20, Tween 80 (Li et al., 2004), hexadecane (Boekema et al., 2007), and synthetic like tributyrin and tripalmitin, are used. Generally, production of lipases increases when the relative percentage of C18:n fatty acid esters in the respective vegetable oil is increased; this indicates the importance of such substances in the synthesis and secretion of the enzyme (Lakshmi et al., 1999). Among vegetable oils, olive oil has also been referred as one of the best inductors of lipase production (Table 1; Sokolovska et al., 1998). To elucidate some aspects of the induction effect of lipid related substrates on lipase production, mixed carbon sources consisting of a soluble compound selected for its growth-promoting capacity and a fatty acid selected as inductor of the enzyme production were used (Dalmau et al., 2000). This strategy allows for the microorganism to use both substrates in a sequential or simultaneous way, depending on its metabolism. Biomass production can be higher in most cases, no increase in lipolytic activity can be observed. This suggested a possible competing effect of some soluble carbon sources or a close relation between extracellular lipase activity production and consumption of fatty acids.

Source of lipase	References
Aspergillus sp.	Cihangir & Sarikaya, 2004; Papanikolaou et al., 2011
Aspergillus niger	Pokorny et al., 1994
Aspergillus niger MYA 135	Colin et al., 2011
Bacillus sp.	Sugihara et al., 1991; Eltaweel et al., 2005
Bacillus subtilis NS 8	Olusean et al., 2011
Burkholderia cepacia LTEB11	Baron et al., 2011
Candida sp.	Annibale et al., 2006; Brozzoli et al., 2009
Candida cylindracea (ATCC 14830)	Salihu et al., 2011
Candida rugosa (DSM 2031)	Lakshmi et al., 1999
Geotrichum candidum 4013	Stránsky et al., 2007; Brabcová et al., 2010
Mycotorula sp.	Peters & Nelson, 1948
Penicillium sp.	Lima et al., 2003; Papanikolaou et al., 2011
Penicillium aurantiogriseum	Lima et al., 2003
Penicillium cyclopium	Chahinian et al., 2000
Pseudomonas aeruginosa KKA-5	Sharon et al., 1998
Rhizopus arrhizus	Elibol & Oyer, 2000
Rhizopus delemar	Acikel et al., 2011
Rhizopus oryzae	Hiol et al., 2000; Salleh et al., 1993; Nunes et al., 2011
Rhodotorula glutinis	Papaparaskevas et al., 1992
Serratia rubidaea	Immanuel et al., 2008
Yarrowia lipolytica	Dominguez et al., 2003; Pigněde et al., 2000; Najjar et al., 2011

Table 1. Sources of lipases induced by olive oil

The induction process can be accomplished by adding edible oils such as butter fat, olive, canola and fish oils to the fermentation medium. It is well known that certain lipids in the culture medium can influence the production and activity of lipases from microorganisms. Generally, the activity of intra and extracellular lipases increases with increasing lipid concentrations, although excessive levels in the growth medium may be cytotoxic. The mechanisms regulating lipase biosynthesis vary widely in different microorganisms.

2.1 *Aspergillus niger*

Lipases from *Aspergillus niger* were induced by solid-state fermentation using, as substrate, agroindustrial residue supplemented with by-products from corn oil refining process or olive oil. Based on the values of lipase activity obtained after 48 hour fermentation by-products from corn oil refining were tested as inductors in the preparation of fermentation medium. The best results were achieved with soapstock and stearin, reaching values of 62.7 and 37.7 U/gds, respectively, which are higher than the value for olive oil (34.1 U/gds). The use of fatty acids residue inhibited lipase production. This kind of inhibition has already been reported by other authors (Corzo & Revah, 1999; Li et al., 2004). The inhibition effect was not observed for low fatty acid concentrations using palmitic and oleic acid during lipase production by *Candida rugosa* (Dalmau et al., 2000) and *Rhyzopus arrhizus* (Li et al., 2006), respectively.

2.2 *Candida rugosa*

The synthesis and secretion of lipases in *C. rugosa* have been studied with carbon sources that are known to affect the production of lipase in two opposite ways: glucose (repressor) and oleic acid (inductor; Ferrer et al., 2001). In these studies, lipase production was monitored both by enzyme activity and by immunodetection with specific antibodies. These studies showed that, according to their regulation, lipase-encoding genes might be grouped in two classes, one of which is constitutively expressed and the other is induced by fatty acids. The synthesis of inducible enzymes is inhibited at the level of transcription by the addition of glucose, and, conversely, oleic acid appears to hinder the synthesis of the constitutive lipase (Lotti et al., 1998).

The studies clearly show that different inductors may change the expression profile of individual lipase genes. A differential transcriptional control of *lip* genes had been previously suggested from several studies on the relationship between culture conditions of *C. rugosa* and the lipase/esterase profiles secreted by this organism (Gordillo et al., 1995; Lotti et al., 1998; Linko & Wu, 1996). *Lip* isoenzymes have differences in their catalytic properties (Rua et al., 1993; Diczfalusy et al., 1997; Tang et al., 2000).

Del Río et al. (Del Río et al., 1990) demonstrated the diauxic growth of *C. rugosa* on olive oil. Two stages could be observed in the consumption of the olive oil: the first one was related to the glycerol depletion without lipase production, and the second one was associated with the fatty acids consumption when the enzyme appeared in the medium. According to this observation, the initial presence of a small quantity of lipase would be sufficient to hydrolyze the triacylglycerol to glycerol and fatty acids. Therefore, production of high levels of lipase would be associated with the consumption of fatty acids. Similar results have been obtained by Sokolovska et al. (Sokolovska et al., 1998), who used olive oil and oleic acid for lipase production. It has been observed that the uptake of oleic acid by *C. rugosa* is favored by the presence of extracellular lipases (Montesinos et al., 1996). Based on the observations and hypothesis just described, Serra et al. (Serra et al., 1992) calibrated and validated a model for lipase production on olive oil and free fatty acids in batch fermentation. Sokolovska et al. (Sokolovska et al., 1998) did not observe significant differences in lipase production using these substrates. Montesinos et al. (Montesinos et al., 1996) developed a simple structured mathematical model for lipase production by *C. rugosa* in batch fermentation. Lipase production is induced by extracellular oleic acid present in the medium. The acid is transported into the cell, where it is consumed, transformed, and stored. Lipase is then excreted to the medium, where it is distributed between the available oil-water interface and the aqueous phase. Cell growth is modulated by the intracellular substráte concentration. Model parameters were determined in a calibration step, and then the whole model was experimentally validated with good results. This model was later modified to be applied from batch to fed-batch and continuous lipase production (Montesinos et al., 1997). Finally, it was exploited in simulations and for the design of new operational conditions as discussed next.

Annibale et al. (Annibale et al., 2006) and Brozzoli et al. (Brozzoli et al., 2009) confirmed that lipase production by *Candida* sp. was found to be completely repressed by the presence of simple sugars and induced by using natural oils.

2.3 *Pseudomonas* sp

An extracellular lipase was isolated and purified from the culture broth of *Pseudomonas aeruginosa* SRT 9 (Borkar et al., 2009). Production medium was prepared containing olive oil (1% w/v) as inductor. Marked stability and activity of induced lipase in organic solvents suggest that this lipase is highly suitable as a biotechnological tool with a variety of applications including organo synthetic reactions and preparation of enantiomerically pure pharmaceuticals. A strain of *Pseudomonas mendocina* producing extracellular lipase was isolated from soil (Dahiya et al., 2010). The bacterium accumulates lipase in culture fluid when grown aerobically at 30 °C for 24 h in a medium composed of olive oil (1%) as substrate. This lipase was capable of hydrolyzing a variety of lipidic substrates and is mainly active towards synthetic triglycerides and fatty acid esters that possess a butyryl group. The medium for lipase production from *Pseudomonas fluorescens* P21 had glucose as carbon source (Cadirci & Yasa, 2010). When glucose was replaced by various lipids, olive oil was the effective lipid for lipase production (3.5 U/l). When glucose in the medium was replaced with olive oil, the lipase yield was increased by 48.9% between 12 and 18 h.

2.4 *Mucor hiemalis*

The influence on lipase induction in *Mucor hiemalis* of different types of triacylglycerols containing mainly oleic acid (olive oil), erucic acid (mustard oil), or saturated fatty acids of 8 to 16 carbons (coconut oil) was studied (Akhtar et al., 1980). The fungus produced a significant amount of lipase in the presence of glucose, but the lipase activity increased markedly when olive oil was added to the medium at the beginning of the fermentation. Among the various sources of triacylglycerols used as the carbon source, olive oil was found to be most effective in inducing the lipase. The lipase of *M. hiemalis* can be considered to be both constitutive and inducible.

2.5 *Penicillium restrictum*

While supplementation with olive oil gave the best lipase results, the highest values of glucoamylase and protease activities (de Azeredo et al., 1999) were achieved with starch enrichment. This indicates that the type of carbon source used as supplementation plays a determinant role in the kind of major enzymes that will be produced by *P. restrictum*. Enriching the babassu cake with different carbon sources favours the synthesis of different enzymes: olive oil supplementation results in high lipase activities, while starch supplementation results in high glucoamylase activities. Therefore, according to the application desired, the basal medium may be differentially enriched to give high yields of the desired enzyme.

2.6 *Rhizopus homothallicus*

Different mixtures of triacylglycerols (Rodriguez et al., 2006): olive, sunflower, corn, peanut, walnut and grape seed oils, were used as energy and carbon sources in addition to lactose, and with urea as nitrogen source. It should be emphasized that the presence of the carbohydrate account for the early growth of the strain *Rhizopus homothallicus* and later growth occurs due to the added oil (Pokorny et al., 1994). This fungal strain was able to produce similar high lipase activities with all studied oils. In the fermentation system, lipase synthesis was not prevented at 4% of triglycerides. To complete these studies, the above medium using olive oil and urea was chosen to evaluate the effect of different carbohydrates on lipase production: glucose, fructose, glycerol, xylose, sucrose and lactose. There were little or no differences with these substrates. This fact suggests that carbohydrate type does not influence lipase production, probably because the carbohydrate concentration is low (5 g/ l) compared to the oil amount (40 g/l) added to culture media and because they are probably utilized before the oil and consequently, before lipase production (Cordova et al., 1998).

2.7 *Rhizopus oryzae*

In the study (Hama et al., 2006) utilizing *Rhizopus oryzae* cells as whole-cel biocatalysts, various substrate-relate compounds such as olive oil, oleic acid, oleyl alkohol, methyl carpate and Tween 80 were tested. It was found that the addition of olive oil on lipase production and localization in suspension cells were therefore investigated (Ban et al., 2001). Olive oil increased intracellular lipase production. However that extracellular hydrolysis activity was much higher in the absence of olive oil. Because the *Rhizopus oryzae* cells used in the study were able to produce lipase constitutively regardless of whether substrate-related compounds were present, it seems likely that these compounds are effective in retaining lipase within the cells.

Nitrogen and carbon sources influencing the growth and production of lipase by *Rhizopus oryzae* were studied by Fadiloglu & Erkmen (1999). High yields of enzyme activity were obtained when protease peptone was the nitrogen source in media with olive oil and without olive oil. Carbon sources increased lipase activity in the media without olive oil, but decreased it slightly in the presence of oil. Lipase activity was significantly higher in the media with olive ail than that without olive ail. Biomass concentration was also higher in the presence of oil (Fadiloglu & Erkmen 1999). Rapid induction of enzymes able to break down foodstuffs appearing in the environment of the micro-organism is clearly of great ecological advantage. This induction process effects a change in the phenotype allowing further production of energy required for metabolism and/or growth (Wiseman, 1975).

2.8 *Geotrichum candidum*

The fungus *Geotrichum candidum* 4013 produces two types of lipases (extracellular and cell-bound; Stránsky et al., 2007). Both enzymes were induced by addition of olive oil. The differences in the abilities of these two enzymes to hydrolyze *p*-nitrophenyl esters were observed. Yan and Yan (2008) tested a combination of different experimental designs to optimize the production conditions of cell-bound lipase from *Geotrichum* sp. A single factorial design showed that the most suitable carbon source was a mixture of olive oil and citric acid and the most suitable nitrogen source was a mixture of corn steep liquor and

NH_4NO_3. Burkert et al. (2004) studied the effects of carbon source (soybean oil, olive oil, and glucose) and nitrogen source concentrations (corn steep liquor and NH_4NO_3) on lipase production by *Geotrichum* sp. using the methodology of response surface reaching a lipase activity of 20 U mL^{-1}.

3. Conclusion

Inducible enzyme systems in micro-organisms display many features of microbiological and biochemical interest. There is no doubt that in the microbial systems investigated, the major induced enzyme (usually a hydrolase) was fonned *de novo*. Enzyme formation occurs from amino acids, rather than from inactive peptide or protein precursor existing prior to the addition of the inductor to the culture (Wiseman, 1975).

Most published experimental data have shown that lipid carbon sources (especially natural oils) stimulate lipase production. Among vegetable oils, olive oil has been referred as one of the best inductors of lipase production. The review showed that olive oil plays significant role in lipase production. It could be concluded that the higher content of unsaturated free fatty acids contained in oil, the higher intracellular and extracellular lipase activity could be obtained with the oil as inductor for cells cultivation.

4. Acknowledgment

The author thanks for funding from the Czech Science Foundation (grant No. 502/10/1734) and the Institute of Organic Chemistry and Biochemistry, research program Z40550506.

5. References

Acikel, U.; Ersan, M. & Acikel ,Y.S. (2011). The effects of the composition of growth medium and fermentation conditions on the production of lipase by *R. delemar*. *Turkish Journal of Biology*, Vol.35, No.1, pp. 35-44, ISSN: 1300-0152

Akhtar, M.W.; Mirza, A.Q. & Chughtai, M.I.D. (1980). Lipase induction in *Mucor hiemalis*. *Applied and Environmental Microbiology*, Vol.40, No.2, pp. 257-263, ISSN 0099-2240

Annibale, A.; Sermanni F.F. & Petruccioli, T. (2006). Olive-mill wastewaters: a promising substrate for microbial lipase production. *Bioresource Technology*, Vol.97, No.15, pp. 1828-1833, ISSN 0960-8524

de Azeredo, L.A.I.; Gomes, P.M.; Sant'Anna, G.L.; Castilho, L.R. & Freire, D.M.G. (2007). Production and regulation of lipase activity from *Penicillium restrictum* in submerged and solid-state fermentations. *Current Microbiology*, Vol.54, No.5, pp. 361-365, ISSN 0343-8651

Babu, I.S. & Rao, G.H. (2007). Optimization of process parameters fort he production of lipase in submerged fermentation by *Yarrowia lipolytica* NCIM 3589. *Research Journal of Microbiology*, Vol.2, No.1, pp. 88-93, ISSN 1816-4935

Ban, K.; Kaieda, M.; Matsumoto, T.; Kondo, A. & Fukuda, H. (2001). Whole cell biocatalyst for biodiesel fuel production utilizing *Rhizopus oryzae* cells immobillized within biomass support particles. *Biochemical Engineering Journal*, Vol.8, No.1, pp. 39-43 ISSN 1369-703X

Baron, A.M.; Zago, E.C.; Mitchell, D.A. & Krieger, N. (2011). SPIL: Simultaneous production and immobilization of lipase from *Burkholderia cepacia* LTEB11. *Biocatalysis and Biotransformation*, Vol.29, No.1, pp. 19-24, ISSN 1024-2422

Boekema, Bouke K. H. L.; Beselin, A.; Breuer, M.; Hauer, B.; Koster, M.; Rosenau, F.; Jaeger, K.E. & Tommassen, J. (2007). Hexadecane and tween 80 stimulate lipase production in *Burkholderia glumae* by different mechanisms. *Applied and Environmental Microbiology*, Vol.73, No.12, pp. 3838-3844, ISSN 0099-2240

Brabcova, J.; Zarevucka, M. & Mackova, M. (2010). Differences in hydrolytic abilities of two crude lipases from Geotrichum candidum 4013. *Yeast*, Vol.27, No.12, pp. 1029-1038, ISSN: 0749-503X

Brozzoli, V.; Crognale, S.; Sampedro, I.; Federici, F.; Annibale. A. & Petruccioli, M. (2009). Assessment of olive-mill wastewater as a growth medium for lipase production by *Candida cylindracea* in bench-top reactor. *Bioresource Technology*, Vol.100, No.13, pp. 3395-3402, ISSN: 0960-8524

Borkar, P.S.; Bodade, R.G.; Rao, S.R. & Khobragade C.N. (2009). Purification and characterization of extracellular lipase from a new strain - *Pseudomonas aeruginosa* SRT 9. Brazilian Journal of Microbiology, Vol. 40, No. 2, pp. 358-366, ISSN 1517-8382

Burkert, J.F.M., Maugeri, F. & Rodrigues, M.I. (2004). Optimization of extracellular lipase production by *Geotrichum* sp. using factorial design. *Bioresource Technology*, Vol. 91, No. 1, pp. 77–84, ISSN 0960-8524

Cadirci, B.H. & Yasa, I. (2010). An organic solvents tolerant and thermotolerant lipase from *Pseudomonas fluorescens* P21. *Journal of Molecular Catalysis B-Enzymatic*, Vol. 64, No. 3-4, pp. 155-161, ISSN 1381-1177

Chahinian, H.; Vanot, G.; Ibrik, A.; Rugani, N.; Sarda L. & Comeau, L.C. (2000). Production of extracellular lipases by *Penicillium cyclopium. Bioscience Biotechnology and Biochemistry*, Vol.64, No.2, pp. 215-222, ISSN 0916-8451

Cihangir, N. & Sarikaya, E. (2004). Investigation of lipase producing by a new isolate of *Aspergillus* sp. *World Journal of Microbiology and Biotechnology*, Vol.20, No.2, pp. 193-197, ISSN: 0959-3993

Colin, V.L.; Baigori, M.D. & Pera, L.M. (2011). Mycelium-bound lipase production from *Aspergillus niger MYA 135*, and its potential applications for the transesterification of ethanol. *Jounal of Basic Microbiology*, Vol.51, No. 3, pp. 236-242, ISSN 0233-111X

Cordova, J.; Nemmaoui, M.; Ismaili-Alaoui, M.; Morin, A.; Roussos, S.; Raimbault, M. & Benjilali, B. (1998). Lipase production by solid state fermentation of olive cake and sugar cane bagasse. *Journal of Molecular Catalysis B-Enzymatic*, Vol.5. No.1-4, pp. 75-78, ISSN 1381-1177

Corzo, G. & Revah, S. (1999). Production and characteristics of the lipase from *Yarrowia lipolytica* 681. *Bioresource Technology*, Vol.70, No.2, pp. 173-180, ISSN 0960-8524

Dahiya, P.; Arora, P.; Chaudhury, A.; Chand, S & Dilbaghi, N. (2010). Characterization of an extracellular alkaline lipase from *Pseudomonas mendocina* M-37. *Journal of Basic Microbiology*, Vol. 50, No. 5, pp. 420-426, ISSN 0233-111X

Dalmau, E.; Montesinos, J. L.; Lotti, M., & Casas, C. (2000). Effect of different carbon sources on lipase production by *Candida rugosa. Enzyme Microbial Technology*, Vol. 26, No.9-10, pp. 657–663, ISSN 0141-0229

Del Río, J.L.; Serra, P.; Valero, F.; Poch, M. & Sola, C. (1990). Reaction scheme of lipase production by *Candida rugosa* growing on olive oil. *Biotechnology Letters*, Vol.12, No.11, pp. 835–838, ISSN 0141-5492

Diczfalusy, M.A.; Hellman, U. & Alexon, S.E.H. (1997). Isolation of carboxylester lipase (CEL) isoenzymes from *Candida rugosa* and identification of the corresponding genes. *Archives of Biochemistry and Biophysics*, Vol. 348, No.1, pp. 1–8, ISSN 0003-9861

Dominguez, A.; Deive, F.J.; Sanroman, M.A. & Longo, M.A. (2003). Effect of lipids and surfactants on extracellular lipase production by Yarrowia lipolytica. *Journal of Chemical Technology and Biotechnology*, Vol.78, No.11, pp. 1166-1170, ISSN 0268-2575

Elibol, M. & Ozer, D. (2000). Lipase production by immobilized *Rhizopus arrhizus*. *Process Biochemistry*, Vol.36, No.3, pp. 219-223, ISSN: 0032-9592

Eltaweel, M.A.; Rahman, R.N.Z.R.A.; Salleh, A.B. & Basri, M. (2005). An organic solvent-stable lipase from Bacillus sp. strain 42. *Annals of Microbiology*, Vol.55, No.3, pp. 187-192, ISSN 1590-4261

Fadiloglu, S. & Erkmen, O. (1999). Lipase production by *Rhizopus oryzae* growing on different carbon and nitrogen sources. *Journal of the Science of Food and Agriculture*, Vol.79, N.13, pp. 1936-1938, ISSN 0022-5142

Fernandes, M.L.M.; Saad, E.B.; Meira, J.A.; Ramos, L.P.; Mitchell, D.A. & Krieger, N. (2007). Esterification and transesterification reactions catalysed by addition of fermented solids to organic reaction media. *Journal of Molecular Catalysis-B Enzymatic*, Vol.44, No.1, pp. 8-13, ISSN 1381-1177

Ferrer, P.; Montesinos, J.L.; Valero, F.; Sola, C. (2001) Production of native and recombinant lipases by *Candida rugosa* - A review. *Applied Biochemistry and Biotechnology*, Vol.95, No.3, pp. 221-255, ISSN 0273-2289

Gombert, A.K.; Pinto A.L., Castilho, L.R. & Freire, D.M.G. (1999). Lipase production by *Penicillium restrictum* in solid-state fermentation using babassu oil cake as substrate. *Process Biochemistry*, Vol.35, No.1-2, pp. 85–90, ISSN 0032-9592

Gordillo, M. A.; Obradors, N.; Montesinos, J. L.; Valero, F.; Lafuente, J. & Sola, C. (1995). Stability studies and effect of the initial oleic-acid concentration on lipase production by *Candida rugosa*. *Applied Microbiology and Biotechnology*, Vol.43, No.1, pp. 38–41, ISSN 0175-7598

Gunstone, F.D. & Qureshi, M.I. (1965). Glyceride studies. Part IV. The component glycerides of ten seed oils containing linoleic acid. *Journal of the American Oil Chemists Society*, Vol.42, No.11, pp. 961-965, ISSN 0003-021X

Gunstone, F.D.; Hamilton, R. J.; Padley, F.B. & Ilyas-Qureshi, M. (1965). Glyceride studies. V. The distribution of unsaturated acyl groups in vegetable triglycerides. *Journal of the American Oil Chemists Society*, Vol.42, No.11, pp. 965-970, ISSN 0003-021X

Hama, S.; Tamalampudi, S.; Fukumizu, T.; Miura, K.; Yamaji, H.; Kondo, A. & Fukuda, H. (2006). Lipase localization in *Rhizopus oryzae* cells immobilized within biomass support particles for use as whole-cell biocatalysts in biodiesel-fuel production. *Journal of Bioscoience Bioingineering*, Vol.101, No.4, pp. 328-333, ISSN 1389-1723

Hasan, F.; Shah A.A. & Hameed, A. (2006). Industrial applications of microbial lipases. *Enzyme Microbial Technology*, Vol.39, No.2, pp. 235-251, ISSN: 0141-0229

Hiol, A.; Jonzo, M.D.; Rugani, N.; Druet, D.; Sardo L. & Comeau, L.C. (2000). Purification and characterization of an extracellular lipase from a thermophilic *Rhizopus oryzae* strain isolated from palm fruit. *Enzyme Microbiology and Technology*, Vol.26, No.5-6, pp. 421-430, ISSN 0141-0229

Hun, C.J.; Rahman, R.N.Z.A.; Salleh A.B. & Barsi, M. (2003). A newly isolated organic solvent tolerant *Bacillus sphaericus* 205y producing organic solvent-stable lipase. *Biochemical Engineering Journal*, Vol.15, No.2, pp. 147-151, ISSN 1369-703X

Immanuel, G.; Esakkiraj, P.; Jebadhas, A.; Iyapparaj, P. & Palavesam, A. (2008). Investigation of lipase production by milk isolate *Serratia rubidaea*. *Food Technology and Biotechnology*, Vol.46, No.1, pp. 60–65, ISSN 1330-9862

Jaeger, K.E .& Eggert, T. (2002). Lipases for biotechnology. *Current Opinion in Biotechnology*, Vol.13, No.4, pp. 390-397, ISSN 0958-1669

Kumar, S.; Kikon, K.; Upadhyay, A.; Kanear, S.S. & Gupta, R. (2005). Production, purification and characterisation of lipase from thermophilic and alkaliphilic *Bacillus coagulans* BTS-3. *Protein Expression and Purification*, Vol.41, No.1, pp. 38-44, ISSN 1046-5928

Lakshmi, B.S.; Kangueane, B.; Abraham, B. & Pennatur, G. (1999). Effect of vegetable oils in the secretion of lipase from *Candida rugosa* (DSM 2031). *Letters in Applied Microbiology*, Vol. 29, No.1, pp. 66–70, ISSN 0266-8254

Lee, D.W.; Kim, H.W.; Lee, K.W.; Kim, B.C.; Choe, E.A.; Lee, HS; Kim D.S. & Pyun, Y.R. (2001). Purification and characterization oft wo distinct termostable lipases from the gram-positive theromphilic bacterium *Bacillus thermoleovorants* ID-1. *Enzyme Microbial Technology*, Vol.29, No.6-7, pp. 363-371, ISSN 0141-0229

Li, C.Y.; Cheng, C.Y. & Chen, T.L. (2004). Fed-batch production of lipase by *Acinetobacter radioresistens* using Tween 80 as carbon source. *Biochemical Engineering Journal*, Vol.19, No.1, pp. 25-31, ISSN 1369-703X

Li, D.; Wang, B. &Tan, T. (2006). Production enhancement of *Rhizopus arrhizus* lipase by feeding oleic acid. *Journal of Molecular Catalysis B-Enzymatic,*, Vol.43, No.1-4, pp. 40-43, ISSN 1381-1177

Lima, V.M.G.; Krieger, N.; Sarquis, M.I.M.; David, D.; Mitchell, D.A.; Ramos L.P. & Fontana, J.D. (2003). Effect of nitrogen and carbon sources on lipase production by *Penicillium aurantiogriseum*. *Food Technology and Biotechnology*, Vol.41, No.2, pp. 105-110, ISSN 1330-9862

Linko,Y. Y. & Wu, X. Y. (1996). Biocatalytic production of useful esters by two forms of lipase from *Candida rugosa*. *Journal of Chemical Technology and Biotechnology*, Vol.65, No.2, pp. 163–170, ISSN 0268-2575

Lotti, M.; Monticelli, S.; Montesinos, J. L.; Brocca, S.; Valero, F. & Lafuente, J. (1998). Physiological control on the expression and secretion of *Candida rugosa* lipase. *Chemistry and Physics Lipids,* Vol.93, No.1-2, pp. 143–148, ISSN 0009-3084

Martinelle, M. & Hult, K. (1995) Kinetics of acyl transfer reactions in organic media catalysed by *Candida antartica* lipase B. *Biochimica et Biophysica Acta*, Vol.1251, No.2, pp. 191-197, ISSN 0167-4838

Montesinos, J.L.; Obradors, N.; Gordillo, M.A.; Valero, F.; Lafuente, J. & Sola, C. (1996). Effect of nitrogen sources in batch and continuous cultures to lipase production by *Candida rugosa*. *Applied Biochemistry and Biotechnology*, Vol.59, No.1, pp. 25–37, ISSN 0273-2289

Montesinos, J.L.; Lafuente, J.; Gordillo, M.A.; Valero, F.; Sola, C.; Charbonnier, S. & Cheruy, A. (1995). Structured modeling and state esti,mation in a fermentation process-lipase production by *Candida rugosa*. *Biotechnology and Bioengineering*, Vol. 46, No.6, pp. 573–584, ISSN 0006-3592

Montesinos, J. L.; Gordillo, M. A.; Valero, F.; Lafuente, J.; Sola, C. & Valdman, B. (1997). Improvement of lipase productivity in bioprocesses using a structured mathematical model *Journal of Biotechnology*, Vol.52, No.3, pp. 207–218, ISSN 0168-1656

Najjar, A.; Robert, S.; Guerin, C.; Violet-Asther, M. & Carriere, F. (2011). Quantitative study of lipase secretion, extracellular lipolysis, and lipid storage in the yeast *Yarrowia lipolytica* grown in the presence of olive oil: analogies with lipolysis in humus. *Applied Microbiology and Biotechnology*, Vol.89, No.6, pp. 1947-1962, ISSN 0175-7598

Nunes, P.A.; Pires-Cabral, P.; Guillen, M.; Valero, F.; Luna, D. & Ferreira-Dias, S. (2011). Production of MLM-Type structured lipids catalyzed by immobilized heterologous *Rhizopus oryzae* lipase. *Journal of the American Oil Chemists Society*, Vol.88, No.4, pp. 473-480, ISSN 0003-021X

Olusesan, A.T.; Azura, L.K.; Abubakar, F.; Mohamed, A.K.S.; Radu, S.; Manap, M.Y.A. & Saari, N. (2011). Enhancement of thermostable lipase production by a genotypically identified extremophilic *Bacillus subtilis* NS 8 in a continuous bioreactor. *Journal of Molecular Microbiology and Biotechnology*, Vol.20, No.2, pp.105-115, ISSN 1464-1801

Papanikolaou, S.; Dimou, A.; Fakas, S.; Diamantopoulou, P.; Philippoussis, A.; Galiotou-Panayotou, M. & Aggelis, G. (2011). Biotechnological conversion of waste cooking olive oil into lipid-rich biomass using *Aspergillus* and *Penicillium* strains. *Journal of Applied Microbiology*, Vol. 110, No.5, pp. 1138-1150, ISSN 1364-5072

Papaparaskevas, D.; Christakopoulos, P.; Kekos, D. & Macris, B.J. (1992). Optimizing production of extracellular lipase from Rhodotorula glutinis. *Biotechnology Letters*, Vol.14, No.5, pp. 397–402. ISSN 0141-5492

Peters, I.I. & Nelson, F.E. (1948). Factors influencing the production of *Mycotorula lipolytica*. *Journal of Bacteriology*, Vol.55, No.5, pp. 581-591, ISSN 0021-9193

Pignĉde, G.; Wang, H.; Fudalej, F.; Gaillardin, C.; Seman, M. & Nicaud, J.M. (2000). Characterization of an extracellular lipase encoded by LIP2 in *Yarrowia lipolytica*. *Journal of Bacteriology*, Vol.182, No.10, pp. 2802–2810, ISSN 0021-9193

Pokorny, D.; Friedrich, J. & Cimerman, A. (1994). Effect of nutritional factors on lipase biosynthesis by *Aspergillus niger*. *Biotechnology Letters*, Vol.16, No.4, pp. 363–6, ISSN 0141-5492

Pokorny, D.; Friedrich, J. & Cimerman, A. (1994). Effect of nutritional factors on lipase biosynthesis by *Aspergillus niger*. *Biotechnology Letters*, Vol.16, No.4, pp. 363–366. ISSN 0141-5492

Rodriguez, J.A.; Mateos, J.C.; Nungaray, J.; Gonza´lez, V.; Bhagnagar, T.; Roussos, S.; Cordova, J. & Baratti, J. (2006), Improving lipase production by nutrient source modification using *Rhizopus homothallicus* cultured in solid state fermentation. *Process Biochemistry*, Vol.41, No.11, pp. 2264–2269, ISSN 1359-5113

Rua, M. L.; Díaz-Mauriño, T.; Fernández, V. M.; Otero, C. & Ballesteros, A. (1993). Purification and characterization of 2 distinct lipases from *Candida cylindracea*. *Biochimica et Biophysica Acta*, Vol.1156, No.2, pp. 181–189, ISSN 0006-3002

Salihu, A.; Alam, M.Z.; AbdulKarim, M.I. & Salleh, H.M. (2011). Suitability of using palm oil mill effluent as a medium for lipase production. *African Journal of Biotechnology*, Vol.10, No.11, pp. 2044-2052, ISSN: 1684-5315

Salleh, A.B.; Musani, R.; Basri, M.; Ampon, K.; Yunus, W.M.Z. & Razac, C.N.A. (1993). Extracellular and intracellular lipases from a thermophilic *Rhizopus oryzae* and factors affecting their production. *Canadian Journal of Microbiology,* Vol.39, No.10, pp. 978–981 ISSN 0008-4166

Serra, P.; del Río, J.L.; Robusté, J.; Poch, M.; Sola, C. & Cheruy, A. (1992). A model for lipase production by *Candida cylindracea. Bioprocess Eng.* Vol.8, No.3–4, pp. 145–150, ISSN 0178-515X

Sharon, C.; Furugoh, S.; Yamakido, T.; Ogawa, H.I. & Kato, Y. (1998). Purification and characterization of a lipase from *Pseudomonas aeruginosa* KKA-5 and its role in castor oil hydrolysis. *Journal of Industrial Microbiology and Biotechnology,* Vol.20, No.5, pp. 304-307, ISSN 1367-5435

Sokolovska, I.; Albasi, C.; Riba, J.P. & Bales, V. (1998). Production of extracellular lipase by *Candida cylindracea* CBS 6330 *Bioprocess Engineering,* Vol.19, No.3, pp. 179–186, ISSN 0178-515X

Stransky, K.; Zarevucka, M.; Kejik, Z.; Wimmer, Z.; Mackova, M. & Demnerova, K. (2007). Substrate specificity, regioselectivity and hydrolytic activity of lipases activated from *Geotrichum* sp. *Biochemical Engineering Journal,* Vol.34. No.3, pp. 209-216, ISSN 1369-703X

Sugihara, A.; Tani, T. & Tominaga, Y. (1991). Purification and characterization of a novel thermostable lipase from *Bacillus* sp. *Journal of Biochemistry,* Vol.109, No.2, pp. 211-215, ISSN 0021-924X

Tan, T.W.; Zhang, M.; Wang, B.W.; Ying, CH. & Deng, L. (2003). Screening of high lipase producing *Candida* sp. and production of lipase by fermentation. *Process Biochemistry,* Vol.39, No.4, pp. 459-465, ISSN 0032-9592

Tang, S.J.; Sun, K.H.; Sun, G.H.; Chang, T.Y. & Lee, G.C. (2000). Recombinant expression of the *Candida rugosa* lip4 lipase in *Escherichia coli. Protein Expression and Purification,* Vol.20, No.2, pp. 308–313, ISSN 1046-5928

Wiseman, A. (1975). Enzyme induction in microbial organisms catalyze the hydrolysis of esters other than acylglycerols. In *Enzyme Induction,* D.V. Parke, (Ed.), Plenum Press. New York, N.Y., ISBN 0470208562

Yan, J. Y., & Yan, Y. I. (2008). Optimization for producing cell-bound lipase from *Geotrichum* sp. and synthesis of methyl oleate in microaqueous solvent. *Applied Microbiology and Biotechnology,* Vol.78, No. 3, pp. 431–439, ISSN 0175-7598

Zhang A.J.; Gao, R.J.; Diao, N.B.; Xie, G.Q.; Gao G. & Cao, S.G. (2009)a. Cloning, expression and characterisation of an organic solvent tolerant lipase from *Pseudomonas fluorescens* JCM5963. *Journal of Molecular Catalysis B. Enzymatic,* Vol.56, No.2-3, pp. 78-84, ISSN: 1381-1177

Zhang, H.Z.; Zhang F.L. & Li, Z.Y. (2009)b. Gene analysis, optimized production and property of marine lipase from *Bacillus pumilus* B106 associated with South China Sea sponge *Halichondria rugosa. World Journal Microbiology Biotechnology,* Vol.25, No.7, pp. 1267-1274, ISSN 0959-3993

13

Olive Oil-Based Delivery of Photosensitizers for Bacterial Eradication

Faina Nakonechny[1,2], Yeshayahu Nitzan[2] and Marina Nisnevitch[1]
[1]Department of Chemical Engineering, Biotechnology and Materials,
Ariel University Center of Samaria, Ariel,
[2]The Mina and Everard Goodman Faculty of Life Sciences,
Bar-Ilan University, Ramat-Gan,
Israel

1. Introduction

Olive oil is a natural product of *Olea europaea*. It contains triacylglycerols of unsaturated and saturated fatty acids as well as free acids and numerous other biologically active components. Modern pharmaceutical industries are turning to natural herbal sources in order to find effective, low allergenic and non-irritating components that can be used in drug delivery systems or as recipients for both hydrophobic and hydrophilic active agents. Combining hydrophobic compounds with olive oil components is not problematic at all. However, this is quite different for hydrophilic compounds. One possible way for overcoming this problem is by mechanochemical treatment. This method has become widespread for preparing powdered solid materials in a large variety of compositions and involves the use of a conventional high-energy ball mill to initiate chemical reactions and structural changes of materials in solid-phase processes. Mechanochemical activation appears to be an environmentally friendly method, since it does not require organic solvents (Grigorieva et al., 2004; Margetić, 2005; Lugovskoy et al., 2008; Lugovskoy et al., 2009). It was shown that the mechanochemical method enabled some olive oil components to covalently attach to talc or to titanium dioxide - the solid ingredients of creams, ointments and powders (Nisnevitch et al., 2011). The remaining components were deeply absorbed by solid phases. New solid-phase composite materials which combined useful properties of various components with a different nature were thus created. Talc combined with olive oil exhibited good antioxidant properties scavenging ca. 40% of free radicals. Olive oil phenols with one or two hydroxyl groups, such as hydroxytyrosol, caffeic acid, photocatechuic acid, syringic acid, derivatives of elenolic acid, derivatives of oleuropein, tyrosol and some others are among the olive oil components responsible for its *in vitro* antioxidative activity (Papadopoulos & Boskou 1991; Briante et al., 2001; Lesage-Meessen et al., 2001; Tovar et al., 2001; Vissers et al., 2004). These compounds retain their antioxidant properties when combined with talc by a mechanochemical method. Furthermore, the possibility of combining water-soluble ascorbic acid (vitamin C) with olive oil on a talc or titanium dioxide support using mechanochemical activation has been reported (Nisnevitch et al., 2011). These triple mixtures (support-olive oil-ascorbic acid) scavenged free radicals instantly and totally due to the presence of ascorbic acid, which is a well-known effective antioxidant (Cathcart, 1985). The scavenging ability in the triple mixtures after mechanochemical treatment was as good as that of the double mixtures of ascorbic acid with the supports. Mechanochemical inclusion of ascorbic acid into composites of olive oil with talc or olive oil with titanium dioxide successfully combined hydrophobic and hydrophilic components and provided high antioxidant properties to the entire system despite the covalent bonding between the components (Nisnevitch et al, 2011).

New olive oil-based composite materials exhibit pronounced bactericidal properties. The antimicrobial activity of the mechanochemically treated triple mixtures which were pressed into pellets was examined against the Gram-positive S. aureus and the Gram-negative E. coli bacteria. Samples containing ascorbic acid on a titanium dioxide support were more effective against both bacteria than a talc support, probably because of weaker bonding of ascorbic acid to titanium dioxide than to talc, which contributed to better diffusion of the ascorbic acid out of the pellets. Gram-positive S. aureus was more sensitive to all the ascorbic acid-containing samples than the Gram-negative E. coli, but E. coli responded to addition of olive oil into both talc-ascorbic acid and titanium dioxide-olive oil mixtures. In the latter case, the inhibitory activity of the triple composites was higher than that of double ascorbic acid-support composites. The antimicrobial activity of all the ascorbic acid-containing samples depended on the ascorbic acid content in the pellets. Olive oil, olive fruit and olive leaf extracts are known to exhibit a broad antimicrobial, antimycoplasmal and antifungal spectrum due to the presence of long chain α,β–unsaturated aldehydes, phenolic glycoside oleuropein and several other phenol compounds (Fleming et al. ,1973; Kubo et al., 1995; Bisignano et al., 2001; Furneri et al., 2002; Medina et al., 2007; Covas et al., 2009; Kampa et al., 2009). Mechanochemical combination of natural antimicrobial agents from olive oil with ascorbic acid, which is a strong bacterial suppressor, enabled the production of highly active solid-phase antibacterial composites.

Hydrophilic and hydrophobic components can also be combined by encapsulating hydrophilic constituents in lipid vesicles called liposomes. Such lipid-based formulations are actually possible carriers for both hydrophobic and hydrophilic active components and can be applied as drug delivery systems. Liposome formulations possess enhanced abilities to penetrate the skin, thus improving the delivery process. Lipid-based drug administration can increase treatment efficiency in cases of skin infections and inflammations caused by bacterial invasion.

2. Olive oil-containing liposomes

Liposomes (nano or micro-scale vesicles) can be obtained using phospholipids' property of self-assembly in the presence of an aqueous phase. Phospholipids spontaneously form a closed spherical phospholipid bilayer such that phosphate groups are in contact with the aqueous phase on the internal and external surfaces, and lipid chains are hidden within the membrane. Such a phospholipid assembly results in large multilamellar liposomes, which are constructed from alternating concentric lipid and aqueous layers. Treatment of multilamellar liposomes by ultrasound, membrane extrusion or other methods leads to the formation of unilamellar liposomes which consist of a single lipid bilayer (Chrai et al., 2001). Liposomes are convenient carriers of both hydrophilic and hydrophobic molecules, where the former can be incorporated into aqueous layers of multilamellar liposomes or encapsulated in the inner space of unilamellar ones, and hydrophobic compounds can be incorporated into the lipid bilayers (Chrai et al., 2002).

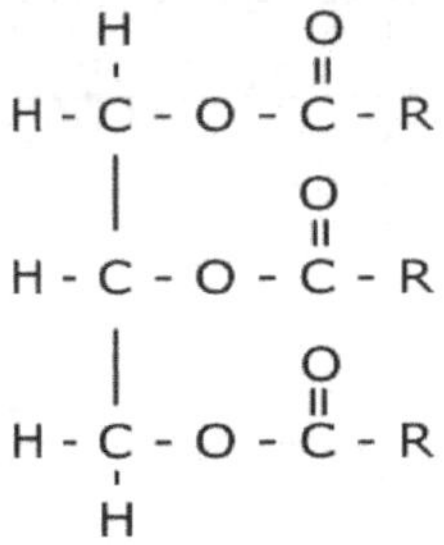

$$
\begin{array}{c}
\text{H} \qquad\quad \text{O} \\
| \qquad\quad\; \| \\
\text{H} - \text{C} - \text{O} - \text{C} - \text{R} \\
| \qquad\quad\; \text{O} \\
\qquad\qquad\;\; \| \\
\text{H} - \text{C} - \text{O} - \text{C} - \text{R} \\
| \qquad\quad\; \text{O} \\
\qquad\qquad\;\; \| \\
\text{H} - \text{C} - \text{O} - \text{C} - \text{R} \\
| \\
\text{H}
\end{array}
$$

Fig. 1. Structure of a triacylglycerol. R - various residues of fatty acids.

Fatty acid	Structure	% in Virgin Olive Oil (Hatzakis et al., 2008)		% in EPC (Ternes, 2002; Sigma aldrich. com)	% in DPPC (northern lipids.com)
		Triacylgly-cerols	Phospho-lipids		
Oleic acid	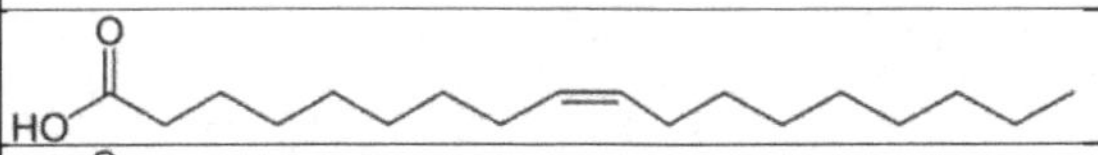	72.0-81.6	72.5-82.9	26-31	-
Linoleic acid	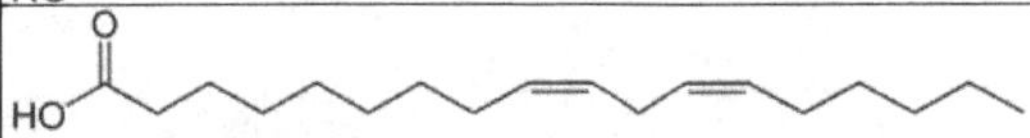	4.6-11.0	2.7-12.0	13-19	-
Palmitic acid	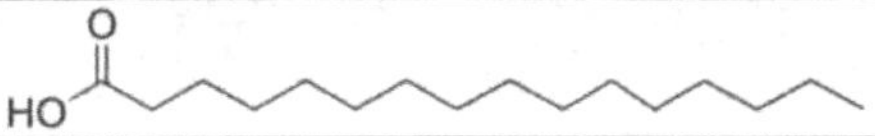	12.3-19.7	11.2-19.4	27-33	100
Stearic acid	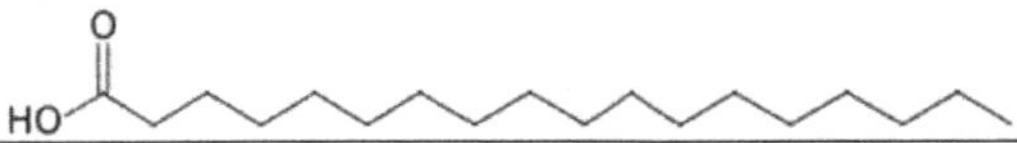			13-15	-
α-Linolenic acid	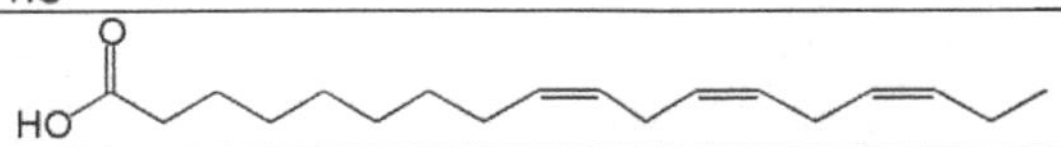	0.08-0.53	0.11-0.47	0-0.2	-

Table 1. Main virgin olive oil, egg phosphatidylcholine and dipalmitoyl phosphatidylcholine fatty acids. (Nichols & Sanderson, 2002; oliveoilsource.com)

Liposomes can be exploited as carriers for controlled drug delivery and targeting to cells. Liposome formulations of drugs have several advantages over the use of drugs in their free form: liposomes guarantee delivery of a highly concentrated drug, liposomes protect the drugs from degradation during the delivery process, liposomes are applicable for polar as well as for nonpolar drugs, and ingredients of the liposomes themselves are nontoxic and biodegradable (Chrai et al., 2002). Liposome components participate in drug delivery, but not in drug function, such that liposomes actually play the role of excipients (Chen, 2008). Additional ingredients can be incorporated into the phospholipid bilayer in order to impart needed properties to liposomes, as indicated by the following examples: negatively charged phosphatidylinositol or positively charged stearylamine can be incorporated into the phospholipid bilayer in order to obtain charged liposomes (Robinson et al., 2001); addition of cholesterol provides rigidity to the liposome structure (New, 1994). The latter example is explained by an increase in the gel-to-liquid crystalline phase transition temperature (T_c) of the lipid liposome layer upon the addition of cholesterol (Beaulac et al., 1998).

Component	Structure
Dipalmitoyl phosphatidylcholine (DPPC)	
Egg phosphatidylcholine (EPC)*	
Cholesterol	

*Alternative fatty acids residues are listed in the Table 1.

Table 2. Compounds used as a basis for liposome preparations.

The major ingredients of olive oil are triacylglycerols (Fig. 1) of unsaturated and saturated fatty acids (Table 1), mainly of oleic acid, but it also contains mixed triacylglycerols of palmitic-oleic-oleic, linoleic-oleic-oleic, palmitic-oleic-linoleic, stearic-oleic-oleic, linolenic-oleic-oleic and other acids (Nichols & Sanderson, 2002; oliveoilsource.com).

Olive oil also contains a small amount of free fatty acids and several minor constituents necessary for health – tyrosol, hydroxytyrosol and their derivatives such as oleuropein, oleuropein aglycone, dialdehydic form of oleuropein aglycone, decarboxymethyl form of oleuropein aglycone and ligstroside aglycone; phenolic acids, for example, 4-hydroxybenzoic acid, protocatechuic acid, syringic acid and 4-hydroxy-phenylacetic acid; flavonoids and lignads, for instance, apigenin, luteolin, pinoresinol and acetopinoresinol; squalene, α-tocopherol, vitamins E and K, pigments chlorophyll, pheophytin, carotenoids and other compounds (Boskou et al., 2006a,b; Boskou 2009a,b). In addition, olive oil includes phospholipids at a concentration range of 11-157 mg/kg in virgin olive oil (Hatzakis et al., 2008) and 21-124 mg/kg in cloudy (veiled) virgin olive oil (Koidis & Boscou, 2006).

Component	DPPC liposomes, % (w/w)	EPC liposomes % (w/w)
Dipalmitoyl phosphatidylcholine	62	-
Egg phosphatidylcholine	-	64
Olive oil	30	28
Cholesterol	8	8

Table 3. Weight compositions of olive oil based liposomes.

The liposomes used in this work were composed of dipalmitoyl phosphatidylcholine (DPPC, Northern Lipids Inc., Canada) or egg yolk phosphatidylcholine (EPC, Sigma, USA) also named L-α-lecithin. These phospholipids are constructed based on the phosphatidylcholine. However, DPPC has a homogeneous composition and contains only saturated palmitic acid residues, contrary to the heterogeneous composition of EPC, which includes several saturated and unsaturated fatty acid residues (Table 1). EPC liposomes are composed of two different fatty acid residues, where one residue is usually saturated and the other is unsaturated, as demonstrated in Table 2 (Kent & Carman, 1999). The most common fatty acids incorporated into the EPC structure are presented in Table 1. Phosphatidylcholine, the major membrane phospholipid in eukaryotic cells, is the source of the bioactive lipids lysophosphatidylcholine, phosphatidic acid, diacylglycerol, lysophosphatidylcholine, platelet activating factor and arachidonic acid. It also plays a role as a reservoir for several lipid messengers (Kent & Carman, 1999).

We incorporated virgin olive oil (Yad Mordechai, Israel) into the lipid bilayer in order to enhance the biocompatibility of liposomes and enrich them with natural salubrious components. For this purpose, organic solutions of DPPC or EPC together with olive oil were prepared, and the organic solvent was evaporated in a round-bottom flask to dryness in a vacuum rotary evaporator to obtain a thin lipid film which was vigorously agitated with buffer solutions with or without water-soluble active agents.

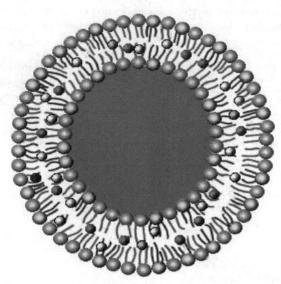

Fig. 2. A schematic representation of unilamellar olive oil including liposome. Phospholipid molecules are orange coloured, olive oil components are green, cholesterol molecules are grey coloured. Internal liposome volume containing water-soluble active components is pink coloured.

As can be seen from Table 1, both used by us phospholipids are built from the same fatty acids as olive oil triacylglycerols and olive oil phospholipids, although in different proportions. This fact points to high compatibility between the used phospholipids and olive oil. Various combinations of phospholipids and olive oil were attempted, and it was found that a homogeneous lipid film could not be obtained with any combination of olive oil and EPC and at a high olive oil content added to DPPC. A small amount of cholesterol (Table 2) was added to the lipid mixture solution in order to increase the lipid film's rigidity. Even and homogeneous films were attained after this addition, which resulted in stable liposomes. The multilamellar liposomes were transformed into unilamellar liposomes by sonication, as described previously (Nisnevitch et al., 2010; Nakonechny et al., 2010). Final compositions found to be appropriate for liposome preparation are presented in Table 3.

A schematic representation of the olive oil-based unilamellar liposomes is presented in Fig. 2. Triacylglycerol olive oil components are organically incorporated into the phospholipid-based structure, hydrophobic olive oil constituents such as polyphenols or vitamins are incorporated into the liposome bilayer and the aqueous solution is located in the inner liposome space. The prepared liposomes were used for encapsulation of active bactericidal factors as described in part 4.

The prepared olive oil-based liposomes were characterized by average size, evaluated by measuring the turbidity spectra. This method is based on the determination of an equation of the turbidity spectra curves, estimation of the power "n" in the equation (1):

$$\log \frac{I_o}{I} = K\lambda^{-n} \tag{1}$$

where $\log \frac{I_o}{I}$ – a measured turbidity value, I_o – initial light intensity, I – light intensity and λ – wavelength, and the liposome average size evaluation with a calibration curve representing "n"-values' dependence on vesicle sizes (Trofimov & Nisnevich, 1990; Nisnevitch et al, 2010). Higher "n"-values correspond to smaller vesicle sizes. Turbidity spectra of DPPC and EPC liposomes with and without addition of cholesterol and olive oil were measured (Fig. 3), and corresponding type (1) equations were found in each case. As can be seen from Table 4, "n" values in these equations and correspondingly, vesicle sizes, are different for DPPC and EPC liposomes, and vary when cholesterol and olive oil are incorporated into the phospholipid layers.

As can be seen from Table 4, the DPPC-based liposomes are smaller than the EPC ones obtained using the same treatment conditions. This phenomenon can be explained by two factors – by the lipid structure and by the lipid phase state of the liposomes. The homogeneous composition of DPPC, which contains only saturated palmitic acid residues, enables dense lipid packing in the liposome bilayers, in contradistinction to the heterogeneous composition of EPC, which includes several saturated and unsaturated fatty acid residues (Table 1). Such a denser package leads to the formation of unilamellar DPPC liposomes with a smaller diameter. At the temperatures of our experiments (from room temperature to 37°C), DPPC exists in a gel phase state (T_c of DPPC is 41°C (avantilipids.com)), whereas EPC is found in a liquid crystal state (T_c of EPC is –10°C (Kahl et al., 1989)). Acyl chains of phospholipids are more disordered and bulky in a fluid state, thus causing an increase in surface area per phospholipid molecule which results in bigger liposomes in the case of EPC liposomes (New, 1994).

Liposome composition	DPPC-based liposomes		EPC-based liposomes	
	"n"-value in equation (1)	Vesicle size, nm	"n"-value in equation (1)	Vesicle size, nm
Phospholipid alone	2.44	200	1.57	> 400
Phospholipid and cholesterol	2.50	190	2.00	> 400
Phospholipid, cholesterol and olive oil	2.03	> 400	1.36	> 400

Table 4. Turbidity spectra parameters and vesicle size for liposomes of various compositions.

Addition of cholesterol to phospholipids resulted in an increase in the "n"-value, which means that the vesicle size decreased upon the addition of cholesterol. The liposome size increased again after olive oil was incorporated into the membrane structure (Table 4). These facts can be explained by taking the correlation between liposome rigidity and size into account. Addition of cholesterol caused the liposome vesicles to become more rigid and respectively smaller, and further addition of olive oil led to disturbance of the lipid layer and to an increase in size (Table 4).

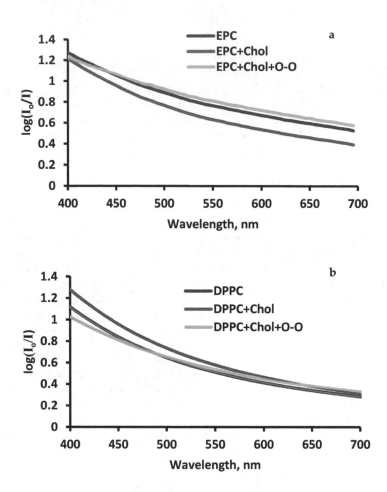

Fig. 3. Turbidity spectra of liposomes with and without additions of cholesterol (Chol) and olive oil (O-O). a – EPC-based liposomes, b – DPPC- based liposomes.

3. Photosensitizers encapsulated in olive oil-containing liposomes

Bacterial resistance to antibiotics has become a serious problem worldwide, causing an urgent need to develop new approaches and ways to overcome the evolution and spread of drug-resistant strains (Patterson, 2006; Maragakis et al., 2008; Moellering et al., 2007). One alternative to treatment of infections by antibiotics is photodynamic antimicrobial chemotherapy (PACT), which is based on the use of non-toxic compounds – photosensitizers, which can be activated by visible light. Excited photosensitizer molecules return to a ground level by transfering their energy to dissolved molecular oxygen with production of reactive oxygen species, which leads to direct damage of cellular components (Macdonald & Dougherty, 2001; Wainwright, 1998). This process is explained in Fig. 4.

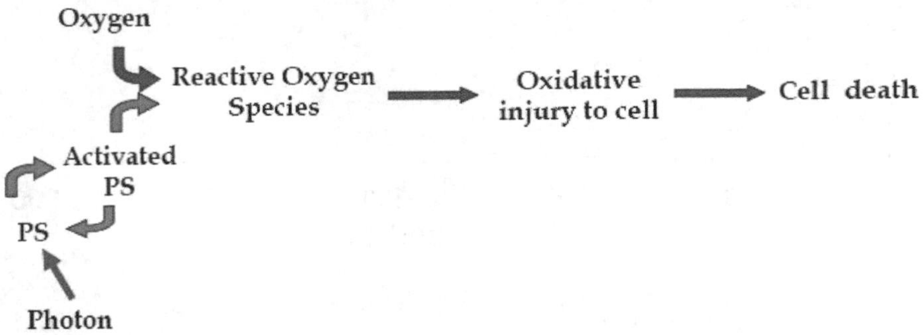

Oxygen

Reactive Oxygen ⟶ Oxidative ⟶ Cell death
Species   injury to cell

Activated
   PS

PS

Photon

Fig. 4. A scheme of photosensitizer (PS) activation upon illumination which visible light and its cytotoxic action.

Photosensitizes refer to several chemical groups - porphyrins, phenothiazinium, phthalocyanines, xanthenes, chlorin derivatives and others. However, a feature common to all of these groups is the presence of conjugated double bonds, which allow effective absorbance of light energy. The history, mechanism of action and biomedical applications of PACT have been reviewed extensively (Nitzan & Pechatnikov, 2011; Malik et al., 2010; Reddy et al., 2009; Randie et al., 2011; Daia et al., 2009). Two photosensitizers, Rose Bengal and Methylene Blue, were used in this work. Rose Bengal relates to a xanthene (halogenated xanthenes) group of photosensitizers, and is negatively charged under physiological conditions. Methylene Blue represents a phenothiaziniums group and exists in cationic form. The structures of these compounds are shown in Fig.5.

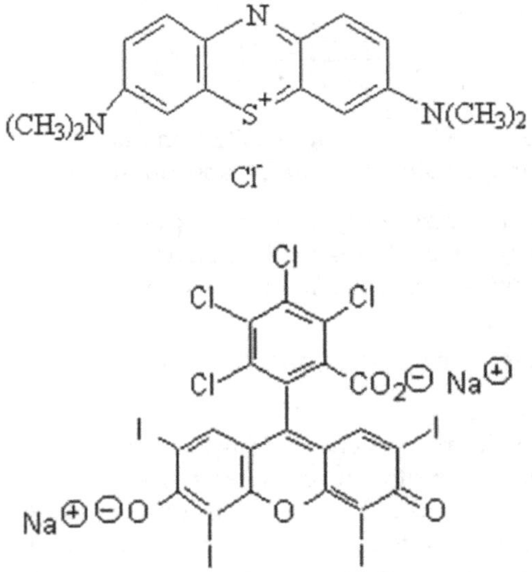

Fig. 5. Structures of photosensitizers Methylene Blue (upper) and Bengal Rose (lower).

Both photosensitizes absorb visible light, and their absorption spectra are presented in Fig. 6.

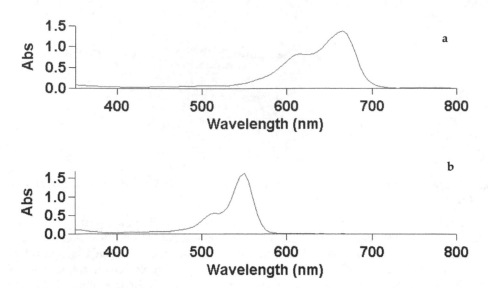

Fig. 6. Absorption spectra of (a) Methyle Blue and (b) Bengal Rose.

The described photosensitizers were encapsulated into DPPC and EPC liposomes with and without addition of olive oil as previously described by us (Nisnevitch et al., 2010). Liposomes with encapsulated photosensitizers were separated from free photosensitizers by centrifugation, and absorption of free photosensitizers was measured at the appropriate wavelengths (665 nm for Methylene Blue and 550 nm for Rose Bengal, Fig. 6).

$$\frac{A_o \cdot V_o - A \cdot V}{A_o \cdot V_o} \cdot 100\% \qquad (2)$$

where - A_0 - absorbance of the initial photosensitizer in the volume V_o and A- absorbance of the free photosensitizer in the volume V. The encapsulation rate reached 50±5% in all cases.

The extent of the photosensitizers encapsulation in liposomes was estimated by formula (2) as the ratio of the encapsulated photosensitizer amount, taken as the difference between initial and free photosensitizer amount, and the initial amount.

4. Bactericidal properties of photosensitizers encapsulated in olive oil-based liposomes

Application of liposomal forms of various drugs is widely used in cases of cancer and bacterial infection treatment. Treatment of tumours by liposomal forms of doxorubicin led to a manifold accumulation of the drug in the malignant cells (Drummond et al., 1999). Entrapment of photosensitizers into liposomes was also successfully applied for eradication of cancer cells (Derycke & de Witte, 2004). Liposome-encapsulated tobramycin, unlike its free form, was demonstrated to be highly effective against chronic pulmonary *P. aeruginosa* infection in rats (Beaulac et al., 1996). Drug administration using liposomes provided a delivery of active components in a more concentrated form and contributed to their enhanced cytotoxicity. A mechanism of drug delivery by liposomes was examined for Gram-negative and Gram-positive bacteria. Gram-negative and Gram-positive bacteria differ in their cell wall structure. Gram-negative cells possess an outer membrane which contains phospholipids, lipoproteins, lipopolysaccharides and proteins, peptidoglycan and cytoplasmic membrane. Gram-positive bacteria do not have an outer membrane, and their cell wall consists of peptidoglycan and an inner cytoplasmic membrane (Baron, 1996).

In Gram-negative bacteria, fusion between drug-containing liposomes and the bacterial outer membranes occurs, which results in the delivery of the liposomal contents into the cytoplasm. This mechanism was verified by scanning electron microscopy (Mugabe et al., 2006; Sachetelli et al., 2000), and it is schematically shown on the Fig. 7a.

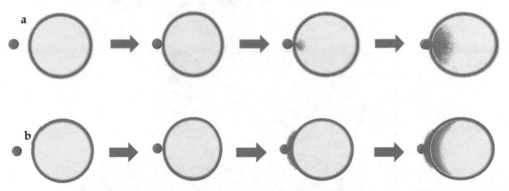

Fig. 7. A schematic representation of liposome-encapsulated drug delivery to (a) Gram-negative and (b) Gram-positive bacteria cells.

In Gram-positive bacteria, liposomes are assumed to release their content after interaction with the external peptidoglycan barrier, enabling passive diffusion through the cell wall (Furneri et al., 2000). This drug delivery mechanism is demonstrated in Fig. 7b. Application of liposomal forms of drugs leads to prolongation of their action in infected tissues and provides sustained release of active components (Storm & Crommelin, 1998).

Gram-positive and Gram-negative bacteria respond differently to PACT, with the former being more susceptible to the treatment. Gram-negative bacteria do not bind anionic photosensitizers (Minnock et al., 2000), unless additional manipulations facilitating membrane transport are used (Nitzan et al., 1992), due to the more complex molecular and physico-chemical structure of their cell wall. PACT is considered to have good perspectives in the control of oral and otherwise localized infections (Meisel & Kocher, 2005; O'Riordan et al., 2005). Local application of liposome-entrapped drugs can prolong their action in infected tissues and provide sustained release of active components (Storm & Crommelin, 1998). It should be mentioned that bacterial resistance to phosphosensitizers has not been reported to date.

Liposome formulations of photosensitizers showed high efficiency in eradication of both Gram-negative and Gram-positive bacteria. Liposome or micelle-entrapped hematoporphyrin and chlorin $e6$ were found to be effective against several Gram-positive bacteria, including methicillin-resistant *S. aureus* (Tsai et al., 2009).

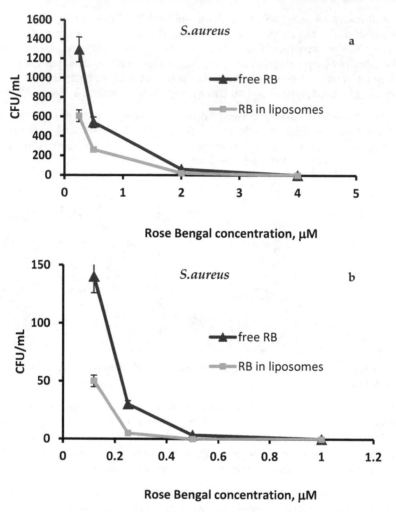

Fig. 8. Eradication of *S. aureus* by various concentrations of Rose Bengal (RB) in a free form and encapsulated into EPC–olive oil liposomes under white light illumination at initial bacteria concentration of (a) $3 \cdot 10^9$ cells/mL and (b) $3 \cdot 10^7$ cells/mL.

Encapsulation of photosensitizers into liposomes does not always result in enhancement compared to the free-form cytotoxic activity. The activity of m-tetrahydroxyphenylchlorin in liposomal form was comparable to the free form activity of PACT inactivation of a methicillin-resistant *S. aureus* strain (Bombelli et al., 2008). When tested against methicillin-resistant *S. aureus*, chlorophyll *a* was reported to be more efficient in free form than in a liposomal formulation, whereas hematoporphyrin as well as a positively charged PS 5-[4-(1-dodecanoylpyridinium)]-10,15,20-triphenyl-porphyrin were less effective in free form than upon encapsulation in liposomes. These results were explained by differences in photosensitizer chemistry which may influence their association with liposomal components, lipid fluidity and localization in liposome vesicles (Ferro et al., 2006; 2007).

We have previously shown that Methylene Blue encapsulated in liposomes composed of DPPC or EPC effectively deactivated several Gram-positive and Gram-negative bacteria, including *S. lutea*, *E. coli*, *S. flexneri*, *S. aureus* and MRSA, and that liposomal Rose Bengal also eradicated *P. aeruginosa* (Nisnevitch et al., 2010; Nakonechny et al., 2010; 2011).

Olive oil-containing liposomes loaded with photosensitizers were tested for their antimicrobial activity under white light illumination against two Gram-positive bacteria of the genus *Staphylococcus* – *S. aureus* and *S. epidermidis*. Although *S. epidermidis* is part of the normal skin flora, it can provoke skin diseases such as folliculitis, and may cause infections of wounded skin, in particular around surgical implants. *S. aureus* is defined as a human opportunistic pathogen and is a causative agent in up to 75% of primary pyodermas, including carbuncle, ecthyma, folliculitis, furunculosis, impetigo and others (Maisch et al., 2004).

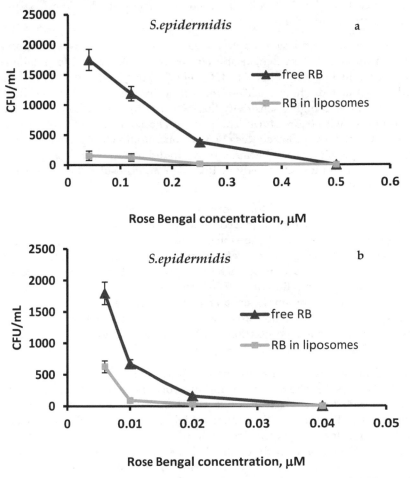

Fig. 9. Eradication of *S. epidermidis* by various concentrations of Rose Bengal (RB) in a free form and encapsulated into EPC–olive oil liposomes under white light illumination at initial bacteria concentration of (a) $3{\cdot}10^8$ cells/mL and (b) $3{\cdot}10^6$ cells/mL.

The water-soluble photosensitizers Rose Bengal and Methylene Blue were encapsulated in the above-described unilamellar liposomes at various concentrations and were examined under white light illumination against various cell concentrations by a viable count method as described previously (Nakonechny et al., 2010) and the number of bacterial colony forming units (CFU) was determined. This number characterized the concentration of bacterial cells which survived after a treatment.

The antimicrobial effect of liposomes incorporated with olive oil and loaded with Rose Bengal was strongly dependent on its concentration (Fig. 8 and 9). As can be seen from Fig. 8a, treatment of S. aureus with EPC-based liposomes caused a million-fold suppression of the bacterial cells at 0.25 μM of Rose Bengal and total eradication at a concentration of 2 μM when tested at an initial cell concentration of $3·10^9$ cells/mL. Total eradication of S. aureus at an initial concentration of $3·10^7$ cells/mL occurred already at a liposome-encapsulated Rose Bengal concentration of 0.5 μM (Fig 8b).

A principal similar trend was observed for S. epidermidis. It was necessary to apply liposome-encapsulated Rose Bengal at a concentration of 0.25 μM for total eradication of bacteria at an initial concentration of $3·10^8$ cells/mL (Fig. 9a), and it was enough to apply 0.02 μM encapsulated photosensitizer for killing bacteria at $3·10^6$ cells/mL (Fig. 9b). S. epidermidis exhibited a higher sensitivity than S. aureus for the liposome formulation of Rose Bengal compared with its free form. For S. aureus, liposomal Rose Bengal was only twice as effective as its free form – at each Rose Bengal concentration its liposomal form caused two-fold higher suppression of the bacteria. In contradistinction, S. epidermidis was suppressed three to twelve times more effectively by Rose Bengal encapsulated in liposomes than by the free photosensitizer.

Bacterial eradicating ability of the encapsulated as well as of the free Rose Bengal was demonstrated to depend on the initial concentration of the bacteria. When tested at the same Rose Bengal concentration, a suppression of both bacteria varied from partial to total. As can be seen from Fig. 10a, a 0.25 μM concentration of Rose Bengal encapsulated in EPC-olive oil liposomes caused a decrease of up to $6·10^2$ cells/mL in the S. aureus concentration when taken at an initial concentration of $3·10^9$ cells/mL (corresponding to 6.7 log_{10} CFU/mL) and up to zero cell concentration when taken at $3·10^7$ or $3·10^6$ cells/mL. In the case of S. epidermidis, 0.01 μM encapsulated Rose Bengal induced bacterial reduction of up to $1.5·10^4$ cells/mL from the initial concentration of 10^8 cells/mL, and to the zero concentration at an initial concentration of $3·10^6$ cells/mL (Fig. 10b).

DPPC-based liposomes were also examined, in addition to EPC-based olive oil-containing liposomes. The results showed high antimicrobial efficiency of the olive oil-containing liposomes in both bases, which was not less than that of the liposomes without olive oil supplements. Fig. 11 relates to the antimicrobial activity of Rose Bengal, applied against S. epidermidis, in free form or encapsulated in olive oil-containing ECP- and DPPC-liposomes, as well as to EPC-liposomes without olive oil. The data presented in Fig. 11 indicate that at each initial concentration, all liposomal forms of Rose Bengal eradicated bacteria more effectively than its free form (P-value 0.015), but there was no statistically significant difference in the photosensitizer activity when encapsulated in various types of liposomes (P-value 0.86).

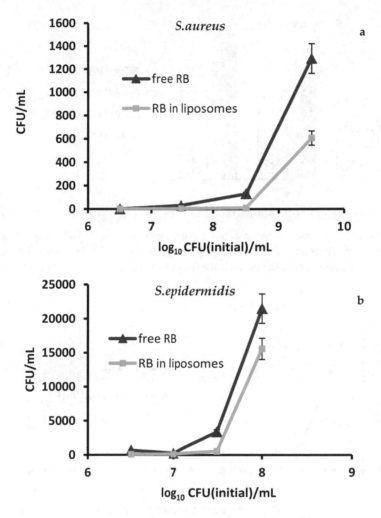

Fig. 10. Eradication of (a) *S. aureus* by 0.25 μM and (b) *S. epidermidis* by 0.01 μM Rose Bengal (RB) in a free form and encapsulated into EPC–olive oil liposomes under white light illumination at various initial bacteria concentrations presented in a logarithmic form.

Olive oil-containing liposomes with encapsulated Methylene Blue were tested against *S. epidermidis*. Bacterial sensitivity to this photosensitizer was much lower than to Rose Bengal in both free and liposomal forms. Thus, at the same initial bacterial concentration of $3 \cdot 10^6$ cells/mL, total eradication of *S. epidermidis* by liposomal Rose Bengal was achieved at 0.02 μM (Fig. 9b), and by liposomal Methylene Blue only at a concentration of 62.5 μM (Fig. 12). As to the general effect of free and liposomal Methylene Blue, it can be said that this photosensitizer exhibits the same trends as Rose Bengal. A liposome-encapsulated form was twice to three times more effective than the free form at all Methylene Blue concentrations (Fig. 12).

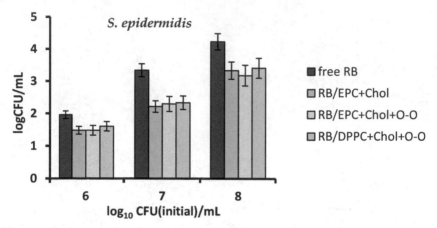

Fig. 11. Eradication of *S. epidermidis* under white light illumination by 0.01μM Rose Bengal (RB) in a free form and when encapsulated into liposomes with or without olive oil (O-O) and cholesterol (Chol) at various initial bacteria concentrations presented in a logarithmic form.

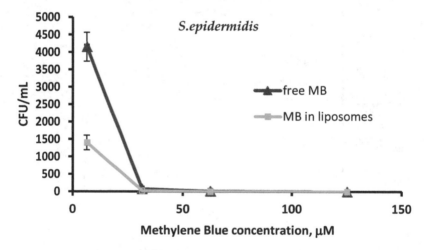

Fig. 12. Eradication of *S. epidermidis* by various concentrations of Methylene Blue in a free form and encapsulated into EPC–olive oil liposomes under white light illumination at initial bacteria concentration of $3 \cdot 10^6$ cells/mL.

It is important to mention that in no case did olive oil incorporation into the membrane of liposomes with encapsulated photosensitizers cause any decrease in their antimicrobial activity.

5. Perspectives for application of olive oil-containing liposomes

Several types of drug delivery systems containing lipids for oral, intravenous or dermal administration are described in the literature (Wasan, 2007). One of them is an oil-in-water emulsion, composed of isotropic mixtures of oil triacylglycerols, surfactant and one or more hydrophilic solvents. The typical particle size of such systems is between 100 and 300 nm (Constantinides, 1995). Another system, called a lipidic self-microemulsifying drug delivery system, represents transparent microemulsions with a particle size of 50-100 nm (Constantinides, 1995; Holm et al., 2003). The described emulsions and microemulsions were based on structural triacylglycerols or sunflower oil. Such systems were proven to appropriately deliver lipophilic drugs such as cyclosporine A, saquinavir, ritonavir and halofantrine (Charman et al., 1992; Holm et al., 2002). A soybean lecithin-based nanoemulsion enriched with triacylglycerols was used for efficient delivery of Amphotericin B (Filippin et al., 2008). An additional example represents solid lipid nanoparticles which were shown to not only deliver glucocorticoids, but also to enhance drug penetration into the skin (Schlupp et al., 2011). Colloid dispersions of solid triacylglycerol 140 nm-sized nanoparticles stabilized with poly(vinyl alcohol) were applied for delivery of the drugs diazepam and ubidecarenone (Rosenblat & Bunje, 2009). Soybean and olive oils were suggested as drug delivery vehicles for the steroids progesterone, estradiol and testosterone (Land et al., 2005). All of the above-mentioned examples illustrate successful use of lipid-based systems for delivery of hydrophobic drugs. However, they are all unsuitable for carrying hydrophilic components.

Liposomes are devoid of this serious disadvantage and are applicable for delivery of both hydrophobic and hydrophilic agents. In case of dermal application, lipid-based drug formulations exhibit enhanced abilities to penetrate into skin, improving the delivery process of active agents, thus enabling an increase in treatment efficiency in cases of skin infections and inflammations caused by bacterial invasion. Liposomes were shown to carry the encapsulated hydrophilic agents into the human stratum corneum and possibly into the deeper layers of the skin (Verma et al., 2003). Packaging of drugs into liposomes enables a more concentrated delivery, enhanced cytotoxicity, improved pharmacokinetic qualities, sustained release and prolonged action of active components.

In this chapter we considered only one type of antimicrobial agents delivered by olive oil-containing liposomes, but the list of active drugs can be continued and expanded. Incorporation of olive oil into the lipid bilayer increases the biocompatibility of liposomes and enriches them with a broad spectrum of natural bioactive compounds. Integration of olive oil into the liposome lipid bilayer enriches the liposome features by new properties. Such enriched liposomes can not only fulfill a passive role in drug delivery, but can also supply active components for post-treatment recovery of skin. It has been proven that daily treatment with olive oil lowered the risk of dermatitis (Kiechl-Kohlendorfer et al., 2008). Olive oil vitamins and antioxidants could help overcome skin damage caused by skin infection and by the active treatment itself. Olive oil-containing liposomes can thus be converted from passive excipients into active supporting means of drug delivery systems. Totally natural and biocompatible olive oil-containing liposomes carrying any of the antimicrobial agents can be administrated in ointments and creams for application on skin areas contaminated with bacteria.

6. Conclusions

Olive oil can be incorporated into the liposome phospholipid bilayer, composed of an egg phosphatidylcholine or a dipalmitoyl phosphatidylcholine bilayer. The photosensitizers Rose Bengal and Methylene Blue encapsulated in olive oil-containing liposomes showed high efficiency in the eradication of Gram-positive *Staphylococcus aureus* and *Staphylococcus epidermidis* bacteria. The effectiveness of the antimicrobial agents was concentration-sensitive and depended on the initial concentration of the bacteria.

Application of olive oil-containing liposomes for drug delivery can change their perception as having a passive role of lipid-based excipients, converting them into a new generation of active and supporting drug carriers, supplying natural bioactive components for post-treatment recovery of skin.

7. Acknowledgment

This research was supported by the Research Authority of the Ariel University Center of Samaria, Israel.

We acknowledge graphical and design assistance of Ms. Julia Nakonechny.

8. References

Baron, S. (1996). *Medical Microbiology. 4th edition*, University of Texas Medical Branch, ISBN 978-0-963117-21-2, Galveston,TX

Beaulac, C.; Clément, S. (major); Hawari, J. & Lagacé, J. (1996). Eradication of mucoid *Pseudomonas aeruginosa* with fluid liposome-encapsulated tobramycin in an animal model of chronic pulmonary infection. *Antimicrobial Agents and chemotherapy*, Vol.40, pp. 665–669

Beaulac, C.; Sachetelli, S. & Lagace, J. (1998). In-vitro bactericidal efficacy of sub-MIC concentrations of liposome- encapsulated antibiotic against Gram-negative and Gram-positive bacteria. *Journal of Antimicrobial Chemotherapy*, Vol.41, pp. 35-41

Bisignano, G.; Lagana, M.G.; Trombetta, D.; Arena, S.; Nostro, A.; Uccella, N. et al. (2001). In vitro antibacterial activity of some aliphatic aldehydes from *Olea europaea* L. *FEMS Microbiology Letters*, Vol.198, pp. 9-13

Bombelli, C.; Bordi, F.; Ferro, S. ; Giansanti, L.; Jori, G.; Mancini, G. et al. (2008). New cationic liposomes as vehicles of *m*-tetrahydroxyphenylchlorin in photodynamic therapy of infectious diseases. *Molecular Pharmaceutics*, Vol 5 ,pp. 672–679

Boskou, D.; Blekas, G. & Tsimidou, M.Z. (2006a). Olive oil composition. In: *Olive Oil, Chemistry and Tehnology*, D. Boskou, (Ed.), 41-72, 2nd Edition, AOCS Press, ISBN 978-1-893997-88-2, Boca Raton, Fl

Boskou, D.; Tsimidoum M.Z. & Blekas, G. (2006b). Polar phenolic compounds. In: *Olive Oil, Chemistry and Tehnology*, D. Boskou, (Ed.), 73-92, 2nd Edition, AOCS Press, ISBN 978-1-893997-88-2, Boca Raton, Fl

Boskou, D. (2009a). Phenolic compounds in olives and olive oil. In: *Olive Oil: Minor Constituents and Health*, D. Boskou, (Ed.), 11-44, CRC Press, ISBN 978-1-4200-5993-9, Boca Raton, Fl

Boskou, D. (2009b). Other important minor constituents. In: *Olive Oil: Minor Constituents and Health*, D. Boskou, (Ed), 45-54, CRC Press, ISBN 978-1-4200-5993-9, Boca Raton, Fl

Cathcart, R.F. 3rd. (1985). Vitamin C: the nontoxic, nonrate-limited, antioxidant free radical scavenger. *Med Hypotheses*, Vol.18, pp. 61-77

Charman, S.A.; Charman, W.N.; Rogge, M.C.; Wilson, T.D.; Dutko, F.J. & Pouton, C.W. (1992). Self-emulsifying drug delivery systems: formulation and biopharmaceutic evaluation of an investigational lipophilic compound. *Pharm Res. Jan*, Vol.9, No.1, pp. 87-93

Chen, M.-L. (2008). Lipid excipients and delivery systems for pharmaceutical development: A regulatory perspective. *Advanced Drug Delivery Reviews*, Vol.60, pp. 768–777

Chrai, S.S.; Murari, R. & Ahmad, I. (2001). Liposomes (a Review) Part One: Manufacturing Issues. *BioPharm*, Vol.11, pp. 10-14

Chrai, S.S.; Murari, R. & Ahmad, I. (2002). Liposomes (a Review) Part Two: Drug Delivery Systems. *BioPharm*,Vol.1, pp.40-43

Constantinides, P.P. (1995). Lipid microemulsions for improving drug dissolution and oral absorption: physical and biopharmaceutical aspects. *Pharm Res.*, Vol.12, No.11, pp. 1561-1572

Covas, M.-I.; Kymenets, O.; Fitó, M. & de la Torre, R. (2009). Bioavailability and antioxidant effect of olive oil phenolic compounds in humans. In: *Olive Oil: Minor Constituents and Health*, D. Boskou, (Ed.), 109-128, CRC Press, ISBN:978-1-4200-5993-9, Boca Raton, Fl

Daia, T.; Huanga, Y-Y. & Hamblin, MR. (2009). Photodynamic therapy for localized infections - State of the art. *Photodiagnosis and Photodynamic Therapy*, Vol.6, pp. 170-188

Derycke, A.S. & de Witte, P.A. (2004). Liposomes for photodynamic therapy. *Adv. Drug Delivery Rev*, Vol.56, pp. 17-30

Drummond, D.C.; Meyer, O.; Hong, K.; Kirpotin, D.B. & Papahadjopoulos, D. (1999). Optimizing liposomes for delivery of chemotherapeutic agents to solid tumors. *Pharmacological Reviews*, Vol51, pp. 691-743

Ferro, S.; Ricchelli, F.; Mancini, G.; Tognon, G. & Jori, G. (2006). Inactivation of methicillin-resistant *Staphylococcus aureus* (MRSA) by liposome-delivered photosensitising agents. *Journal of Photochemistry and Photobiology B.: Biology*, Vol.39, pp. 98-104

Ferro, S.; Ricchelli, F.; Monti, D.; Mancini, G. & Jori, G. (2007). Efficient photoinactivation of methicillin-resistant *Staphylococcus aureus* by a novel porphyrin incorporated into a poly-cationic liposome. *The International Journal of Biochemistry and Cell Biology*, Vol.83, pp. 1026-1034

Filippin, F.B.; C.Souza, L. & Maranhão R.C. (2008). Amphotericin B associated with triacylglycerol-rich nanoemulsion: stability studies and in vitro antifungal activity. *Quím. Nova*, Vol.31, No.3 , pp. 591-594

Fleming, H.P.; Walter Jr., W.M. &. Etchells, J.L. (1973). Antimicrobial properties of oleuropein and products of its hydrolysis from green olive. *Applied Microbiology*, Vol.26, pp. 777-782

Furneri, P.M.; Fresta, M.; Puglisi, G. & Tempera, G. (2000). Ofloxacin-loaded liposomes: in vitro activity and drug accumulation in bacteria. *Antimicrob. Agents Chemother*, Vol.44, pp. 2458-2464

Furneri, P.M.; Marino, A., Saija, A.; Uccella, N. & Bisignano, G. (2002). In vitro antimycoplasmal activity of oleuropein. *International Journal of Antimicrobial Agents*, Vol.20, pp. 293-296

Grigorieva, T.F; Vorsina, I.A.; Barinova, A.P. & Lyakhov, N.Z. (2004). Mechanocomposites as new materials for solid-phase cosmetics. *Chemistry for Sustainable Development*, Vol.12, pp. 139-146

Hatzakis, E.; Koidis, A.; Boskou, D. & Dais, Ph. (2008). Determination of phospholipids in olive oil by ^{31}P NMR spectroscopy. *J. Agric. Food Chem.*, Vol.56, pp. 6232–6240

Holm, R.; Porter, C.J.; Edwards, G.A.; Müllertz, A.; Kristensen, H.G. & Charman, W.N. (2003). Examination of oral absorption and lymphatic transport of halofantrine in a triple-cannulated canine model after administration in self-microemulsifying drug delivery systems (SMEDDS) containing structured triacylglycerols. *Eur J Pharm Sci.*, Vol.20, No.1, pp. 91-97

Holm, R.; Porter, C.J.; Müllertz, A.; Kristensen, H.G. & Charman, W.N. (2002). Structured triacylglycerol vehicles for oral delivery of halofantrine: examination of intestinal lymphatic transport and bioavailability in conscious rats. *Pharm Res.*, Vol.19, No. 9, pp. 1354-1361

http://www.northernlipids.com/products/documents/List%20of%20Lipids%202007.pdf

http://www.oliveoilsource.com/page/chemical-characteristics

http://www.sigmaaldrich.com/etc/medialib/docs/Sigma/Product_Information_Sheet/1/p2772pis.Par.0001.File.tmp/p2772pis.pdf

http://avantilipids.com/index.php?option=com_content&view=article&id=216&Itemid=206&catnumber=850355

Kahl, L.P.; Scott, C.A.; Lelchuk, R.; Gregoriadis, G. & Liew, F.Y. (1989). Vaccination against murine cutaneous leishmaniasis by using Leishmania major antigen/liposomes. Optimization and assessment of the requirement for intravenous immunization. *J. Immunol.*, Vol.142, pp. 4441-4449

Kampa, M.; Pelekanou, V.; Notas, G. & Castanas, E. (2009) Olive oil phenols, basic cell mechanisms and cancer. In: *Olive Oil: Minor Constituents and Health*, D. Boskou, (Ed.), 129-172, CRC Press, ISBN:978-1-4200-5993-9, Boca Raton, Fl

Kent, C. & Carman, G. M. (1999). Interactions among pathways for phosphatidylcholine metabolism, CTP synthesis and secretion through the Golgi apparatus. *Trends in Biochemical Sciences*, Vol.24, No4, pp. 146-150

Kiechl-Kohlendorfer, U.; Berger, C. & Inzinger, R. (2008). The effect of daily treatment with an olive oil/lanolin emollient on skin integrity in preterm infants: a randomized controlled trial. *Pediatr Dermatol*, Vol.25, No.2, pp. 174-178

Koidis, A. & Boskou, D. (2006). The contents of proteins and phospholipids in cloudy (veiled) virgin olive oils. *Eur. J. Lipid Sci. Technol.*, Vol.108, pp. 323–328

Kubo, A.; Lunde, C.S. & Kubo I. (1995). Antimicrobial activity of the olive oil flavor compounds. *J. Agric. Food Chem.*, Vol.43, pp. 1629–1633

Land, L.M.; Li, P. & Bummer, P.M. (2005). The influence of water content of triacylglycerol oils on the solubility of steroids. *Pharm Res.*, Vol.22, No.5, pp. 784-788

Lesage-Meessen, L.; Navarro, S.; Maunier, D.; Sigoillot, J.C.; Lorquin, J.; Delattre, M. et al. (2001). Simple phenolic content in olive oil residues as a function of extraction systems. *Food Chemistry*, Vol.75, pp. 501-507

Lugovskoy, S.; Nisnevitch, M.; Lugovskoy, A. & Zinigrad, M. (2009). Mechanochemical synthesis of dispersed layer composites on the basis of talc and a series of biological active species. *Clean Techn Environ Policy*, Vol.11, pp. 277-282

Lugovskoy, S.; Nisnevitch, M.; Zinigrad, M & Wolf, D. (2008). Mechanochemical synthesis of salicylic acid—formaldehyde chelating co-polymer. *Clean Techn Environ Policy*, Vol.10, pp. 279–285

Macdonald, I. & Dougherty, T. (2001). Basic principles of photodynamic therapy. *J. Porphyrins Phthalocyanines*, Vol.5, pp. 105-129

Maisch, T.; Szeimies, R.-M.; Jori, G. & Abels, Ch. (2004). Antibacterial photodynamic therapy in dermatology. Photochem. Photobiol. Sci ., Vol.3, pp. 907 – 917

Malik, R.; Manocha, A. & Suresh, D.K. (2010). Photodynamic therapy - a strategic review. *Indian J of Dental Research*, Vol.21, pp. 285-291

Maragakis, L.L.; Perencevich, E.N. & Cosgrove, S.E. (2008). Clinical and economic burden of antimicrobial resistance. *ExpertRev. Anti Infect. Ther.*, Vol.6, pp. 751–763

Margetić, D. (2005). Mehanokemijske organske bez koristenja otapala. *Kemija u industriji (Zagreb)*, Vol.54, pp. 351-358.

Medina, E.; Romero, C.; Brenes, M. & De Castro, A. (2007). Antimicrobial activity of olive oil, vinegar, and various beverages against foodborne pathogens. *J Food Prot.*, Vol.70, pp. 1194-1199

Meisel, P.; & Kocher, T. (2005). Photodynamic therapy for periodontal diseases: State of the art. *J. Photochem. Photobiol. B: Biology*, Vol.79, pp. 159–170

Minnock, A.; Vernon, DI.; Schofield, J.; Griffiths, J.; Parish, JH. & Brown, SB. (2000). Mechanism of uptake of cationioc water-soluble pyridium zinc phthalocynaine across the outer membrane of *Escherichia coli*. *Antimicrob. Agents Chemother*, Vol44, pp. 522-527

Moellering, R.C.Jr.; Graybill, J.R.; McGowan J.E.Jr. & Corey, L. (2007). Antimicrobial resistance prevention initiative—an update: Proceedings of an expert panel on resistance. *Am J Infect Control*, Vol.35, pp. S1-S23

Mugabe, C.; Halwani, M.; Azghani, A.O.; Lafrenie, R.M. & Omri, A. (2006). Mechanism of enhanced activity of liposome-entrapped aminoglycosides against resistant strains of *Pseudomonas aeruginosa*. *Antimicrob. Agents Chemother*, Vol.50, pp. 2016-2022

Nakonechny, F.; Firer, M.A.; Nitzan, Y.& Nisnevitch, M. (2010). Intracellular antimicrobial photodynamic therapy: a novel technique for efficient eradication of pathogenic bacteria. *Photochem Photobiol.*, Vol.86, pp.1350-1355

Nakonechny, F.; Nisnevitch, M.; Nitzan, Y. & Firer, M. (2011). New techniques in antimicrobial photodynamic therapy: scope of application and overcoming drug resistance in nosocomial infections. In: *Science against microbial pathogens: communicating current research and technological advances*, Microbioloby Book Series, A. Méndez-Vilas, (Ed.), FORMATEX, ISBN 978-84-611-9421-6, Badajoz, Spain, in press

New R.R.C. (1994). Influence of liposome characteristics on their properties and fate. In: *Liposomes as tools in basic research and industry*, J.R. Philippol & F. Schuber, (Ed.), 3-20, CRC Press Inc., ISBN 0-8493-4569-3, Boca Raton, USA

Nichols, D.S. & Sanderson, K. (2002). The nomenclatiure, structure and properties o food lipids, In: *Chemical and Functional Properties of Food Lipids*, Z.E. Sikorski & A.Kolakowska, (Ed.), 18-47, CRC Press, ISBN 9781587161056, BocaRaton, London, NewYork, Washington, D.C.

Nisnevitch, M.; Lugovskoy, S.; Gabidulin, E. & Shestak, O. (2011). Mechanochemical production of olive oil based composites, In: *Olive Oil and Health*, J.D. Corrigan, (Ed.), 569-580, Nova Science Publishers Inc, ISBN 9781617286537, NY, USA

Nisnevitch, M.; Nakonechny, F. & Nitzan, Y. (2010). Photodynamic antimicrobial chemotherapy by liposome-encapsulated water-soluble photosensitizers. *Russian Journal of Bioorganic Chemistry*, Vol.36, pp. 363-369

Nitzan, Y. & Pechatnikov, I. (2011). Approaches to kill gram-negative bacteria by photosensitized process. In: *Photodynamic inactivation of microbial pathogens: medical and environmental applications*, M.R. Hamblin, (Ed.), 47-67, Royal Society of Chemistry, ISBN 978-1-84973-144-7

Nitzan, Y.; Gutterman, M.; Malik, Z. & Ehrenberg B. (1992). Inactivation of Gram-negative bacteria by photosensitized porphyrins. *Photochem. Photobiol.*, Vol.55, pp.89-96

O'Riordan K., Akilov, O.E. & Hasan, T. (2005). The potential for photodynamic therapy in the treatment of localized infections. *Photodiag. Photodynamic Therapy*, Vol. 2, pp. 247-262

Papadopoulos, G. & Boskou, D. (1991). Antioxidant effect of natural phenols on olive oil. *Journal of the American Oil Chemists Society*, Vol.68, pp. 669-671

Patterson, J.E. (2006). Multidrug-resistant gram-negative pathogens: multiple approaches and measures for prevention. *Infect Control Hosp Epidemiol.*, Vol.27, pp. 889-892

Randie, H.; Kim, Ph.D.; April, W. & Armstrong M.D. (2011). Current state of acne treatment: highlighting lasers, photodynamic therapy and chemical peels. *Dermatology Online Journal*, Vol.17, No.2. Available on: http://dermatology.cdlib.org/1703/2_reviews/2_11-00063/article.html

Reddy, V.N., Rani, K.R.; Chandana, G. & Sehrawat, S. (2009). Photodynamic therapy. *Indian J of Dental Advancements*, Vol.1, pp. 46-50

Robinson, A.; Bannister, M.; Creeth, J. & Jones, M. (2001). The interaction of phospholipid liposomes with mixed bacterial biofilms and their use in the delivery of bactericide. *Colloids and Surfaces A: Physicochemical and Engineering*, Vol.186, pp. 43-53

Rosenblatt, K.M. & Bunjes, H. (2009). Poly(vinyl alcohol) as emulsifier stabilizes solid triacylglycerol drug carrier nanoparticles in the alpha-modification. *Mol Pharm.*; Vol.6, No.1, pp. 105-120

Sachetelli, S.; Khalil, H.; Chen, T.; Beaulac, C.; Senechal, S. & Lagace J. (2000). Demonstration of a fusion mechanism between a fluid bactericidal liposomal formulation and bacterial cells. *Biochimica et Biophysica Acta*, Vol.1463, pp. 254-266

Schlupp, P.; Blaschke, T.; Kramer, K.D.; Höltje, H-D.; Mehnert, W. & Schäfer-Korting, M. (2011). Drug Release and Skin Penetration from Solid Lipid Nanoparticles and a Base Cream: A Systematic Approach from a Comparison of Three Glucocorticoids. *Skin Pharmacol Physiol*, Vol.24, pp. 199-209

Storm, G. & Crommelin, D.J.A. (1998). Liposomes: quo vadis? *Pharmaceutical Science & Technology Today*, Vol.1, pp. 19-31

Ternes, W. (2002). Egg lipids, In: *Chemical and Functional Properties of Food Lipids*, Z.E. Sikorski & A.Kolakowska, (Ed.), 270-292, CRC Press, ISBN: 9781587161056, BocaRaton, London, NewYork, Washington,D.C.

Tovar, M.J.; Motilva, M.J. & Paz Romero, M. (2001). Changes in the phenolic composition of virgin olive oil from young trees (Olea europaea L. cv. Arbequina) grown under linear irrigation strategies. *Journal of Agricultural and Food Chemistry*, Vol.49, pp. 5502-5508

Trofimov V.I. & Nisnevich M.M. (1990). Production of sterile drug-containing liposomes and control over their properties. *Vestnik Akademil Meditsinskikh Nauk SSSR*, Vol.6, pp. 28-32

Tsai, T.; Yang, Y.T.; Wang, T-H.; Chien, H-F. & Chen, C-T. (2009). Improved photodynamic inactivation of Gram-positive bacteria using hematoporphyrin encapsulated in liposomes and micelles. *Lasers in Surgery and Medicine*, Vol.41, pp. 316–322

Verma, D.D.; Verma, S.; Blume, G. & Fahr, A. (2003). Liposomes increase skin penetration of entrapped and non-entrapped hydrophilic substances into human skin: a skin penetration and confocal laser scanning microscopy study. *Eur J Pharm Biopharm.*, Vol.55, No3, pp. 271-277

Vissers, M. N.; Zock, P. L. & Katan M. B. (2004). Bioavailability and antioxidant effects of olive oil phenols in humans: a review. *European Journal of Clinical Nutrition*, Vol.58, pp.955-965

Wainwright, M. (1998). Photodynamic antimicrobial chemotherapy (PACT), *Journal of Antimicrobial Chemotherapy*, Vol.42, pp. 13–28

Wasan, K.M. (2007). *Role of Lipid Excipients in Modifying Oral and Parenteral Drug Delivery*, John Wiley & Sons, Inc., ISBN 9780471739524, Hoboken, New Jersey, US

14

Biocatalyzed Production of Structured Olive Oil Triacylglycerols

Laura J. Pham[1] and Patrisha J. Pham[2]
[1]BIOTECH, University of the Philippines at Los Baños College, Los Baños, Laguna,
[2]Dave C. Swalm School of Chemical Engineering Mississippi State University,
[1]Philippines
[2]USA

1. Introduction

Functional properties of fats and oils do not depend only on their fatty acid composition but also on the distribution of these fatty acids in the three positions of the glycerol backbone. This gives the fat or oil its commercial value. (Zhao,2005) There is a growing demand for lipids with desired characteristics ,thus researches have given way to these demands by the development of structured lipids with triacylglycerols that have predetermined composition and distribution of fatty acids. Structured lipids are now considered as alternatives to conventional fats not on the basis of saturate/polyunsaturate ratios but rather on their impact on cholesterol deposition. With the advances in the biotechnology and chemistry of fats and oils it is now possible to design fats and oils with properties that are desired. Recent years have seen great interest in the biotechnological modification and synthesis of structured triacylglycerols. Modification of fats and oil triacylglycerols to improve functionality have been carried out with various oils including olive oil. Olive oil enjoys a privileged position amongst edible oils and is still a buoyant commerce because of the large consumption of Mediterranean inhabitants (Oh et al,2009). It is one of the most expensive vegetable oils and of all the vegetable oils, olive oil is the best source of the monounsaturated fatty acid, oleic acid (72-83%)Risk factors for cardiovascular disease such as the level of homocysteine and total and low density lipoprotein (LDL) cholesterol in plasma have been reported to be reduced by oleic acid (Baro et al,2003). Olive oil is more than just oleic acid and because of its properties and qualities, it is used almost entirely in dietary consumption and even new markets have been created for this oil.

2. Triacyglycerol structure and characteristics

Triacylglycerols are by far the most abundant single lipid class and virtually all important fats and oils of plant or fat origin and most animal depot fats consist almost entirely of this lipid.

2.1 Triacylglycerol structure

Glycerol is a trihydric alcohol (containing three -OH hydroxyl groups) that can combine with up to three fatty acids to form monoacylglycerols, diacylglycerols, and triacylglycerols.

Fatty acids may combine with any of the three hydroxyl groups to create a wide diversity of compounds. A triacylglycerol (TAG)(Fig.1) consists of three fatty acids(R) to one glycerol molecule.

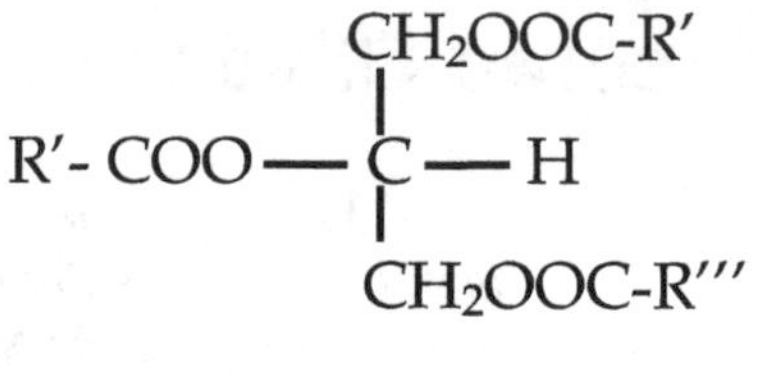

$$CH_2OOC\text{-}R'$$
$$R'\text{-}COO\text{---}\overset{|}{C}\text{---}H$$
$$CH_2OOC\text{-}R'''$$

Fig. 1. Stucture of a triacylglycerol

If all three fatty acids are identical,it is a simple triacylglycerol. The more common forms however are the "mixed" triacylglycerols in which two or three kinds of fatty acids are present in the molecule. The positions occupied by these fatty acids are numbered relative to their stereospecificity or stereospecific numbering (sn) as sn-1,sn-2 and sn-3.The orientation of the triacylglycerol structure specificity is as follows: if the fatty acid esterified to the middle carbon of the glycerol backbone is considered to the left (on the plane of the page) , then the top carbon is sn-1,the bottom carbon is numbered sn-3 (below or behind the plane of the page) and the middle carbon is subsequently numbered as sn-2. The fatty acids in the triacylglycerol define the characteristics and properties of the triacylglycerol molecule. Both the physical and chemical characteristics of fats are influenced greatly by the kinds and proportions of the component fatty acids and the way in which these are positioned in the glycerol molecule (Breckenridge, W.C. 1978; Christie, W.W.1982; Karupiah, T. & Sundram, K.2007).

2.2 Triacylglycerol species of olive oil

The triacylglycerol composition is a relevant information for the restructuring of lipids. Most often this defines the properties being sought to make them more suitable for their end use. These are mainly nutritional or physical. Nutritional properties are important in structured lipids as there is a growing appreciation for this information because metabolism is intimately linked to triacylglycerol composition. Triacylglycerol composition by HPLC of olive oil as reported in literature is given in Table 1(Christie,1982;Uzzan,1996;Aranda et al,2004). Most prevalent triacylglycerols in olive oil is the oleic-oleic-oleic (OOO) triacylglycerol, followed, in order of incidence, by palmitic-oleic-oleic (POO), then oleic-oleic-linoleic (OOL), then palmitic-oleic-linoleic (POL), then by stearic-oleic-oleic (SOO).The triacylglycerol species show a small degree of asymmetry in the distribution of fatty acids among the three positions of the glycerol moiety.

However, a single symmetric triacylglycerol specie (OOO) represents almost half of the total triacylglycerols. New developments in analytical methodology have allowed the evaluation of the degree of asymmetry in other fractions. The information on the individual triacylglycerols would be very useful in the structured lipid production.

Triacylglycerol (TG)specie	% of Total TG (Range)
LLL	0 - 0.8
OLL	0.3 - 5.8
OLLn	0.9 – 0.6
OOLn	1.0 – 1.5
PLL	0.5 – 2.8
POLn	0.3 - 1.1
OOL	10.4 - 18.2
PoOO	0 - 1.1
POL	4.5 - 12.3
PPoO	0.4 - 1.2
PPL	0.7 - 2.1
OOO	21.8– 43.1
POO	20 - 23.1
PPO	2.9 - 5.3
PSPo	0 - 0.8
PPP	0 - 0.5
SOO	3.6 - 3.7
PSO	0.4 - 1.2
PPS	0 - 0.6

Table 1. Triacylglycerol Composition as Analyzed by HPLC

2.3 Fatty acid profile and distribution intriacylglycerols

The fatty acid composition of olive oil varies widely depending on the cultivar, maturity of the fruit, altitude, climate, and several other factors(Galtier et al,2008). The major fatty acids in olive oil triacylglycerols are oleic acid ($C_{18:1}$), a monounsaturated omega-9 fatty acid which makes up 55 to 83% of olive oil. Another fatty acid is linoleic acid ($C_{18:2}$), a polyunsaturated omega-6 fatty acid that makes up about 3.5 to 21% of olive oil. Palmitic Acid ($C_{16:0}$), a saturated fatty acid that makes up 7.5 to 20% of olive oil, stearic Acid (C18:0), a saturated fatty acid that makes up 0.5 to 5% of olive oil and linolenic acid ($C_{18:3}$) (specifically alpha-Linolenic Acid), a polyunsaturated omega-3 fatty acid that makes up 0 to 1.5% of olive oil .Olive oil contains more oleic acid and less linoleic and linolenic acids than other vegetable oils, that is, more monounsaturated than polyunsaturated fatty acids . This renders olive oil more resistant to oxidation. The different fatty acids have stereospecific distribution on the glycerol backbone rather than a completely random or "restricted random" distribution. In most vegetable oils either 18:1 or 18:2 are exclusively at the sn-2 position in the triacylglycerol species like OOO,LLL,POL and LLO. Linolenic acid (C18:3) occurs less commonly,but when present , is at the sn-3 position as seen for OOLn in canola oil. Oleic acid is commonly at the sn-2 position of the olive oil triacylglycerols. (Karupiah & Sundram, 2007).Table 2 shows the fatty acid distribution in the three positions of the glycerol molecule as reported by Uzzan (1996). In esterified olive oil, the content of saturated acids palmitic and stearic in position 2 is higher, with values of approximately 13-15% compared to the normal 1.5-2%.

Nature of FA	% total FA in 2	% total FA in 1+3
C14:0		
C16:	1.4	15.0
C18:0	-	3.4
C18:1	82.9	72.8
C18.2	14.0	7.4
C18:3	0.8	0.9

Table 2. Fatty Acid Distribution in the three positions of olive oil triacylglycerol

3. Structured triacylglycerols (sTAGS)

Structured lipids may be defined as triacylglycerols restructured or modified to change the fatty acid composition or their positional distribution in glycerol molecules by a chemical or enzymatic process The term "structured triacylglycerol" was first introduced by Babayan (1987) to describe fats and oils that have been modified to change the fatty acid composition and the structure of triacylglycerols after the application of modification technologies. According to Hoy and Xu (2003) structured triacylglycerols (ST) generally are any fats that are modified or restructured from natural oils and fats, or fatty acids there from, having functionalities or nutritional properties for edible or pharmaceutical purposes. This definition covers any fats produced by either chemical or enzymatic methods for special functionality or nutritional use, including cocoa butter equivalents, breast milkfat substitutes,some low calorie fats, oils enriched in essential fatty acids γ-linolenic, arachidic, α-linolenic, eicosapentaenoic (EPA) and docosahexaenoic (DHA) acids, margarines or plastic fats and structured triacylglycerols containing both long chain (essential) and medium/short chain fatty acids. Adamczak (2004) defined structured triacylglycerols as triacylglycerols with a precisely defined composition and position of fatty acids esterified with glycerol possible only with the use of lipases or enzymatic modification. Structured triacylglycerols are often referred to as a new generation of lipids that are considered as nutraceuticals or functional foods or functional lipids. Regardless of the definition restructuring of triacylglycerols can be designed for use as medical or functional food as well as nutraceuticals depending on the type of application.

3.1 Synthesis and production of sTAGS

Biotechnology has experienced considerable advances in the past years via the use of fats and oils. The enzymatic process of modification is one of the advantages of fats and oils biotechnology which gives additional levels of flexibility in controlling and designing structured triacylglycerols.

3.1.1 Lipases

In recent years, the use of lipases to modify the properties of triacylglycerols has received considerable interest and has been the subject of extensive research worldwide. Lipases catalyze three types of reactions and the catalytic action of lipases is reversible. They catalyze hydrolysis in an aqueous system, but also esterification (reverse reaction of hydrolysis) in a microaqueous system, where water content is very low. Transesterification is categorized into four subclasses according to the chemical species which react with the ester. Alcoholysis is the reaction with an ester and an alcohol, while acidolysis is the one with an ester and an acid. Interesterification is a reaction between two different esters, where alcohol and acid moiety is swapped. Lipases can be classified according to their positional specificity into two groups: 1,3-positional-specific and non-positional-specific. Usually, pancreatic and fungal lipases are 1,3-positional-specific, while yeast and bacterial ones are non-positional specific or weakly 1,3-positional-specific. It should however be noted that the positional specificity of lipases is not strictly divided into the two categories, but it varies widely in the range of very distinctly 1,3-positional-specific to very weakly specific or completely non-positional-specific By exploitation of the specificity of lipases it is possible to produce acylglycerol mixtures which cannot be obtained by conventional chemical modification processes. Specificity of lipases can be utilized to produce products that cannot be produced otherwise which means that with 1,3 specific lipases, reactions involving triacylglycerol changes are confined to the sn-1 and sn-3 positions and the sn-2 acyl groups remain unaltered. There are several advantages connected to the use of lipases. The relatively mild reaction conditions for lipases reduce the amount of by products formed in a reaction. The use of lipases also renders it possible to process substances such as polyunsaturated fatty acids which cannot be processed by the conventional high temperature/high pressure processes.(Kennedy,1991Adamczack,2004) With the application of new biotechnological techniques ,companies are now able to produce lipases at lower costs. This will make the enzymatic processes far more competitive to the existing processes for the production of structured triacylglycerols.

3.1.2 Enzymatic processes of modification

Structured triacylglycerols may be prepared by hydrolysis of fatty acyl groups from a mixture of TAGs and random re-esterification follows onto the glycerol backbone. Depending on the desired metabolic effect , a variety of fatty acids are used in this process, including different classes of saturated, monounsaturated, and polyunsaturated fatty acids.Thus, a mixture of fatty acids is incorporated onto the same glycerol molecule. These manufactured lipids are structurally and different metabolically from the more simple, random physical mixtures of medium-chain triacylglycerols (MCTs) and long-chain triacylglycerol (LCT). Six possible fatty acid combinations could result for structured triacylglycerols prepared with an MCT and LCT and these are two MCFAs and one LCFA; one MCFA and two LCFAs; the two positional isomers; and small amounts of the starting MCT and LCT (Fig.2).Based on their high regiospecificity, lipases are effective biocatalysts for the manufacture of structured lipids that have a predetermined composition and distribution of fatty acids on the glycerol backbone. Structured lipids resembling TAGs of human milk have been produced by trans-esterification of tripalmitin, depending on the desired metabolic effect from plant oil, with oleic acid or PUFAs, obtained from plant oils, using sn-1,3-specific lipases as biocatalysts . Such TAGs were found to closely mimic the fatty acid distribution of human milk and may be used in infant food formulations. Apart from imitating the human milk more closely, the occurrence of palmitic acid lipase catalyzed esterification has been used in fat modification to improve absorption properties and the nutritional value of lipids. The most commonly used method is acidolysis for the production of MLM type (M-medium chain fatty acid; L-long chain fatty acid using a regiospecific lipase to incorporate the medium chain fatty acids into the sn-1 and sn-3 positions of the triacyglycerol molecule. Currently interesterification is viewed as an alternate process to the partial hydrogenation of oils and fats. The process involves randomization among all three stereospecific positions of fatty acids in native edible oils and fats by enzymatic catalysis at low temperatures. The positional distribution of the fatty acids on the glycerol backbone is altered either through fatty acids switching positions within a triacylglycerol molecule or between triacylglycerols. If interesterification involves triacylglycerol species within the dietary fat, the fatty acid composition remains the same. There are many applications of interesterification. It is not only the management of the fatty acid mixtures which could lead to the improvement of physical properties such as in the case of cocoa butter equivalents or substitutes but it is used in the production of structured lipids which can provide specific metabolic effects for nutritive and therapeutic purposes (Kennedy,1991; Marangoni, 1993; Klemann,1994).

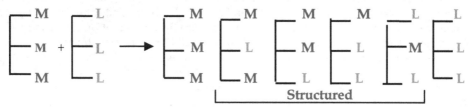

Fig. 2. Production of Structured Triacylglycerols

4. Structured olive oil triacylglycerols

Several researches have reported of the use of lipase catalyzed modification as the process for the production of structured olive oil triacylglycerols. In a study by Lee et al., 2006, olive oil triacylglycerols was used as a delivery medium for enrichment of conjugated linoleic acid in a dietary oil. Conjugated linoleic acid (CLA) are positional isomers of conjugated octadienoic acids two of which are cis-9,trans 11 and trans 10 cis-12 and are known to possess biological activity. The consumption of dietary CLA has decreased in recent years due to the replacement of animal lipids that contain little CLA .To consume more CLA and derive more health benefits, the enrichment of CLA in food has been attempted through modification of lipids in which synthesis of structured lipids is the most desirable method. Commercially produced CLA isomers were incorporated into extra virgin olive oil through a I,3 specific lipase from *Rhizomucor miehie* that catalyzed acidolysis to produce the structured olive oil triacylglycerols. The olive oil synthesized structured olive oil contained reduced content of oleic acid which was 43.1 mol % from the original value of 75.7 mol% of the total fatty acids. The decrease was compensated by the increase of CLA content at 42.5 mol %.Major CLA isomers incorporated into the triacylglycerol molecules were cis-9,trans-

11 at 16.9 mol % and trans-10, cis-12 at 24.2%.The study suggests that restructuring olive oil may be a suitable way to incorporate or deliver CLA into human diets.

Structured triacylglycerols synthesized by the acidolysis of olive oil and capric acid was carried out with immobilized lipase derived from *Thermomyces lanuginosus* to produce olive oil triacylglycerols with medium chain fatty acids in its glycerol moiety.(Oh et al,2009).

Medium chain triacylglycerols (MCTs) also offer numerous health benefits and have been widely studied for medical, nutritional and food applications. Structured lipids containing medium chain fatty acids at sn 1,3 positions and long chain fatty acids at the sn-2 position of triacylglycerols are more readily absorbed and oxidized for energy as compared to long chain triacylglycerols (LCT). Results of the study showed that the fatty acid composition of the olive oil triacylglycerols was significantly changed. The major fatty acid in the triacylglycerols was oleic acid originally, but after restructuring capric and oleic acids became the major fatty acids of the triacylglycerols. The study carried out by Fomuso and AKoh (2002) performed the lipase (1,3 specific lipase from *Rhizomucor miehie*) catalyzed acidolysis of olive oil in a bench scale packed bed reactor. Findings showed olive oil to be characterized by four major clusters of triacylglycerol species with Equivalent carbon number (ECN) ,C_{44},C_{46}, C_{48}, and C_{50}.Three monosubstituted products and two disubstituted products were detected after the reaction Monosubstituted products have ECN of C_{36}, C_{38}, and C_{40}. And the disubstituted products had ECN of, C_{30}, and C_{32}. Fatty acid distribution of the sn-2 position of olive oil was .74.8% oleic acid and 25.2 % linoleic acid .The structured olive oil had 7.2% capryllic acid, 69.6% oleic acid, 21.7% linoleic acid and 1.5% palmitic acid at the sn-2 position. The results showed a structured olive oil that would have improved properties and nutritional value.

The production of cocoa butter equivalents is a promising application of the biotechnological production of structured lipids. Due to the high cost and fluctuations in supply and demand of cocoa butter, the industry has looked into the production of cocoa butter equivalents from other oil sources. Cocoa butter equivalents can be produced by the enzymatic acidolysis using sn 1,3 specific lipases that can catalyze the incorporation of palmitic acid (C_{16}) and stearic acid (C_{18}) to the sn-1,3 positions of a source containing oleic acid at sn-2 position until a similar composition of cocoa butter is obtained. The three main triacylglycerols are the 1,3 dipalmitoyl-2- oleoyl-glycerol (POP); 1(3)-palmitoyl-3(1)stearoyl-2-oleoyl glycerol (POS) and 1,3-distearoyl-2-oleoyl glycerol (SOS) with oleic acid at the sn-2 position of the glycerol backbone (Lipp,et al,2001).Using olive pomace oil these three major triacylglycerols can be achieved to produce a cocoa butter like fat. Olive pomace oil's chemical composition does not differ from refined olive oil. It has the same triacylglycerol profile of olive oil because it is olive oil extracted via solvent. In a study by Ciftci and Fadiloglu(2009) utilizing the olive pomace oil for the production of a cocoa butter like fat, findings showed that the triacylglycerol composition of the prepared product was similar to that of the commercial cocoa butter which contained 18.9% POP, 33.1% POS, and 24.7% SOS..The triacylglycerol composition of refined olive pomace oils was redesigned so that properties such as the melting point, solid fat content and fat crystal network microstructures of the structured olive pomace oil and cocoa butter were very much similar.

In another study of Olive oil triacylglycerol restructuring, olive oil was blended with coconut oil to get a balanced proportion of saturated to unsaturated fatty acids and was subjected to lipase catalyzed interesterification to rearrange the fatty acids in the triacylglycerol molecule that would have both a short chain fatty acid and long chain fatty acid in one triacylglycerol molecule (Nagaraju & Lokesh, 2007). Results showed there were no significant differences between the blended and interesterified oils in terms of the fatty acid composition but HPLC analysis showed that there were new triacylglycerol molecular species formed. Studies have shown that structured lipids have a unique metabolism and exhibit better benefits when compared with the blended oils having similar fatty acid composition. The study of Nagaraju and Lokesh (2007) showed a reduction of serum cholesterol levels by 25% as compared to the oil blend. Cholesterol levels in rat liver was also reduced by 32% as compared with results using physical blending The effect was certainly significant.

5. Conclusion

Through enzyme biotechnology, olive oil triacylglycerols can be structured to contain medium chain fatty acids or functional fatty acids like the conjugated linoleic acid for the production of a structured olive oil that would have improved biological and nutritional properties. Structured lipids provide attributes that consumers will find valuable whether for demands of healthier oils or for physical requirements to give appropriate properties. There is also a need for more researches with olive oil restructuring that will allow for better understanding and more control over the various interesterification processes and reduction in costs associated with large-scale production of structured olive oil triacylglycerols.

6. References

Adamczak, M.(2004).The application of Lipases in Modifying the Composition, Structure and Properties of lipids-A Review. *Polish Journal Of Food and Nutrition Sciences*. Vol. 13/54, No.1 pp. 3-10

Aranda, F.; Gomez-Alonso, S.; Rivera del Alamo, R.M.; Salvador, M.D. & Fregapane, G.(2004)Triglyceride, total and 2-position fatty acid composition of Cornicabra virgin olive oil: Comparison with other Spanish cultivars. *Food Chemistry*.Vol.86.No.4, pp.485-492

Babayan, V.K. (1987).Medium chain Triglycerides and Structured Lipids.*Lipids*, 22: 417-420

Baro L.; Fonolla, J.; Pena, JL.; Martinez-Ferez, A.; Lucena, A.; Jimenez,.; Boza, JJ.& Lopez-Huertas, E.(2003). n-3 fatty acids plus oleic acid and vitamin-supplemented milk consumption reduces total and LDL cholesterol, homocysteine and levelsof endothelial adhesion molecules in healthy humans. Clin Nutr 22(2):175–182

Breckenridge, W.C. Stereospecific analysis of triacylglycerols. in 'Handbook of Lipid Research. Vol. 1. Fatty Acids and Glycerides, pp. 197-232 (edited by A. Kuksis, Plenum Press, New York) (1978).

Ciftci, O.N., Fadiloglu, S, & Gogus, F.(2009). Utilization of Olive Pomace Oil for Enzymatic Production of Cocoa Butter-like Fat. *J Am Oil Chem Soc.*, Vol.86: pp119-125

Christie, W.W.(1982).*Lipid Analysis*.2nd Edition.PergamonPress.ISBN 0-08-023792-4.U.K.

Galtier, O.;LeDreau, Y.;Ollivier, D.;Krister, J.;Artaud, J & Dupuy, N. 2008.Lipid Compositions and French Registered Designations of Origins of Virgin Olive Oils Predicted by Chemometric Analysis of Mid Infrared Spectra. *Applied Spectroscopy.* Vol. 6, No.5Hoy, C.E. & Xu.X. (2001). Structured Triacylglycerols , In:*Structured and Modified Llipids,* Gunstone, F. (Ed), pp.209-239, Marcel Dekker, Inc.ISBN:0-8247-0253-0, New York

Karupiah, T., & Sundram, K. (July 2007).Effects of Stereospecific positioning of fatty acids in triacylglycerol structures in native and randomized fats: A review of their nutritional implications: Nutrition & Metabolism, 10.1186/1743-7075-4-6.Available from /http./creative commons.or/licenses/by/2.0)

Kennedy J.P.(1991) Structured lipids: fats of the future. FoodTechnol., 11: 76–83.

Klemann, L.P.; Aji K., Chrysam M.M.; D'Amelia, R.P., Henderson J.M.; Huang A.S.& Otterburn, M.S.(1994) produced by the catalysed interesterification of short- and long-chain fatty acid triglyceride. J. Agric.Food Chem., 42: 443–446.

Lee, J.H.; Kim, M.R.; Kim, I.H.; Kim, H.; Shin, J.A. & Lee, K.T. (2004).Physical and Volatile Characterization of Structured Lipids from Olive Oil Produced in a Stirred tank Batch Reactor.*Journal of Food Science, Vol.69, Nr. 2, pp.*89-95

Lee, J.;H., Lee;K.T.; Akoh, C.C.; Chung, S.K. & Kim, M.R. (2006).Antioxidant Evaluation and Oxidative Stability of Structured Lipids from Extravirgin Olive Oil and conjugated Linoleic Acid. J. *Agric Food Chem., Vol* 54, No.15, pp. 5416-5421.

Lipp, M.; Simoneau, C.; Ulberth, F.; Anklam, E.; Crews, C.; Brereton, P.; deGrey, W.; Schwacks, W. & Wiedmaiers, C. (2001). Composition of Genuine Cocoa Butter and Cocoa Butter Equivalents. *Journal of food Composition and analysis. Vol. 14, pp.399-408*

Marangoni, A.G.; McCurdy, R.D. &Brown E.D.(1993). Enzymatic Interesterification of Triolein with Tripalmitin in Canola Lecithin-Hexane Reverse Micelles. *J Am Oil Chem Soc.* Vol. 70, No.8, pp. 737-744.

Moreda, W.; Perez-Camino, M.C & Cert A. (2003).Improved method for the determination of triacylglycerols in olive oils by High performance liquid chromatography. *Grasas y Aceites.* Vol 54.No.2, 175-179

Nagaraju, A. & Lokesh,. R. (2007). Interesterified coconut oil blends with groundnut oil or olive oil exhibit greater hypochlesterolemic effect compared with their physical blends in rats. *Nutrition Research.*Vol.27: pp. 580-586

OH, J. E.; Lee, K.; Park, W.; H.K., Kim, J.Y.; Kwang, I.K.; Kim, J.W., Kim, H.R. & Kim, I.H.(2009). Lipase Catalyzed Acidolysis of Olive oil with Capric Acid: Effect of Water Activity on incorporation and Acyl Migration. J. *Agric Food Chem.,* Vol 57, No.19, pp9280-9283

Uzzan, A. (1996). Oleaginous Fruits and their oils, In: *Oils and Fats Manual, A Comprehensive Treatise, Vol.1.*Karleskind, A. (Ed), pp. 225-233.Lavoisier Publishing, ISBN:1-898298-08-4.Paris, France

Zhao, H.; Bie, X;, Lu, F. & Liu, Z.(2005). Lipase Catalyzed Acidolysis of Lard with Capric Acid in Organic Solvent. *J Food Eng.*Vol.78:41-46

Part 4

Regional Studies

15

Olive Oil Sector in Albania and Its Perspective

Ana Mane Kapaj[1] and Ilir Kapaj[2,3]

[1]Agriculture University of Tirana, Faculty of Economy and Agribusiness,
Department of Economy and Agrarian Policy Tirana,
[2]Agriculture University of Tirana, Faculty of Economy and Agribusiness,
Department of Agribusiness Management Tirana,
[3]Hohenheim University, Institute of Agribusiness Management and
Computer Applications in Agriculture, Stuttgart,
[1,2]Albania
[3]Germany

1. Introduction

Albania, situated on the eastern shore of the Adriatic Sea, may be divided into two major regions: a mountainous highland region (north, east, and south) constituting 70% of the land area, and a western coastal lowland region that contains nearly all of the country's agricultural lands and is the most densely populated part of Albania. Due to the mountains landscape and especially because of its many divisions, the climate varies from region to region. It is warmer in the western part of the country which is affected by the warm air masses from the sea (the Adriatic costal region has a typical Mediterranean climate). This climate makes Albania an important producer of olives and olive oil for the region.

The transition of Albanian economy from a centrally planned to a market economy is associated with the implementation of a considerable number of structural and institutional reforms necessary for a sustainable market economy. Trade liberalization policies were implemented associated with elimination of price controls as the economy was decentralized to balance the supply and demand of goods and services.

Despite the progress made, especially in terms of macroeconomic and financial stability, Albania continues to have one of the lowest levels of income per capita in. In addition, there is a big income gap between rural and urban areas, since the agricultural sector comprise about 58% of total labour force and count for 25% of Albania Gross Domestic Product (INSTAT, 2010). Albania's economic growth can be achieved primarily through strengthening the agricultural sector. The current macroeconomic situation along with the climatic, geographic, and cultural advantages as comparable to neighbouring countries provide the opportunity for a fast and sustainable growth of the agricultural sector. Even though the olive production does not take a large share in the total agricultural production, it is an industry with huge potentials that has been steadily growing during the years.

Like many of the other agricultural products, the major supply of oil (vegetable and olive) in Albania comes from imports. This is because of the inconsistent and unreliable supply of local raw material needed for the oil processing industry. In addition, the distribution infrastructure linking to the markets is also poor. With current prices and expected yield, the farmers do not have the incentives to grow oil-bearing plants because of the low economic returns. Furthermore, many processing plants had been destroyed after the 1990s. However, if Albania reaches an average yield, similar to that of its neighbouring countries (Greece and Italy), there will be a great potential for Albania to develop an olive industry comparable to its neighbours with similar climatic and soil conditions. To make this a reality olive productivity has to increase along with a favourable marketing situation conducive to exports. The surface plant with olives is 42 thousand hectares, with a total number of olive trees of around 5 million. Because of the insufficient services olive tree have low growth rates with a very high yield fluctuation. The result is mall quantity and low quality olive oil. Almost 10.3 million US $ have been invested in the olive oil processing industry since 1992. The major part of the processing machinery in use is obsolete.

The olive and olive oil sector is an important segment of Albanian primary production and agro industry. Primary production of olives accounts for approximately 16% of total fruit output in value, including grapes. The number of planted trees is nearly 5 million and is rapidly increasing, as a response to sustained demand, good prices and government subsidies for expanding the production base.

Official data on olive oil production show an output ranging between 6,400 Mt (Million ton) in bad harvest years and 11,900 Mt in good harvest years. There is a structural production deficit of approximately 1,000 Mt per year, mostly covered by imports of bottled olive oil from Italy and other EU countries. Main production areas of olives for olive oil are in the center and south of the country. In these areas, 90% to 95% of cultivars are for olive oil production. (Leonetti et al, 2009)

Processing industry has a specializing and modernizing trend, producing mostly olive oil and table olives (15-20% of total olive production). Official data for 2009 show that there are 108 enterprises processing all edible oils including olive oil, and 16 enterprises processing table olives. The structural deficit of table olives is covered mainly by imports from Greece.

2. Olive cultivation in Albania

Albania is one of the few countries in Europe and the only country in the Central-East Europe that has the favourable climatic and geographical conditions for olive cultivation. The olive cultivation story in Albania is very old. The people of the rural areas are used with the cultivation of olives, and a good tradition has been heritage from one generation to the other.

The demand for olive oil and table olives in the domestic market is very high. On the other hand, with an adequate technological improvement in the olive processing industry, this product could be traded in the international market.

Olives are among the most important fruit tree crops grown in Albania, covering an estimated 8% of the arable land. As shown in Figure 1, the Albanian olive production zone covers the entire coast from Saranda (South) to Shkodra (North) and inland river valleys in the districts of Peqin/Elbasan, Berat/Skrapar, and Tepelene/Permet.

Olive tree in Albania is cultivated in the regions along the western costal lowland. Geographically 3.3% is cultivated in the plain zone and 96.7% in the hilly zone. In 77% of the farms olives are cultivated in organized plantations whereas in the remaining 23% of the farms this culture is found in a not organized form. The olive concentration in plantations gives the possibility for more careful services and the use of adequate technologies. According to the data taken from INSTAT (Institute of Statistics, 2008), the dynamic of the surface and the number of the olives during the years is as follows.

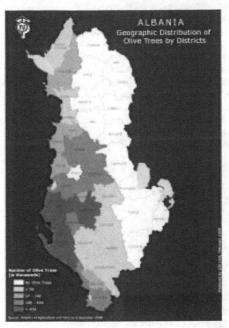

Fig. 1. Map of Albania showing olive cultivation area (USAID, 2011)

According to Figure 2 the surface of olive plantation and the number of olive trees has increased by four times in the year 1990 compared with the year 1938. After the 1990s, as the result of the late processing of the Land Agrarian Reform in this sector, the olive production industry has suffered a lot of considerable damages. As many other sectors of the country's economy, this sector was characterized by a visible depreciation in the main indicators. Huge olive blocks like those in Fier, Mallakaster, Berat and Lushnje were burned and destroyed. The transformation of the State Farms into private economies in this sector of the economy has been very slow. Even today, there are regions where the reform changes have not yet been completed. Table 1 shows olive production and yield in the main regions of the country and Table 2 describes in numbers the overall country situation.

Although there has been a considerable investment in the new olive plantations, the production investments and the services for this culture have been minimal. Today the olive production has low and fluctuating yields. The extensive character of the olive cultivation and the insufficient treatments that are usually done to the olives are the cause of this phenomenon. The yield fluctuation in the olive production has been and still is a serious phenomenon for our country. According to statistical data, the ratio between an "empty" year (year with very low production) and the year with a good production is very high.

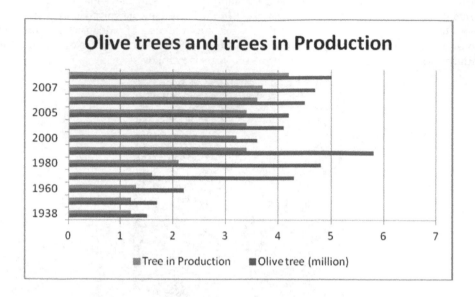

Fig. 2. Olive trees and trees in production for the period 1938-2008 (INSTAT, 2010)

Nr	Region	Number of olives (000 trees)		Yield (Kg/tree)	Production (Ton)
		Total	In production		
1	Berat	628	492	22,0	10841
2	Vlorë	532	495	13,0	6436
3	Elbasan	364	331	10,0	3315
4	Fier	347	311	12,7	3955
5	Tiranë	318	294	9,1	2664
6	Sarandë	312	310	6,6	2048
	TOTAL	**2501**	**2233**	**13,1**	**29259**
	REPUBLIC	**3564**	**3200**	**13,1**	**42012**

Table 1. Olive production data for the main regions 2009 (Ministry of Agriculture, Food and Consumer Protection, 2010)

Nr.	Region	Number of olives (000 trees)		Yield (Kg/tree)	Production (Ton)
		Total	In production		
1	Berat	628	492	22,0	10841
2	Delvinë	127	126	3,3	419
3	Durrës	57	52	19,3	1004
4	Elbasan	364	331	10,0	3315
5	Fier	347	311	12,7	3955
6	Gramsh	2	2	20,3	31
7	Gjirokastër	5	4	34,4	150
8	Kavajë	75	75	13,4	998
9	Kruje	104	87	4,5	393
10	Kuçovë	39	37	20,4	753
11	Laç	10	10	13,0	126
12	Lezhë	18	15	9,9	148
13	Lushnjë	227	209	19,5	4070
14	Mallakastër	197	161	20,9	3362
15	Peqin	65	64	6,5	412
16	Përmet	2	1	12,8	15
17	Sarandë	312	310	6,6	2048
18	Skrapar	1	1	11,6	14
19	Shkodër	93	81	6,0	485
20	Tepelenë	43	43	8,6	373
21	Tiranë	318	294	9,1	2664
22	Vlorë	532	495	13,0	6436
	TOTAL	**3564**	**3200**	**13,1**	**42012**

Table 2. Number of heads, yields and olive production according to the regions, 2009 (Ministry of Agriculture, Food and Consumer Protection, 2010)

3. Olive age and cultivars in Albania

According to the age of the olives there is a visible distinction that divides the olive plantations into two groups;

1. Centennial olive plantations are mainly found in the urban areas of Sarandë, Vlorë, Berat, and Elbasan. These are native varieties with high economic values that consist of the main part of olive production of the country.
2. Olive plantations planted after the 1960s, which are found by the sea and in the central part of the country.

Based on the statistical data the proportion of the olives according to their age result as follows: Olive plantations above 100 years old (30% of the total olive trees), Olive plantations from 30-40 years old (45%) and Olive plantations from 10-20 years old (25%).

One of the most important factors affecting productivity of the olive tree is its cultivar. Albania is rich with more than 28 varieties grown throughout the country. The nine most cultivated are listed in Table 3. With the exception of the Frantoio variety introduced from Italy, the other eight most commonly grown varieties are native to Albania. The two leaders are "Kalinjot", which covers about 40% of the total plantations for oil and table use; and "Kokermadh i Beratit", representing approximately 21% of table olives. The interaction of the Albanian varieties with the local environment (soil, climate, altitude) and cultural practices results in the special characteristics and tastes distinctive to the oils produced in various regions of throughout Albania.

Varieties	Number of Trees	Surface(Ha)	Maximum Oil Yield (% of weight)	Main Use
Kalinjot	2,335,000	17,700	27	Table & Oil
Kokërrmadh i Beratit	1,000,000	7,700	18	Table
Frantoio	470,000	2,600	19	Oil
Kokërrmadh Elbasani	450,000	4,000	20	Table & Oil
Mixan	430,000	3,770	25	Oil
Ulli i Bardhë Tiranes	200,000	1,500	28	Oil
Nisiot	120,000	900	12	Oil
Ulli i Hollë I Himares	70,000	800	15	Oil

Table 3. Olive cultivars in Albania (Ministry of Agriculture, Food and Consumer Protection, 2009)

4. Olive harvesting and collecting

Olive collection in Albania starts at the beginning of October and goes on until February. The harvesting is mostly done manually, and no modern equipment is used. During harvesting no selection between olives is done. Farmers use combined harvesting of olives that fall from the wind or as the effect of diseases and olives that are taken from the trees. This way of harvesting has a big influence on the manufactured oil quality.

The Albanian distribution system is traditional and extremely fragmented, without a real wholesaling sector. Especially for olive oil, distribution to retailers is mainly performed by the bottlers themselves. Wholesalers play a more important role in distributing table olives. More in general, food processing companies are distributing directly to retail outlets bypassing or relaying less on wholesalers. Two major changes occurred in the last three years, which will induce major changes in the distribution system. The establishment of a network of wholesale markets, facilitating wholesale trading and gradually introducing more transparency in price formation and on the other hand the development of organized distribution, with the entrance of two foreign-owned supermarket chains and the parallel growth of some domestic larger retailers into supermarket chains.

More organized logistics are necessary to cope with such evolution. Total mark up in the post-production section of the food chain is also likely to increase, as prices are already high. This is likely to put more pressure on producers to reduce sales prices. For olive oil and table olives, such evolution is likely to induce the following changes:

- Organized distribution needs regular supplies of relatively large quantities of products. The role of bottlers will further increase and medium producers will be forced to upgrade their distribution system or to reduce the share of olive oil sold with their own brand. This evolution is also representing a challenge for the small modern processors which will be forced to increase the resources devoted to marketing, as increasing number of wealthy customers will make their purchases in supermarkets.
- An increasing role will be played by wholesale markets in distribution of table olives, thus facilitating in the short term a further increase in the number of small wholesalers/processors. Generally, wholesalers and importers will become more important players in the table olive trading.

Most urban dwellers buy olive oil in mini-markets and traditional retail outlets whereas imported olive oil is almost exclusively sold through supermarkets. Organized distribution is catching an increasing share of customers. These outlets do not represent any more the higher end of retailing business. Supermarkets are adjusting their prices to those ones of traditional retailers, aiming at widening the range of customers beyond the middle income consumers' segment. Restaurants and other catering outlets are buying, with few exceptions, the cheapest qualities of olive oil. Limited purchasing of higher quality olive oil is made by high-end restaurants. Apart from self-consumption, olive oil in rural areas is mostly informally traded and purchased from local oil mills. A smaller share, estimated in 30% of the total or less, is sold usually by the liter (i.e. not bottled), in traditional retail outlets. Retail shops and green markets are the prevalent market channels in rural areas where there is no olive oil production.

Until the end of the 1970s the olive oil processing was done in traditional primitive ways by the peasants themselves. Gradually with the increases in yield, some plants were built. These were very old technology fashioned plants. Only at the beginning of the 1980s some presses were imported from Italy, and this was the start of innovations in the oil manufacturing plants. Actually almost half of olive oil existing processing plants use the "Pieralisi" type presses for the olive oil production (Figure 3). Second popular kind of press is Alfa Laval with 15% and the next significant types are Eno Rossi (11%) and Mix (5 %). The situation shows that the processing olive oil technology is dominated by three phase decanters.

5. Financing the olive oil sector

After the 1990s, a lot of investments were done in the olive oil processing industry. According to a study done by IFDC in 2002, the total amount of investments in this sector is 1442 million Lekë (or 10.686.230,92 euro). The regions with the highest amount of investment are Vlora with 25.0% of the total, Tirana with 17.6%, Saranda with 17.5%, and Fier with 13.8% of the total investments.

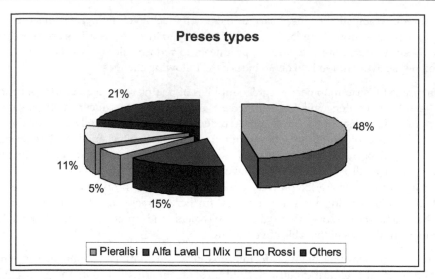

Fig. 3. Olive oil processing presses used in Albania (Ministry of Agriculture, Food and Consumer Protection, 2009)

There are three main investment sources in Albania, as far as the agricultural sector is concerned, own financial sources, bank credits and other funds. The investments are mainly done by the private financial sources of the entrepreneurs. This is followed by a smaller part of those that have taken some bank credits. Figure 4 below, shows schematically the share that each of these forms holds in the total investment structure.

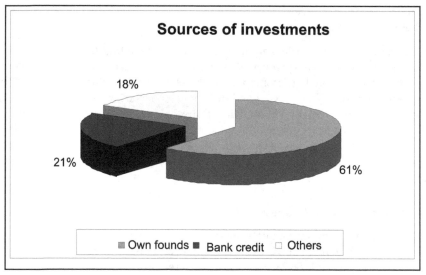

Fig. 4. Sources of financial invetments (Ministry of Agriculture, Food and Consumer Protection, 2009)

6. International trade

Albanian olive oil exports are very encouraging as the industry is maturing and achieving all attributes required for the olive oil quality. The figures however remain modest: 22 tons were exported in 2004; 16 tons in 2005; 54 tons in 2006; 15 tons in 2007 and 4 tons were exported in 2008. The first success was the export of "Shkalla enterprise" certified organic and extra-virgin olive oil to the niche market in Switzerland. This represents a small, but stable export and with potential to increase. This was the first sign of the "recovery" of Albanian olive oil export to the neighboring countries since 1996. The transaction was particularly important because, for the first time, the processing plant was certified. Furthermore, the payment was delivered by the letter of credit, in contrast to cash, that had been the practice until then.

Albanian imports on the other side are significant and range between 850 – 1100 Mt per year, of which almost 90% is supplied by Italy and Greece. Large part of the imported oil is in bulk to be than bottled in Albania. Albanian import of olive oil has increased since year 2000. In 2005 and 2006, due to major increase of EU olive oil prices and higher levels of domestic olive production, imports of olive oil dropped. In 2008, imports of olive oil were considerably higher than the same period of the previous years, due to the low olive oil production in 2007, caused by low olive production. This evolution of imports shows how the olive oil demand in Albania is price sensitive. The olive oil price increased by almost 40% from year 2004 to 2005 and was associated with almost 20% reduction in imports. Simultaneously, the continuous increase of domestic production of olives and olive oil has partially compensated the increasing demand, and contributed to lowering demand for imports. Imports usually increase in the last three months of each year, when consumption is higher and the olive oil of the new crop is not yet ready. Imports reach a minimum in summer. In general, the yearly peak of imports of olive oil follows by one or two months that one of table olives. In 2008 imports remained high also in January and February, due to the scarcity of domestic production.

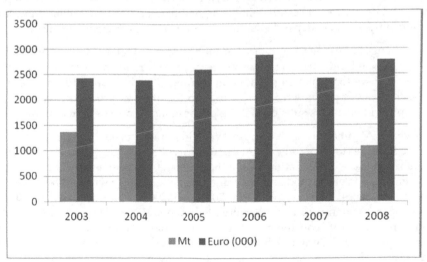

Fig. 5. Olive oil import trends from EU in terms of quantity and value (INSTAT, 2009)

At present, imports cover about 10% of market-based demand, but imports from the highly competitive Mediterranean producers could lead to an increase in market share of imported olive oils in the next years. Increasing domestic demand, years of relatively lower productivity and a general scarce competitiveness of the Albanian products are among the main factors that caused such a situation.

7. Economic issues behind olive oil processing industry

According to the results of the studies that have been undertaken for the olive oil industry indicated that olive oil production is privately profitable. The Private Cost Ratio has been measured and the coefficient was evaluated at 0,703, meaning that this production is profitable for the private enterprises. Still according to these results indicate that Albania has not a comparative advantage in olive oil producing industry in the production year 2008-2009. The calculations resulted in DRC = 2.2. This means, it is not socially desirable to produce and expand olive oil production in Albania, as the use of domestic factors is not efficient in economic terms. Comparing the two values of the above ratios, in current situation, olive oil production can be appreciated as competitive within the country with private prices, but it cannot be appreciated as competitive with social prices.

A sensitivity analysis was done to see how the DRC ratio reacts to different changes in different parameters of the model. Changes were done in the parameters like world market prices of olives and olive oil, exchange rate and labor force price. As a result of the sensitivity analysis it was seen that the olive oil production is very sensitive to changes in input (olives) and output (olive oil) prices of the world market and also to changes in the exchange rate. But it is not sensitive towards changes in the labor force prices. The major determinants of the Albanian olive oil comparative advantage are the favorable world price of olive oil, the exchange rate and the price of olives as input factors for the olive oil manufacturing. The explored values of private profitability and the DRC suggest that that olive oil production is privately and socially profitable, however two important conclusions are to be emphasized particularly: firstly, the private profitability is higher than the social profitability, and secondly, social profitability is largely depended on the situation at the international market.

Due to the changes in the sensitivity analysis it can be seen that a reduction on the olives as input in the olive oil industry, the domestic resource cost ratio enters in the interval values in which we can say that olive oil production is competitive in Albania. In this stage there is need of state policy intervention in order to help the olive oil producers for having lower prices of inputs. Policies like subsidies of the prices of inputs are suggested in this case.

In recent years, Albania has seen a rapid evolution in its citizens' consumption behaviors and life styles due to economic growth, improvement in the standard of living, fast urbanization and trade liberalization within the country. One consequence of this has been the gradual segmentation of the food and beverages market, similar to what has been seen in other transitioning countries (Berisha and Mara, 2005: WB, 2007). The transition from a centrally planned socialist economy to a market oriented economy has also given rise to an urban middle-income class of consumers. Another important study on olive oil consumer preferences conducted in Tirana/Albania 2010 has resulted with some other result on the olive oil (Chan Halbrendt et al, 2010). According to this study 6 consumers' segments and profiles were identified, based on set of preferences and willingness to pay. The fact that it is now possible to clearly identify several segments of consumers marks a milestone in the process of evolution of agri-food marketing, with major consequences on development policies. Origin is a key choice factor for 82% of respondents, in three out of six consumers' segments.

The confidence on quality and safety of domestic product is low. This conclusion emerges from the analysis of several factors: i) imports are growing notwithstanding the consumers' preference for Albanian olive oil; ii) consumers have little confidence on reliability of domestic industrial producers and controls made by competent authorities, so they prefer to buy olive oil directly from trusted farmers, or from the oil mills or to buy imported products; iii) during the analysis there was a scarce correspondence between low income and preference for low prices, as high prices are considered one of the few reliable proxies for quality.

The majority of purchased olive oil is still traded as not bottled product, being sourced either directly from farmers and oil mills or as by quantity in traditional shops. 44% of the interviewed consumers in Tirana confirm that they buy directly from farmer and olive oil mill respectively. This percentage should be much higher in smaller cities or rural areas characterized by olive and olive oil production and consumption. When considering also self-consumption of farmers in production areas, it is possible to conclude that most probably, more than 70% of the olive oil consumed in the country is sold as a non-bottled product and is subject to little quality control.

Under the current extensive inefficient conditions in which the olive culture is cultivated in Albania, there is however a profitability for farmers to produce. This profitability and comparative advantage can be improved if the olive culture is cultivated more intensively. If the farmers are sure that the processing industry will act as a reliable market for their products, they will increase the production. On the other hand the increased olive cultivation will provide more raw materials for the processing industry, assuring its functioning with full capacity. The better utilisation of existing capacities in processing industry will allow favouring from the low of economies of scale and at the end effect result in lower production cost.

8. The future of olives and olive oil in Albania

In the last years olives and olive oil has become one of the priorities of the Albanian Government policies. Recently Albanian Government is undertaking an extremely ambitious policy for expanding the production base, targeting a fivefold increase of the total number of olive trees, i.e. up to 25 ml trees. For this purpose, most subsidies provided from 2007 to the agricultures sector from the State (scoring about 10 m Euro in 2008) are addressed to the olive cultivation and olive oil production. Focusing investments in increasing yields (production per tree), stabilizing output from one year to another and improving harvesting and pest management practices would be at present a more cost/effective option for ensuring a sustainable development of the sector. Priority actions include: i) improvement of value chain governance tools, including harmonization of laws to EC, ii) increased technical assistance to farmers to increase productivity and stabilize output; iii) support to value chain operators for facilitating access to services, iv) supporting establishment and strengthening of farmers' associations and cooperatives and; v) optimize the use of effluents and by-product in olive oil industry, to mitigate environmental impact of olive oil production and increase profitability in olive oil processing.

The Government of Albania lunched since early 2009 the idea of supporting the plantation of 20 million olive trees, which would eventually transform Albania into a world level competitor. At present, domestic demand of olive oil scores around 12,200 Mt and that one of table olives 14,000 Mt per year (excluding self-consumption). Yearly yields and output are highly variable, as Albanian olive orchards receive poor or no services and are highly vulnerable to weather conditions. According to previous surveys and according to the evaluation of specialists, average yield of olive trees in Albania is circa 15 kg/tree.

Under these assumptions and estimations, there is a deficit of 1,500 tons of olive oil which is not very different from the recorded official imports of olive oil – circa 1,000 tons of olive oil (the current yields may be even a bit more than 15 kg/tree, i.e. if assumed 15.5 kg/tree, than we obtain a deficit corresponding exactly to the recorded imports). Improving average yield to 17 kg/tree (+13%) to the current 5,011 million of trees (thus excluding in these calculations the expected increased number of trees in the coming years) there a surplus of production will be already achieved. Considering that many trees will enter in full production in the next years, this objective seems easy to achieve. According to expert (agronomist) evaluations, under irrigation and proper treatment, it is possible to achieve average yield of 25 kg/tree (conservative assessment). At this level, suficit is of equivalent 7,550 tons.

As a conclusion, Albania can meet its demand for olives and olive oil, and even achieve surplus, by simply improving services to the current olives; moreover, even without further support for new plantings, the expansion of the production base will continue, even if at a slower pace: before the introduction of subsidies for new plantings, the average growth of the production base was of 166,000 new trees/year; in 2007-2008, after the introduction of subsidies, this amount increased to 257,000 per year.

In their study, Leonetti et al, 2009, introduced 5 different scenarios considering several investment and related implications.

- Scenario 1 – Average future plantings in accordance to the trend recorded before the introduction of subsidies.
- Scenario 2 - Average future plantings in accordance to the trend recorded after the introduction of subsidies (2007- 2008).
- Scenario 3 - 20 million trees are planted within 5 years, starting in year 2009, at a pace of 4 million trees per year.
- Scenario 4 - 20 million trees are planted within 10 years, starting in year 2009, 2 million trees per year
- Scenario 5 - 20 million trees are planted within 15 years, starting in year 2009, 1.33 million trees per year.

The number of trees according to each scenario is reflected respectively in Figure 6. For the production, based on expert assessment, they assume that old trees (planted till 2008) have a yield of 15 kg/tree, whereas the new ones, 25 kg/tree. In the second year, the new trees achieve 3 kg/tree, third year 8 kg/trees, fourth 20kg/tree, fifth 25 kg/tree.

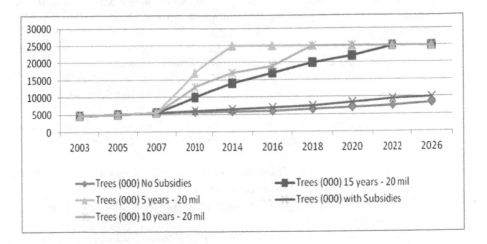

Fig. 6. Dynamics of Olive trees under different scenarios (000 trees), (Adapted from Leonetti et al, 2009)

9. Concluding remarks

Olive oil production can become a very important aspect in the Albanian agriculture economy. Due to the favourable climatic conditions the main input, olives, can be cultivated in a more intensive form, despite the fact, that the areas under olive cultivation in Albania compared with the areas in Greece or Italy are very insignificant.

After a relatively long hiatus related with the democratic political changes in of early '90-s, the Albanian olive and olive oil industry is showing signs of a healthy recovery. Since the country restored political and economic stability in 1999, the olive sector has attracted notable attention from the government and private investors, giving rise to a considerable growth of the sector. The planting of new trees has increased tremendously as a result of the government's supporting programs and private initiatives. New plantings have been established with modern practices and good management. The oil processing industry has also experience significant growth. Small processing plants have become more efficient in producing high quality oil for domestic and foreign markets. Still, the sector faces important challenges to overcome such as high cost of production, dominance of low quality olive oil production, shortage of raw materials, weak contracting relationship between the growers and the processors, and deficiencies in marketing.

10. References

Adelman, I and Taylor, J. E., (1990), "Changing comparative advantage in food and agriculture: lessons from Mexico", Development Centre studies} / Organisation for Economic Co-operation and Development, Paris, France

Agolli, Sh., Cipi, A., and Mance,M., (2000), "Policy Analysis Matrix: Evaluation of Comparative Advantage in the Greenhouse and Field Production", IFDC, Tirana, Albania

Agolli, Sh., Velica. R., and Mance, M., (2002), "The Olive Oil Industry: Marketing, Technology and its Competitiveness", IFDC, Tirana, Albania

Berisha, A. & Mara.V. 2005 "Role of diets in Balkan countries, as part of Mediterranean diets, in people's health" Albanian Journal of Agricultural Sciences, Vol 1, No 1, Tirane, Albania

Chan-Halbrendt, C., Zhllima, E., Sisior, G., Imami, D., Leonetti, L., 2010, "Consumer Preference for Olive Oil: The Case of Albania", IAMA World Symposium, Boston, USA

Chungsoo, K., (1983), "Evolution of Comparative Advantage: the factor proportions theory in a dynamic perspective", Tübingen, Germany

Civici, A., (2003), "The Situation and Competitiveness Level of the Agro-food Sector in Albania", Tirana, Albania

Cungu, A.,and Swinnen, J.F. M., (1998), "Albanian's Radical Agrarian Reform", Policy Research Group, Working Paper No. 15, April 1998

Edward, E. L., (1984), "Sources of International Comparative Advantage: theory and evidence", Cambridge

Ethier, W.J., (1988), "Modern International Economics", University of Pennsylvania, USA

Felderer, B., (2001), "Growth and trade in the international economy: papers for the conference "Dynamics in economic growth and international trade", Vienna, 22 - 23 June 2001

Gandolfo, G., (1987), "International Economics I: The pure theory of International Trade", University of Rome "La Sapienza", Rome, Italy

Gray, H. P., (1987), "International economic problems and policies", New York: St. Martin's Pr., XIV, 486 S

Greenaway, D., (1988), "Economic Development and International Trade", St. Martin's Press, New York, USA

Huang, J., Song, J., and Fuglie, K., (2002), "Competitiveness of Sweet Potato as animal feed in China", CCAP, Working Paper-E12

Kabursi, A. A., "Lebanon's Agricultural Potential: a Policy Analysis Approach", McMasterUniversity and Econometric Research

Kallio, P. K. S., "Old and new challenges in international agricultural trade", European review of agricultural economics; 29, 1 : Special section, Oxford : Oxford Univ. Press

Kemp, M. C., (2001), "International trade and national welfare", Routledge frontiers of political economy, London

Keuschnigg, M., (1999), "Comparative advantage in international trade: theory and evidence", Studies in empirical economics, Florenz, European Univ. Inst., Diss.

Khachatryan, N., (2002), "Assessing the market potential of brandy produced in Armenia", University of Hohenheim, Stuttgart, Germany

Krogman, P.R. and Obstfeld, M., (2003), "International Economics: Theory and Policy", World student series, USA

Leoneti, L., Imami, D., Zhllima, E. and Stefanllari, A.; 2009, "Report on food chain analysis of olive oil and table olives in Albania", prepared for USAID-AAC (Albanian Agriculture Competitiveness) project

Mane, A; KAPAJ, I; Chan-HAlbrendt, C; Totojani, O; "Assessing the Comparative Advantage of Albanian Olive Oil Production" IFAMR (International Food and Agribusiness Management Review); Volume13, Issue 1, 2010. ISSN #: 1559-2448,College Station, TX 77841-4145, USA

Mane, A. (2004), "Assessing the comparative advantage of olive oil production in Albania", Master Thesis, Stuttgart, Germany

Monke, E. A., and Pearson, S. R. (1989), "The policy analysis matrix for agricultural development", IthacaNY: Cornell University Press

Nguyen, M. H. and Heidhues, F., (2004), "Comparative Advantage of Vietnam's Rice Sector under Different Liberalization Scenarios", University of Hohenheim, Stuttgart, Germany

Nguyen, M. H., (2002), "Changing Comparative Advantage of Rice Production under Transformation and Trade Liberalization: a policy analysis matrix study of Vietnam's rice sector", University of Hohenheim, Stuttgart, Germany

Osmani, R., (2000) "The manual for olive cultivation", Tirana, Albania

Pearson, S., Gotsch, C. and Bahri, S., (2003), "Application of the Policy Analysis Matrix in Indonesian Agriculture", May 2003.

Robson, C., (1993), "Real world research; A resource for social scientists and practitioners researcher", Blackwell, Oxford

Sánchez, J. ed., (2002), "Olive oil", European journal of lipid science and technology; 104, 9/10: Special issue, Weinheim: Wiley-VCH

Takayama, T. and Judge, G. G., (1971), "Spatial and temporal price and allocation models/ Takashi Takayama", George G. Judge, Amsterdam: North-Holland Publ. Comp.

Themelko,H., (2001), "Olive Situation in Albania and Measures for Increasing the Olive and Olive Oil Production", Centre for Rural Studies, Tirana, Albania

Tracy, M., (ed), (1998), "CAP reform: the Southern products: wine, olive oil, fruit and vegetables, bananas, cotton, tobacco", Papers by Southern European Experts, Agricultural Policy Studies

Van Marrewijk, C., (2002), "International trade and the world economy", Oxford: Oxford University Press

Vossen, P., (2000), "Olive Oil Production in Italy", Technical report on the olive oil production tour

Zaloshnja, E.,(1997), "Analysis of Agricultural Production in Albania: Prospects for Policy Improvements", Dissertation work, Blocksburg, Virginia, USA

MoAFCP, (Ministry of Agriculture, Food and Consumer protection), 2010, "Annual Report 2009", Tirana, Albania

MoAFCP, (Ministry of Agriculture, Food and Consumer protection), 2009, "Annual Statistics 2008", Tirana, Albania

MoAFCP, (Ministry of Agriculture, Food and Consumer protection), 2001, "Annual Report 2000", Tirana, Albania

MOF, (Ministry of Finance), 2003, "Annual Report", Tirana, Albania
INSTAT, (Institute of Statistics), "Statistical Yearbook 2010", Tirana, Albania
INSTAT, (Institute of Statistics),"Statistical Yearbook 2009", Tirana, Albania

Permissions

The contributors of this book come from diverse backgrounds, making this book a truly international effort. This book will bring forth new frontiers with its revolutionizing research information and detailed analysis of the nascent developments around the world.

We would like to thank Boskou Dimitrios, for lending his expertise to make the book truly unique. He has played a crucial role in the development of this book. Without his invaluable contribution this book wouldn't have been possible. He has made vital efforts to compile up to date information on the varied aspects of this subject to make this book a valuable addition to the collection of many professionals and students.

This book was conceptualized with the vision of imparting up-to-date information and advanced data in this field. To ensure the same, a matchless editorial board was set up. Every individual on the board went through rigorous rounds of assessment to prove their worth. After which they invested a large part of their time researching and compiling the most relevant data for our readers. Conferences and sessions were held from time to time between the editorial board and the contributing authors to present the data in the most comprehensible form. The editorial team has worked tirelessly to provide valuable and valid information to help people across the globe.

Every chapter published in this book has been scrutinized by our experts. Their significance has been extensively debated. The topics covered herein carry significant findings which will fuel the growth of the discipline. They may even be implemented as practical applications or may be referred to as a beginning point for another development. Chapters in this book were first published by InTech; hereby published with permission under the Creative Commons Attribution License or equivalent.

The editorial board has been involved in producing this book since its inception. They have spent rigorous hours researching and exploring the diverse topics which have resulted in the successful publishing of this book. They have passed on their knowledge of decades through this book. To expedite this challenging task, the publisher supported the team at every step. A small team of assistant editors was also appointed to further simplify the editing procedure and attain best results for the readers.

Our editorial team has been hand-picked from every corner of the world. Their multi-ethnicity adds dynamic inputs to the discussions which result in innovative outcomes. These outcomes are then further discussed with the researchers and contributors who give their valuable feedback and opinion regarding the same. The feedback is then collaborated with the researches and they are edited in a comprehensive manner to aid the understanding of the subject.

Apart from the editorial board, the designing team has also invested a significant amount of their time in understanding the subject and creating the most relevant covers. They scrutinized every image to scout for the most suitable representation of the subject and create an appropriate cover for the book.

The publishing team has been involved in this book since its early stages. They were actively engaged in every process, be it collecting the data, connecting with the contributors or procuring relevant information. The team has been an ardent support to the editorial, designing and production team. Their endless efforts to recruit the best for this project, has resulted in the accomplishment of this book. They are a veteran in the field of academics and their pool of knowledge is as vast as their experience in printing. Their expertise and guidance has proved useful at every step. Their uncompromising quality standards have made this book an exceptional effort. Their encouragement from time to time has been an inspiration for everyone.

The publisher and the editorial board hope that this book will prove to be a valuable piece of knowledge for researchers, students, practitioners and scholars across the globe.

List of Contributors

G. Rodríguez-Gutiérrez, A. Lama-Muñoz, M.V. Ruiz-Méndez, F. Rubio-Senent and J. Fernández-Bolaños
Departamento de Biotecnología de los Alimentos, Instituto de la Grasa (CSIC), Avda, Seville, Spain
Instituto al Campus de Excelencia Internacional Agroalimentario, ceiA3, Spain

Bárbara Rincón, Fernando G. Fermoso and Rafael Borja
Instituto de la Grasa (Consejo Superior de Investigaciones Científicas), Sevilla, Spain

Farshad Darvishi
Department of Microbiology, Faculty of Science, University of Maragheh, Iran

Fabiano Jares Contesini, Camilo Barroso Teixeira, Paula Speranza, Danielle Branta Lopes, Hélia Harumi Sato and Gabriela Alves Macedo
Laboratory of Food Biochemistry, Department of Food Science, College of Food Engineering, State University of Campinas (UNICAMP), Campinas, SP, Brazil

Patrícia de Oliveira Carvalho
Multidisciplinar Laboratory, University São Francisco, Bragança Paulista, SP, Brazil

Ozan Nazim Ciftci, Deniz Ciftci and Ehsan Jenab
University of Alberta, Canada

S. Cicerale, L. J. Lucas and R. S. J. Keast
School of Exercise and Nutrition Sciences, Centre for Physical Activity and Nutrition (CPAN), Sensory Science Group, Deakin University, Melbourne, Australia

María Gómez-Romero, Alegría Carrasco-Pancorbo and Alberto Fernández-Gutiérrez
University of Granada, Spain

Rocío García-Villalba
CEBAS CSIC of Murcia, Spain

José G. Fernández-Bolaños, Óscar López, M. Ángeles López-García and Azucena Marset
University of Seville, Spain

Christophe Magnan, Hervé Le Stunff and Stéphanie Migrenne
Université Paris Diderot, Sorbonne Paris Cité, Biologie Fonctionnelle et Adaptative, Equipe d'accueil conventionnée Centre National de la Recherche Scientifique, Paris, France

Basem Mohammed Al-Abdullah1, Khalid M. Al-Ismail and Khaled Al-Mrazeeq
Department of Nutrition and Food Technology, Faculty of Agriculture, University of Jordan, Amman, Jordan

Malak Angor and Radwan Ajo
Al- Huson University Collage, Al- Balqa Applied University, Al-Huson, Jordan

S.N. Kang and I.S. Kim
Gyeongnam National University of Science and Technology, Korea

S.S. Moon and Y.T.Kim
Sunjin Meat Research Center, Korea

C. Jo
Chungnam National University, Korea

D.U. Ahn
Iowa State University, USA

Marie Zarevúcka
Institute of Organic Chemistry and Biochemistry, Czech Republic

Faina Nakonechny
Department of Chemical Engineering, Biotechnology and Materials, Ariel University Center of Samaria, Ariel, Israel
The Mina and Everard Goodman Faculty of Life Sciences, Bar-Ilan University, Ramat-Gan, Israel

Marina Nisnevitch
Department of Chemical Engineering, Biotechnology and Materials, Ariel University Center of Samaria, Ariel, Israel

Yeshayahu Nitzan
The Mina and Everard Goodman Faculty of Life Sciences, Bar-Ilan University, Ramat-Gan, Israel

Laura J. Pham
BIOTECH, University of the Philippines at Los Baños College, Los Baños, Laguna, Philippines

Patrisha J. Pham
Dave C. Swalm School of Chemical Engineering Mississippi State University, USA

Ilir Kapaj
Agriculture University of Tirana, Faculty of Economy and Agribusiness, Department of Agribusiness Management Tirana, Albania
Hohenheim University, Institute of Agribusiness Management and Computer Applications in Agriculture, Stuttgart, Germany

Ana Mane Kapaj
Agriculture University of Tirana, Faculty of Economy and Agribusiness, Department of Economy and Agrarian Policy Tirana, Albania